Butterflies and Moths of Costa Rica

Butterflies and Moths of Costa Rica

Luis Ricardo Murillo-Hiller

Antlion Media
A Zona Tropical Publication

Comstock Publishing Associates
an imprint of
Cornell University Press
Ithaca and London

Copyright © 2025 by Luis Ricardo Murillo-Hiller and
John McCuen

All rights reserved. Except for brief quotations in
a review, this book, or parts thereof, must not be
reproduced in any form without permission in writing
from the publishers. For information within Costa Rica,
visit Zona Tropical at zonatropical.net. For information
in the rest of the world, address Cornell University Press,
Sage House, 512 East State Street, Ithaca, New York
14850, or visit cornellpress.cornell.edu.

First published 2025 by Cornell University Press

Printed in China

Librarians: A CIP catalog record for this book is available
from the Library of Congress.

ISBN 978-1-5017-8160-5 (paperback)

Zona Tropical ISBN 978-1-949469-33-2

Book design: Gabriela Wattson

for Leonardo and Julián

CONTENTS

ACKNOWLEDGMENTS

I am exceedingly grateful to a number of people who helped make this book happen. Particularly important were specialists in certain families, among them Jorge Corrales, José Montero, Kenji Nishida, Dariel Sanabria, Eugenie Phillips, and Bernardo Espinoza. I also received valuable information and advice from Kirby Wolfe, Rolando Cubero, and Jim Córdoba. And I must express my gratitude to botanist Fabián Araya for reviewing plant names and providing important comments about my treatment of plant taxonomy. And a special thanks goes to Tom Fox, Pablo Venegas, German Vega, and Noemi Canet, all of whom provided many valuable biological observations.

Obtaining photographs of specific butterfly species is no easy task, and therefore I'd like to thank all the people who shared their beautiful pictures with me–their names are listed in the Credits. I also want to thank the good people at the Museo Nacional de Costa Rica (MNCR), the School of Biology at the University of Costa Rica, the Costa Rican Entomological Supply, and the Centro de Investigación en Biolodiversidad y Ecología Tropical (CIBET). A special thanks goes to professor Paul Hanson for his encouragement at the outset of this project, to Amy Hughes and John McCuen for their marvelous work in editing this book, and to Gaby Wattson, who created a beautiful design.

Finally, my warmest thanks go to my wife, Daniela, for her motivation, patience, and emotional support, which she gave in great abundance.

INTRODUCTION

Butterflies and moths are among the most charismatic and familiar of wildlife groups, and most people know of or have had contact with a few of these mega-diverse and conspicuous insects. As a group, they are important flower pollinators, co-responsible with bees, other animals, and other processes for the fruit production and sustainability of many of the plants humans eat and use. They play fundamental roles in the food web, as food for other animals and as natural controllers of plant populations. They inspire collectors and educators and play a role in connecting people with nature.

One of the most remarkable features of butterflies and moths, which form the insect order Lepidoptera, is their enormous species diversity. In Costa Rica, they can be found year-round and in various habitats. This field guide presents new and accurate information concerning natural history, ecology, behavior, and taxonomy of Costa Rican lepidopterans. It is meant for naturalists, field biologists, tour guides, science students, nature lovers, gardeners, artists, homeowners, and locals and visitors interested in learning about Costa Rican natural history and biodiversity. Some of the species selected are of great importance as agricultural pests, making this guide a valuable tool for field agronomists, farmers, land managers, and gardeners.

The book is divided into two main parts. First is an introductory section, which treats general aspects of butterflies and moths and their natural history, as well as Costa Rica's geography, climate, and habitats. The second section comprises species accounts of 170 Costa Rican butterflies and moths. The species accounts provide color photographs, a distribution map, and general and specific data for each lepidopteran about where to see it, how to identify it and distinguish it from others, the host plants for its larvae and the preferred food of adults, and information on its behavior, distribution, and more. As some technical terms are needed to explain aspects of morphology and biology, a glossary is included at the end of the book. An appendix provides a list of the plants mentioned in this book, with scientific names and English and Spanish common names. The Index of Plant Names (p. 308) includes scientific names for all plants.

TAXONOMY

From the Greek words *taxis* (arrangement) and *nomos* (law), the word "taxonomy" concerns the rules of the organization of things. In biology, it refers to the hierarchical organization of species, and may be expressed in a system of "boxes inside boxes." Generally, in such a system, all the small boxes inside a bigger box must share the same common ancestor. Configuring the taxonomy of earth's life-forms presents an enormous challenge, given the vast number of living organisms and the many

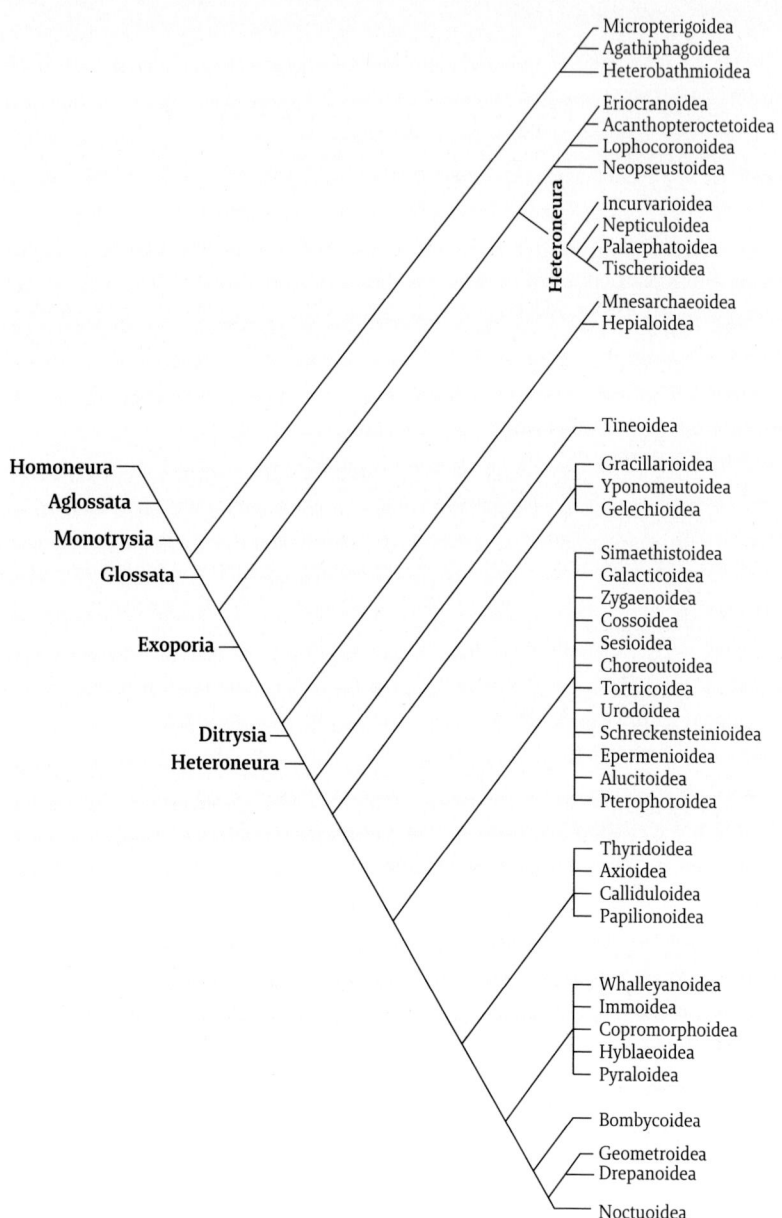

Fig. 1. Phylogeny of main superfamilies of Lepidoptera. Constructed by author by combining results from Scoble 1992; Kristensen et al. 2007; Kawahara and Breinholt 2014; and Espeland et al. 2018.

characteristics to consider. The puzzle begins with the fundamental unit of taxonomy: the species. Even today, there is no globally accepted concept of the species, leading to continuous debate about species richness, conservation, and taxonomy. Emerging technology has complicated the picture. For example, the Neotropical butterfly species *Telegonus fulgerator* was described in 1775, but a 2004 study (Hebert et al.) obtained new evidence from molecular techniques that revealed these skippers represent ten different species. The ten species' natural history features, ecology, and morphology in early stages supported the molecular data.

Similar changes are seen at higher levels, and the subfamilies, families, and superfamilies are in constant transformation as researchers seek true (natural) classifications (Fig. 1). Butterflies were once classified in two superfamilies, Hesperioidea and Papilionoidea (Scoble 1992), but recently the Hesperiidae (skippers) together with the Hedylidae (American moth-butterflies) have been put into the Papilionoidea, converting it into a group that includes butterflies, skippers, and some moths (Heikkilä et al. 2011, Kawahara and Breinholt 2014, Espeland et al. 2018).

Butterflies and moths belong to the insect order Lepidoptera, a name derived from Greek that means "scaled wings." The order comprises four suborders—Zeugloptera, Aglossata, Heterobathmiina, and Glossata—plus five families with plesiomorphic (ancestral) characteristics that have not been satisfactorily classified yet. Zeugloptera, Aglossata, Heterobathmiina each consist of a single family; their species possess mandibles instead of a proboscis. The rest of the known lepidopterans are in the suborder Glossata; among their characteristics (with some exceptions) is the presence of a functional proboscis (Van Nieukerken et al. 2011).

Glossata is divided in five infraorders: four of them, comprising about ten families, are separated by characteristics concerning the female genitalia and wing-vein arrangements. The fifth infraorder, Heteroneura, is the most diverse, containing the remaining 90% of the butterfly and moth families (Van Nieukerken et al. 2011). Heteroneura comprises thirty superfamilies, one of which is Papilionoidea, containing all the true butterflies (Van Nieukerken et al. 2011). The remaining twenty-nine superfamilies include the vast majority of moth families and species; some moths fly in the daytime, which actually makes them butterflies and indicates that butterflies appeared many independent times in the evolution of Lepidoptera.

COSTA RICA'S BUTTERFLIES AND MOTHS

Costa Rica has about 14,000 known species of moths and close to 1541 species of butterflies. The estimate for moths is much less precise than it is for butterflies because there are so few taxonomists systematically collecting and describing moth species. Butterflies have long attracted more attention from researchers, thanks to their bright coloration and because observing and collecting in daylight is easier than at night.

The lepidopteran species of Costa Rica are distributed among seventy-three families. The members of many of these families are rarely seen by people because of their tiny size or because the family is represented by few species in the country, or they live inside forests and are active at night, which makes them hard to find without formal sampling methodologies. A mere handful of families contain species of interest to the general reader, with individuals of large size, great beauty, or abundance, or that share their habitat with humans.

The 170 species selected and described in this guide were chosen because they are common, typical, or representative of a specific habitat; are colorful or otherwise relatively easy to identify; are of economic importance; or are simply too beautiful to be excluded. The selection encompasses a sampling of the species that normally can be seen on a trip around Costa Rica.

Costa Rican Biodiversity Law 7788 (1998) defines biodiversity as the variability of live organisms and the variations that each can present, in any part of the country (sea, air, or land). In this sense, a single species represents several aspects of biodiversity. It may exhibit, for example, sexual dimorphism, polymorphism, geographical variations, different growing stages, use of many habitats through its life history, complex relationships with other species that compromise its existence, and many other traits. Although many ecologists and conservation biologists therefore do not equate biodiversity with species richness (or number), decisions about conservation planning are often based on species richness data (Maclauring and Sterrenly 2008).

Biodiversity is a mathematical index that takes into consideration such factors as the size of the sample, abundance, and the evenness (or relative numbers) among the species; it is possible also to assess biodiversity in the sense of species richness in a given area (Hurlbert 1971). For instance, Costa Rica holds about 1541 species of butterflies (9.5% of the world's species) in the superfamily Papilionoidea excluding the family Hedylidae (Sandoval et al. 2019); while Peru, Colombia, and Brazil have more than double that number. But when we factor geographical area in the comparative analysis, Costa Rica has the highest butterfly species density, or number of species per square kilometer (Fig. 2; Table 1).

In comparisons with countries at the same latitude on other continents, Costa Rica maintains its pattern of higher butterfly diversity (DeVries 1987). The same pattern is found in species diversity of birds and of plants (Prance 1994), which are the lepidopterans' main food source. It is accepted that greater diversity generates greater diversity, and the beginning of that greater diversity in the Neotropics can be

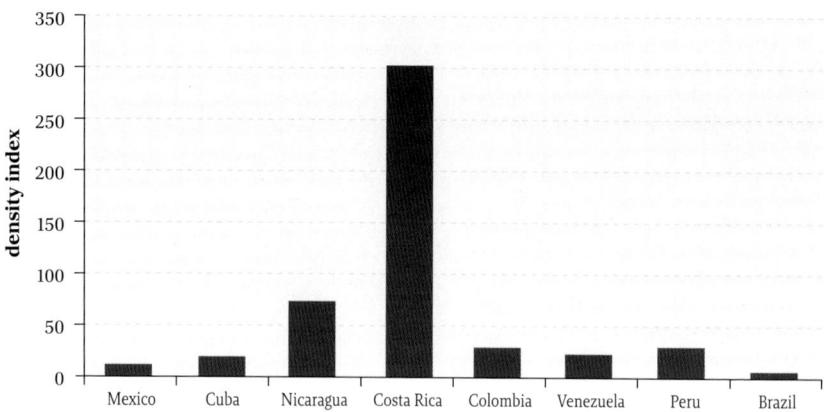

Fig. 2. Species richness (density index) of the butterfly superfamily Papilionoidea (excluding Hedylidae) per 10,000 km² in eight countries of the Neotropical region.

Table. 1: Species richness of the butterfly superfamily Papilionoidea (excluding Hedylidae) per 10,000 km² in eight countries of the Neotropical region.

Country	Area (km²)	Papilionoidea species richness	Density index (Hurlbert 1971): species per 10,000 km²
Mexico	1,964,000	1825 (Llorente-Bousquets et al. 2014)	9
Cuba	109,884	191 (Barro and Núñez 2011)	18
Nicaragua	130,373	925 (J. M. Maes, pers. comm.)	70
Costa Rica	51,100	1541 (Sandoval et al. 2019)	302
Colombia	1,141,748	3877 (Garwood et al. 2022)	34
Venezuela	916,445	1907 (Viloria 2000)	21
Peru	1,285,000	3710 (Lamas 1997)	29
Brazil	8,516,000	3288 (Lewinsohn et al. 2005)	4

explained by the abrupt landscape changes in the Neogene, when the uplift of the Andes Mountains, the formation of the Amazon and associated rivers, and the closure of the Isthmus of Panama, combined with natural organisms' abilities to persist and disperse, led to greater diversity over time (B. Smith et al. 2014). For instance, the land bridge between the North and South American continents that formed in Central America with the closure of the isthmus permitted the passage of species from one landscape to another, allowing colonization of new habitats and giving rise to endemic species.

Around 157,424 butterfly and moth species of 129 families have been described in the world (Van Nieukerken et al. 2011); of those, 73 families, representing 8.9% of global butterfly and moth diversity, inhabit Costa Rica. Two of the moth superfamilies occurring in the country exceed all others in species number: Geometroidea, with 23,748 species (1554 in Costa Rica); and Noctuoidea, with 70,000 species (3424 in Costa Rica).

Butterfly diversity is not equally distributed in Costa Rica; there is a general tendency of decreasing biodiversity with increasing elevation (Henderson 2002). That is not the case in all groups—the glasswing butterflies (tribe Ithomiini), for example, show great diversity in mid-elevation mountains between 600 and 1500 m (DeVries 1987). The butterfly genera *Vanessa*, *Oxeoschistus*, and *Catasticta*, and the moth genus *Neopreptos* are confined to mid-elevations.

Another tendency concerns organism size; families with larger butterflies and moths are less diverse than families with small species (DeVries et al. 1992). For instance, in the swallowtail butterfly family (Papilionidae), all species are medium-size to large, and the family is less diverse than the metalmark butterfly family (Riodinidae), in which all species are small. Likewise, the sphinx moth family (Sphingidae), composed entirely of large species, is far less diverse than the pyralid moth family (Pyralidae), with small species. Finally, some ecosystems hold fewer butterfly species; mangroves and páramo, for example, which are in opposite extremes of the country's elevation range, are two of the least diverse butterfly habitats in Costa Rica.

COSTA RICAN GEOGRAPHY, CLIMATE, AND HABITATS

Between its Caribbean Sea and Pacific Ocean coastlines, Costa Rica contains an enormous diversity of landscapes, including active volcanoes up to 3400 m tall, rainforest-clad flatlands, evergreen swamps, cloud forests, high-elevation páramo, semi-deciduous forests, dry forests, mangroves, and savannas. Each of these environments supports a characteristic composition of species.

The Caribbean and Pacific water masses are major contributors to Costa Rican weather patterns. They bring temperature stability, generate inshore winds, and transport large amounts of water vapor inland. These phenomena, combined with tides, generate the development of mountain rain shadows, high solar-radiation areas, and mangroves, swamps, and other microhabitats.

Topography also plays a crucial role in the nation's weather patterns. Two mountain types are found in Costa Rica: the orogenic mountain chain in the south of the country, called the Cordillera de Talamanca (elevation to 3820 m); and the lower-elevation volcanic mountain chain running longitudinally from the country's center (Cordillera Central) to its northwest (Cordillera de Guanacaste). These ranges result in two distinct drainages, the Pacific slope and the Caribbean slope, each containing a complex of smaller mountain systems forming valleys, passes, and secondary drainages.

The combination of mountains and seas produces a variety of weather conditions. In general terms, on the Pacific side of Costa Rica, the dry season lasts from December to April, and the rainy season from May to November; these seasons are most clearly marked in the Guanacaste area. On the Caribbean side, it is rainy year-round, but there are two periods with less rain, one in February–March and the other in September.

The Reventazón river valley, on the Caribbean slope, is one of the rainiest places on earth, receiving 8000 mm of precipitation annually (Coen 1983; Sánchez 2002), while tropical dry forest occurs in localized patches on the northern Pacific slope (Bolaños et al. 2005). Areas of Guanacaste present rainforest characteristics half of the year, receiving 2000–3000 mm of rain, and emulate tropical dry forest the other half, when no precipitation falls.

The Guanacaste cordillera in northern Costa Rica has lower elevations than the other ranges, and wider, low-altitude passes between the volcanoes allow the passage of the trade winds from the Caribbean to the Pacific slope. From December to April (the dry season), these winds release all their humidity on the Caribbean slope, and when they cross to the Pacific side, they push the onshore Pacific winds away from land, preventing the arrival of humid air from the Pacific and creating a dry rain shadow on this side. In addition, the absence of high-elevation mountains within several kilometers of the Pacific reduces condensation of the humid air coming from the Pacific. During the rainy season (May–November), the intertropical convergence zone generates a low-pressure area in the Caribbean that attracts huge water-charged air masses from the Pacific, which produce heavy rains when entering the hot Guanacaste lands. In the south of the country, on the other hand, the high mountains with abrupt slopes generate a vortex with a horizontal axis on both slopes that pulls the humid air masses from the sea, increasing the amount of rain mostly at 900 m elevation (Coen 1983).

Costa Rica is located in the tropics, between 8° and 11° north latitude. In consequence, the sun is highest in the sky at midday; the mean duration of daylight is 12 hours; and the number of daylight hours is about the same year-round. These conditions influence habitats, as the consistency of precipitation and sunlight (Méndez and Monge-Nájera 2010) maximizes photosynthesis levels, thus providing food and resources for animals constantly, all year long. Animals and plants (especially insects) take advantage of such conditions by producing more generations per year; this, in a certain way, enlarges their capacity for habitat specialization. All these factors, together with a constant water supply and volcanic soils that are very rich in nutrients, are key ingredients to drive diversification of life. The microhabitats formed by weather and topography drive local evolutionary processes, enabling species to adapt to humidity, wind, heat, and other conditions; as animals and plants adapt, they evolve into new species, leading, in the relative short term, to endemism—which describes a species that is native to one particular place and not found in any other place.

Bioregions of Costa Rica

In accordance with its topographic and biological characteristics, Costa Rica can be divided in four bioregions (Fig. 3): Guanacaste, in the northwest; the central mountain ranges forming a spine down the middle of the country; the Caribbean slope; and the rainy Pacific.

Guanacaste. At the northern limit of this bioregion are the volcanoes of the Cordillera Volcánica de Guanacaste. The southern limit is a narrow lowland on the Pacific coastline north of the mouth of the Río Grande de Tárcoles, where mangrove is the predominant element of the landscape and acts as an ecological barrier. The western limit is the Pacific Ocean. The eastern/inland limit is the Guanacaste mountain range. A unique feature of this bioregion is the presence of five large patches of tropical dry forest (in Bagaces; east of Liberia; around Junquillal Bay; on Isla Chira; and around Abangaritos) and about three smaller ones. Some additional tropical dry forest to moist forest transitional areas are dispersed in the region's tropical moist forest. Deforestation has amplified the dry forest characteristics in many moist forest areas, through a process called dry atmospheric association (Bolaños et al. 2005). Savannas, deciduous forests, evergreen riparian forests, and mangroves are common in this bioregion.

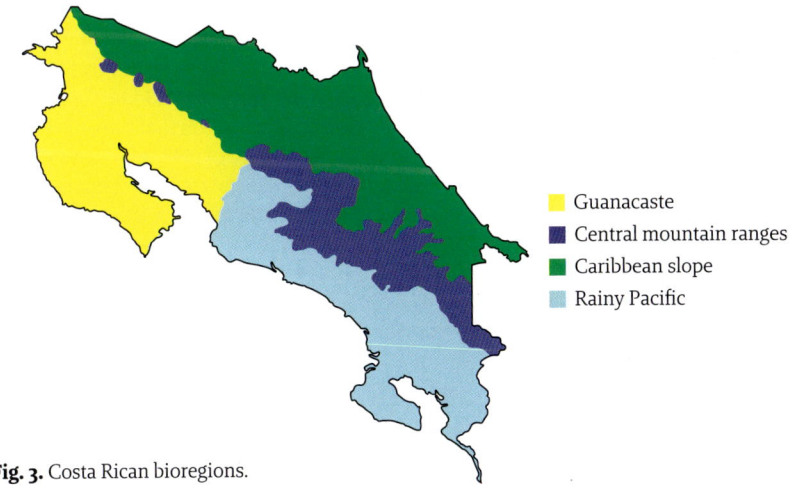

Guanacaste
Central mountain ranges
Caribbean slope
Rainy Pacific

Fig. 3. Costa Rican bioregions.

Central mountain ranges. The Cordillera de Guanacaste, Cordillera Central, and Cordillera de Talamanca, a spine of mountain ranges running from the north to the south of the country, form the boundary between the Pacific and Caribbean slopes. In these mountain highlands, where the elevation reaches 3820 m at the highest point (Cerro Chirripó), are cloud forests (starting at 800 m above sea level), oak forests, and páramo—found on the Poás, Irazú, and Barva volcanoes and the high peaks of Cerro de la Muerte, Cerro Chirripó, and Cerro Buena Vista. Low temperatures, strong winds, and low and succulent vegetation are common in these habitats. Faunal and floral endemism is high because populations developed in isolation after glacial periods.

Caribbean slope. This bioregion is characterized by large expanses of flat evergreen lands in Costa Rica's northeast. On the southern Caribbean seaboard, the flatlands are replaced by the steep hills of the Talamanca mountains, which almost reach the coast. From north to south, this coastline is strongly affected by the Caribbean Sea, which brings inland, by the force of wind, a large amount of moisture. Part of the moisture falls in the lowlands, and the rest on mid-elevation mountain slopes. These rains are reinforced by trade winds during the time of year when the rest of the country is in dry season, giving the Caribbean slope its evergreen appearance. Some areas are constantly flooded (e.g., Barra del Colorado, Tortuguero, and Parismina), and human transportation is by boat. Instead of mangroves, flooded palm forests flourish along the northern (lowland) half of Costa Rica's Caribbean coast.

Rainy Pacific. This bioregion, in west-central and southwestern Costa Rica, is shaped by a narrow coastline followed immediately by the steep mountains of the Pacific side of Cerro de la Muerte and other peaks in the Talamanca cordillera and its secondary mountains. Here the General River flows through low-elevation valley to meet the Térraba River and its basin. In this area, the humidity blown inland from the Pacific precipitates abruptly, generating an evergreen forest. Some of Costa Rica's rainiest places are found in this bioregion, such as Corcovado National Park, on the Osa Peninsula, where it can rain up to 6000 mm per year.

Habitats and Microhabitats

The combination of climate and topography results in a very diverse palette of habitats and microhabitats, even in a country as small as Costa Rica (51,100 km^2). The main habitats of butterflies and moths are briefly described here; within these broad categories are numerous microhabitats. In many areas, different tree species flower and fruit at varying times, and since they are important sources of nectar and fruits for adult feeding, they influence butterfly population dynamics, prompting migrations as well as diapause (a period of inactivity or low metabolic activity) in different stages: egg, larva (or caterpillar), pupa (or chrysalis), and adult. A river canyon in Guanacaste, for example, becomes a microhabitat for Blomfild's Beauty (*Smyrna blomfildia*), which shelters there in reproductive diapause during the dry season for the water supply and cool temperatures and then, during the rains, disperses for reproduction.

Dry–moist forest (Fig. 4): In the northern Pacific lowlands of Guanacaste, dry–moist forest is found in areas characterized by a long dry period. Its many small rivers run dry every year. The trees are an average of 30 m tall, and almost all vegetative species lose

Fig. 4. Dry-moist forest in Sardinal, Guanacaste, during rainy season.

Fig. 5. Savanna in Guanacaste during rainy season.

their leaves during the dry period (Bolaños et al. 2005). Butterflies and moths in this habitat undergo diapause during the dry season (e.g., *Protographium epidaus*, *Abaeis albula*,
Rothschildia lebeau, *Caio championi*, and others) or migrate to higher elevations where
evergreen forest persists (*Phoebis argante*, *Manataria maculata*, *Aellopos titan*) (Janzen
1983, DeVries 1987, Murillo-Hiller and Nishida 2003). When the rains begin, all plants
produce leaves that are a valuable food source for the larvae, and many herbaceous

plants bloom with nectar-rich flowers; at this point, Guanacaste dry–moist forest shows its mega-diversity of butterflies and moths in very high abundance. Some of the common species include the moths *Caio championi*, *Xylophanes tersa*, and *Copaxa moinieri*; and the butterflies *Caligo telamonius menus*, *Opsiphanes fabricii*, *Smyrna blomfildia*, *Protographium epidaus*, and many others. Extensive cattle ranching in this area transformed most of the landscape to savanna more than 100 years ago (Fig. 5). These savannas are covered mainly by grasses and low herbaceous plants, which provide flowers and places for butterflies to lay eggs. Species such as *Junonia genoveva*, *Anartia jatrophae*, and *Euptoieta hegesia* are common residents of the savannas. While the cattle ranches are artificial savannas, areas in the low Pacific slopes of Rincón de la Vieja, Orosí, Tenorio, and Miravalles volcanoes are covered by natural savannas (Jiménez 2016), the species of which disperse to cattle pastures.

Rain- and wet forest (Fig. 6): The most widely represented habitat in Costa Rica, rain- and wet forest covers all the Caribbean flatlands from sea level up to around 750 m, mostly in the provinces of Alajuela, Heredia, and Limón; and occurs on the Pacific side in a small area north of the Guanacaste volcanoes and in a narrow band from the mouth of the Tárcoles River in Puntarenas all the way south to the Osa Peninsula. This habitat is characterized by heavy rains distributed year-round, with annual totals from 1500 to 7000 mm (Bolaños et al. 2005). The forest height generally reaches 50 m, but some canopy-emergent trees reach up to 60 m. This habitat is always wet, and the rivers never run dry. The forest understory is very dark and dominated by short palms and many species of *Heliconia* and similar plants (Fig. 7). Common butterflies include *Eurybia lycisca*, *Caligo atreus*, *Cithaerias pireta*, and *Mechanitis polymnia*; and the moths include *Manduca albiplaga*, *Megalopyge lanata*, and the day-flying *Urania fulgens*.

Most of the butterfly and moth diversity of this habitat is found in the forest canopy, where the sun's light and heat are more constant. There, tree flowers are an important

Fig. 6. Rain- and wet forest of the Caribbean flatlands.

Fig. 7. Understory in Rincón de la Vieja National Park (600 m).

source of food, as are fruits and sap. Very fast and powerful fliers live in the canopy. Males establish territories around large tree branches and chase each other. Among the common butterfly species are *Marpesia petreus*, *Catonephele numilia*, *Archaeoprepona amphimachus*, and *Memphis oenomais*; and among the moths are *Pachylia ficus*, *Adeloneivaia centrojason*, *Syssphinx molina*, *Titaea tamerlan*, and *Didugua argentilinea*.

Similar conditions occur around rivers and roads where the sunlight penetrates all the way to the ground. Some canopy butterflies may fly at lower altitudes here, such as *Phoebis philea* and *Adelpha cocala*. Other species fly in river canyons and over roads because their males drink dissolved nutrients from the mud and sand, such as the moth *Urania fulgens* and the butterflies *Doxocopa laurentia*, *Heraclides thoas*, and *Abaeis xantochlora*. Males of some species, such as *Morpho helenor*, *Morpho menelaus*, and *Adelpha cocala*, establish territories around rivers and roads. One remarkable and temporal forest microhabitat is the light gap created by a landslide or a fallen tree. These gaps in the canopy bring direct sunshine into the otherwise dark forest understory. *Consul fabius*, *Myscelia cyaniris*, *Heliconius* and *Parides* butterfly species, and day-flying moths such as *Telchin atymnius* set up territories in light gaps, and many eat from the flowers of herbaceous plants that grow there.

Pre-montane and montane rain- and wet forest (Fig. 8): These wet mountain forests of tropical areas (popularly, if inaccurately, called "cloud forests") occur in a range of different elevations, from 750 to 2300 m, and receive between 3000 and 8000 mm of precipitation per year. They are characterized by trees from 10 to 35 m tall, a high abundance of perennial epiphytes, and persistent clouds most of the time. Butterfly activity is generally restricted to a few hours a day (at most) in this habitat, but the diversity is very high. Typical butterflies of these forests are *Pterourus garamas*, *Memphis proserpina*, *Cyllopsis hedemanni*, *Oxeoschistus tauropolis*, and many Ithomiini and *Catasticta* species; and the moths include *Neopreptos marathusa*, *Ardonea morio*, *Copaxa rufinans*, and *Amastus aconia*.

Fig. 8. Pre-montane wet forest of Braulio Carrillo National Park.

Fig. 9. Páramo. AC

Páramo (Fig. 9): In Costa Rica, this habitat occurs only in the highest peaks of Cerro Chirripó, Cerro de la Muerte, and Cerro Kamuk, in the Talamanca range. The elevation is about 3500 m, and the average year-round temperature is 5 °C. Instead of forests, thickets of *Chusquea* bamboo dominate, and low evergreen vegetation grows between rocks and swampy ground. Butterflies and moths of the páramo have not been well studied, and only isolated data from collections are available. Both moths (*Aellopos titan*, *Amastus aconia*, and *Neopreptos marathusa*) and butterflies (*Catasticta teutila*, *C. cerberus*, and *Calephelis schausi*) are found here.

Mangroves (Fig. 10): This coastal habitat is at sea level and always has high temperatures. Mangrove habitat is composed of a small number of different tree species adapted to salt water and flooding. There are only a few records of butterfly or moth species using mangrove trees as host plants. In Costa Rica, butterflies and moths reported to feed on the mangrove *Avicennia germinans* include *Lygropia erythrobathrum* (Janzen and Hallwachs 2009, Villalobos 2015) and *Junonia* species. Other lepidopterans seen in mangrove habitats are related to plants that grow in the periphery or are adults attracted by decomposing organic matter.

Cropland. Agricultural land is usually poor in diversity because of the use of pesticides and herbicides. Nonetheless, lepidopteran species that use a specific crop as a host plant will very likely be found there. Moths tend to be more generalist than butterflies in choosing host plants. For instance, *Rothschildia lebeau* can be found on jocote (mombin) and peach; and *Phobetron hipparchia* on avocado, cocoa, coffee, mango, oil palm, and others (Coto and Saunders 2004). Butterflies, however, can be extremely specific about host plants, as in the cases of *Eumaeus godarti*, found on cycads; and *Leptophobia aripa*, on arugula and cabbage.

Parks and gardens. If the right plant species are planted, parks and gardens can support considerable diversity of butterflies and moths. The species composition will vary with the locality, but if there is sunshine, butterflies will appear to eat and/or lay eggs. As long as milkweed is present, the Monarch (*Danaus plexippus*) will undergo its life cycle. The same is true of *Opsiphanes fabricii* and palms, and *Dione juno* and passionflowers. Moths are seen less often because of their night-flying behavior, but their presence can be assessed by finding larvae on plants. *Xylophanes tersa* larvae feed on *Pentas*, and *Dicentria rustica* on roses.

Early successional habitat and thickets (*charral*). In Costa Rica, the word *charral* refers to unoccupied land on which natural vegetation has grown uncontrolled but no

Fig. 10. Mangroves, Bejuco Beach, central Pacific coast.

trees are present. Generally, *charral* is 1–2 m tall and composed of grasses, shrubs, bushes, or thickets of different species. It is characterized by very high sunlight levels, since no trees grow very tall. Butterfly and moth diversity and abundance tend to be high because of the predominance of nectar-rich flowering plants such as *Lantana* species; *Stachytarpheta* species; various Asteraceae species, including *Bidens pilosa*; and others. Among the common butterflies of *charral* are *Danaus plexippus*, *Anartia jatrophae*, *Urbanus proteus*, and *Eurema*, *Abaeis*, and *Phoebis* species. The moths *Melanchroia chephise*, *Aellopos titan*, and *Spodoptera dolichos* are regular inhabitants of these habitats.

BIOLOGY OF BUTTERFLIES AND MOTHS

Butterflies and moths are broadly categorized into three groups based on the configuration of the female reproductive system; basal moths comprise the groups Monotrysia and Exoporia, and the higher moths and butterflies comprise Ditrysia. In Monotrysia, the female has a single genital opening used for both copulation and oviposition. In Exoporia, the female has two pores, one for copulation and the other for oviposition. These basal groups represent less than 5% of Lepidoptera species and are not treated in this book. The remaining 95% of Lepidoptera species are placed in Ditrysia, in which the female has two pores, one for copulation and the other for oviposition. All the species covered in this field guide are in the group Ditrysia.

Day-flying butterflies represent around 12% of the order Lepidoptera; the remaining 88% are moths and micro-moths. Moths first appeared on earth around 195 million years ago (Wolfe et al. 2016) and rapidly diversified, developing a wide variety of morphological adaptations, from mouthparts enabling them to chew pollen and homoneurous wing venation (with the system of veins on the forewings and hind wings alike) to highly specialized syndromes such as parasitism (Epipyropidae) and sphingophyly (the pollination of flowers by hawk moths, family Sphingidae). The modern moths that evolved from earlier ancestors are treated in this book as the "Lower Ditrysia Moths." True butterflies (superfamily Papilionoidea) emerged around 100 million years ago from a highly derived group of moths (Heikkilä et al. 2011) that looked similar to the modern moths in the superfamily Calliduloidea. Another group of moths also evolved from that advanced group of moths, treated in this guide as the "Higher Ditrysia Moths."

Life Cycle

All butterflies and moths go through four life stages (egg, larva, pupa, and adult) (Fig. 11), and their wings develop inside the body at immature stages. This type of life cycle is known as complete metamorphosis; post-embryonic development (i.e., of the larva or caterpillar) is divided into stages called "instars" (Borror et al. 1992).

Egg. The egg develops in the abdomen of the female lepidopteran as a single cell, with its cytoplasm surrounded by a thin vitelline (egg-yolk-like) membrane and externally by a durable shell called the chorion. In the top, the egg has one or more small pores, called micropyles, through which the sperm enters and fertilizes the egg. Once the

Fig. 11. Stages in the life cycle of a butterfly (*Adelpha leucopthalma*, Nymphalidae): (a) egg, (b) larva, (c) pupa, (d) adult.

egg is laid, gas interchange occurs through the micropyle. Unlike all other insects, lepidopteran females are, with few exceptions, heterogametic, and the males are homogametic; this means the female ovum determines the sex of the offspring (Borror et al. 1992).

Copulation is a slow process that can take from an hour to a couple of days. The male attaches to the female, with the help of special structures that are part of the male genitalia, and through an intromittent organ (the aedeagus; see fig. 17) transfers a little sperm package (the spermatophore) into a pouch in the female body called the bursa copulatrix (Vane-Wright 2015). There, the sperm will be viable for the lifetime of the female and will be used to fertilize each egg the butterfly lays. Generally, a single copulation provides all the sperm the female needs; in some genera, such as *Parides*, the male will deposit a structure called a sphragis, which plugs the female's genital pore, preventing other males from mating with her (DeVries 1987). The sphragis can dissolve with time, and the female can then mate again. The spermatophore is often accompanied by packets of fats, carbohydrates, or protective alkaloids, which act as incentive for the female to mate again.

The mature egg passes through a duct and receives a small amount of sperm, therefore becoming fertilized, and then is laid, generally on a host plant—the plant the caterpillar feeds on (Vane-Wright 2015). Oviposition, or egg laying, takes place from 2–3 days to weeks after copulation. Generally, a female will actively search for the appropriate host plant, first by using sensory receptors on the antennae and then by touching every plant leaf with its legs, antennae, and abdomen. In most butterflies,

Fig. 12. Four eggs laid by the butterfly *Hamadryas fornax* (Nymphalidae).

the legs are olfactory organs. The female's forelegs are specialized to recognize volatile compounds on the leaf's cuticle, ensuring she has chosen the correct host plant. She uses the legs, which possess a very sensitive chemoreceptor equipped with sharp spines, to scratch the leaves to release their volatile compounds; the butterfly appears to "drum" very fast before laying an egg–behavior that can be easily observed and sometimes even heard by humans.

Moth and butterfly eggs (Fig. 12) may be laid in huge clusters of hundreds of eggs or singly; they are deposited on the upper side or underside of leaves, on tree trunks or branches, or on flowers, on the host plant or on plants surrounding the host plant. Some generalist feeders drop the eggs on the ground, and the young larvae must find a plant to eat, while the Giant Bagworm (*Oiketicus kirbyi*) lays its eggs without leaving its cocoon, and when hatched the larvae float away on a silk string to feed on any plant they land on. Lepidopteran eggs can range in size from less than 0.1 mm long, as in many micro-moths, such as *Phereoeca allutella*, to 2.0 mm long in the cases of the huge saturniid moths, such as *Rothschildia triloba* and *Adeloneivaia centrojason*. They may be circular, bottle-like, or disk- or barrel-shaped, with a smooth chorion or with a crest or ribs, and come in a spectrum of colors.

Some eggs hatch within a few days of being laid, such as those of *Heliconius* species, which last around 4 days under warm weather conditions; but others, such as some *Automeris* moths, can take up to 1 month. To hatch, the young larvae must break through the egg chorion; some have a head spine for doing so, while others use their mandibles.

Larva. The larva is basically an eating stage. Its work is to eat as much as possible to obtain the nutrients to transform into an adult. Butterfly and moth larvae feed mainly on plant tissue, mostly on leaves. The plant (or plants) a species feeds on as a larva is known as its host plant(s). Some species are monophagous, which means the larva can eat from only a single species of plant; others are oligophagous and can eat from a small group of related plant species, or from unrelated plants that contain a similar chemical composition. Finally, some moths are polyphagous, able to feed from a wide variety of plant species of many families.

In many butterfly and moth species, the coevolution with its host plants has molded its behavior, shape, and colors. Many micro-moth larvae, called leaf miners, are so small they eat the tissue inside the host plant's leaves; larvae of the day-flying sun-moths of the family Castniidae bore inside their host plant's roots or stalks of herbaceous plants. Still, most butterfly and moth larvae live exposed among the host plant's foliage. Larval development can take from 2 weeks, as in the small white butterfly *Abaeis albula*, to 1 year, as in the Costa Rican White Morpho (*Morpho polyphemus catalina*).

The larva, like all insects, has a tough exoskeleton. This thick skin provides protection against desiccation and strikes from predators but is inconvenient at the time of growth. Although in the larval stage the exoskeleton is quite flexible and elastic, there is a limit beyond which it cannot stretch any more, and the larva must molt. Butterfly larvae generally pass through four molts, and the fifth time they molt into a chrysalis, or pupa. In contrast, some moth larvae can go through ten or more molts, and if there is not enough food, they can raise the number of molts until they have obtained the nutrients necessary to become adults.

Before each molt from one instar to the next, the larva spins a silk bed on a branch or leaf, then holds onto this surface. After about 24 hours, the new and larger head capsule pushes forward, expelling the old head capsule, which detaches, releasing the new head; the new prolegs then walk forward, emerging from the old skin and leaving it behind.

Pupa. When the time comes for pupation, the butterfly larva spins a little silk pad underneath a leaf, stem, branch, rock, fence, or roof overhang, and then holds fast with the posterior prolegs (A_{10}; see "Morphology," p. 18). Close to 24 hours later, it releases the rest of the prolegs, which leaves it in a hanging position. It stays like that for around 24 hours more, and then the dorsal part of the first two thorax segments (T_1 and T_2) tears apart, and the pupa starts emerging. With strong circular movements, the head capsule breaks through, and the larval skin is pushed aside, revealing the pupal skin. Finally, the pupa's most posterior part, a sticky, hooked structure called the cremaster, breaks free of the larval skin and in less than 10 seconds must reach the silk pad to attach the pupa; until that moment, the pupa is held by only a few millimeters of old, fragile larval skin. If the cremaster does not reach the silk, the pupa falls and dies.

The process is safer for moths; most of them build a cocoon by putting together leaves, little plant stalks, dirt, or any other available material with silk. Safe inside the cocoon, the molting process takes place.

The lepidopteran pupa (Fig. 13) is held within an enclosure with all the appendages glued to the body forming a single unit. But if you look carefully, it is possible to distinguish the adult butterfly's eyes, antennae, proboscis, legs, wings, and body segments. The pupa is a relatively inactive stage but can produce vigorous movement in the abdominal segments. Despite its idle appearance, a major structural reorganization is occurring inside.

The pupal stage is when metamorphosis is clearest; the abrupt change from caterpillar to moth or butterfly takes place when undifferentiated cell aggregations called imaginal buds, which were inactive during the larval stage, suddenly start dividing. The cell parts to build the new structures are obtained from a "cell soup" issuing from the caterpillar's cells as they break down. Some of the caterpillar's structures, such as the gut, are kept but modified for the adult stage.

Fig. 13. Pupa of the moth *Adeloneivaia centrojason* (Saturniidae).

Because of their limited ability to move, pupae and cocoons depend on camouflage to survive. Generally, they are very well hidden and hard to find among foliage or organic matter on the ground. The pupae of some butterflies are positioned with the head up and held by a silk girdle around the thorax; to emerge, the butterfly uses its legs to pull itself out. Others merely hang, and the butterfly emerges by gravity.

The length of time before a butterfly or moth emerges from the pupa is variable. Some pupal stages last just a couple of weeks, as with the glasswing butterflies (Ithomiini); while others can remain in pupal diapause for up to 4 years, as in the case of the kite swallowtail butterflies (Leptocircini). In general terms, a moth pupa lasts an average of 1–2 months, and a butterfly pupa an average of 3 weeks.

Adult. Butterflies emerge from the pupa during daytime, moths at nighttime. After emerging, all new adults must spread their wings, a process that takes around 15 minutes, depending on the size of the insect. To expand the wings, they pump hemolymph (circulatory fluid) through the wing veins, with the help of gravity—therefore, they need to find a place to hang. After that process, the insect must wait around 4–5 hours to become ready to take flight; in the first day, it will avoid flying unless absolutely necessary.

Adult females are ready to mate from the first day after emerging from the pupa, but adult males generally need 2–3 days. In some cases, as with *Heliconius* butterflies, males will seek female pupae that are almost ready to emerge, even some hours before they hatch, in order to copulate with the new adult female (Vane-Wright 2015).

Morphology: Parts and Structures

The following text and illustrations present the terminology used to describe the parts and structures of butterflies, moths, and their larvae. In describing these creatures, we also use the following general terms: *anterior* (toward the front or head), *posterior* (toward the rear or tail), *dorsal* (referring to the back or upper side), *ventral* (referring to the belly or underside), *lateral* (referring to the side), *basal* (near the base; for wings, the area close to their attachment to the body), *proximal* (toward or near the center or the base), *distal* (away from the center or base, toward the tip or end).

Larva. The larva of a butterfly or moth (Fig. 14) has a prominent head, filled mainly with muscles, attached to a pair of mandibles that cut leaves when feeding. On each side of the head are six simple eyes, called stemmata, which can generate a rudimentary but well-focused image. Two small, three- or four-segmented antennae are located between each mandible and the stemmata. The rest of the larval body is segmented. The three segments right after the head are the thoracic segments (T_1, T_2, T_3); each one has a pair of true segmented legs. These legs are important for holding food while eating and to avoid falling from the plant. After the three thoracic segments are ten abdominal segments (A_1 to A_{10}). In most species, segments A_3, A_4, A_5, A_6, and A_{10} possess fleshy appendages called prolegs. Each proleg has a series of small hooks called crochets, the arrangement of which is an important tool for identifying moths at family level. The number of prolegs is variable among families. The Costa Rican moths known as *tortoloqüilos* (Megalopygidae) have seven pairs of prolegs (A_2–A_7 and A_{10}). Prolegs are completely absent in the moth family Eriocraniidae. The larvae (known as inchworms or loopers) of the family Geometridae have only two pairs of prolegs (A_6 and A_{10}) and move by arching the body with each step as the sets of prolegs meet with the true legs, giving rise to the scientific family name, which means "earth-measurer."

Respiration in larvae takes place through spiracles, small pores at the sides of the larval thorax and abdomen. The spiracles connect inside the body with a system of tracheae and tracheoles (tubes in the breathing system), branching from thicker to thinner and reaching all body tissues. Spiracles are in segments T_1 and A_1 to A_8 (Stehr 1987).

Caterpillar bodies are sometimes armed with variable structures that help in defense or camouflage. The most common structures are setae (singular: seta), which are soft or hard hairs; and pubescence, dense short hairs covering wide areas. Scoli (singular: scolus) are fleshy, branched structures that hold arrangements of setae or hardened spines, sometimes charged with irritant compounds that sting on contact. Others just have fleshy, wartlike tubercles. The arrangement of these structures in the caterpillar body are helpful for species, genus, and family identification.

Adult. All adult lepidopterans have scales covering most of the body and wings. The scales protect the insect against water, with an oily hydrophobic layer; help in thermoregulation (insulation); and are responsible for the insect's coloration (Scoble

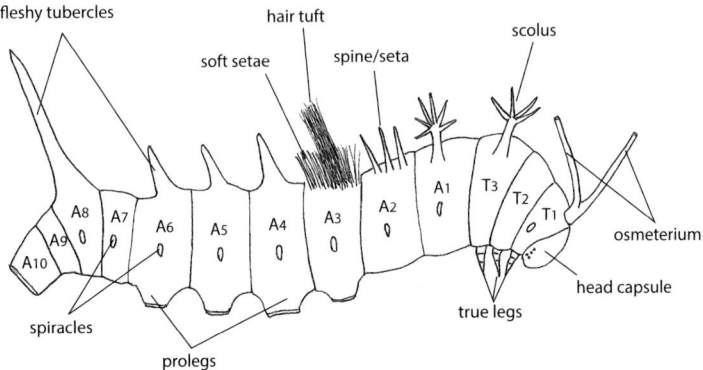

Fig. 14. The parts of a caterpillar (composite of several species).

1992). Some specialized scales on males, called androconia (or androconial scales), are loaded with pheromones and help in mate attraction and species recognition among individuals; they are found on different parts of males' wings or abdomen. Scales do not regenerate, and as a butterfly or moth loses them with time, many of their capacities are lost.

The main body segments—head, thorax, and abdomen—are quite visible in butterflies and moths. The head (Fig. 15) possesses a pair of prominent compound eyes that perceive movement and form in full color. Two ocelli occur between the compound eyes in most groups; ocelli are simple eyes unable to focus, so their function most likely has to do with orientation (up and down) and detecting light and dark. There are two antennae, which are used as sensory organs. The proboscis (feeding organ) is formed from a pair of galeas, which stick together, creating an internal duct. A pair of appendages called labial palps protect the proboscis and also are used to clean the compound eyes from pollen, fruit debris, or other matter.

Behind the head is the thorax, composed of three segments: prothorax, mesothorax, and metathorax. Each segment has a pair of walking legs. In Nymphalidae, the legs on the prothorax evolved as sensory organs that allow the female to recognize the correct host plant on which to lay its eggs; this set of legs is much smaller than the walking legs of segments T_2 and T_3. For scientists and naturalists, this set is useful in determining the sex of individuals when primary reproductive structures are not visible. In some families, such as Papilionidae and Erebidae, the tibia of the foreleg (T_1) has an epiphysis, a comblike structure used for cleaning its antennae, proboscis, and other body parts. The mesothorax holds a pair of wings called the forewings (FW), and the metathorax holds another pair, the hind wings (HW). The wings are thin exoskeletal membranes with a tubular vein system.

The main function of the wings is flight. Most moths have a wing-locking system to keep fore- and hind wings attached to each other so they behave as a single wing;

Fig. 15. Some parts of the head and thorax of a lepidopteran (*Ascalapha odorata*, Erebidae).

this locking system is composed of a rigid spine, the frenulum, in the hind wing, which is inserted into a pocket, the retinaculum, in the forewing. This structure is absent in butterflies.

Wings are also very important because of their color patterns. In order to describe the wings, we use a "map" of the venation (vein pattern), color areas, and other patterns (Fig. 16). Their colors give butterflies and moths the ability to communicate with other species and with other individuals of the same species. For instance, a bright color informs a bird that the butterfly is distasteful; a morpho butterfly male's bright blue coloration informs a female that he is healthy and ready to mate; colors also help in thermoregulation and as camouflage, allowing butterflies and moths to hide from predators (Scoble 1992).

Behind the thorax is the abdomen, composed of ten segments (as in larvae). Its main functions are respiration, reproduction, digestion, and excretion. Copulation

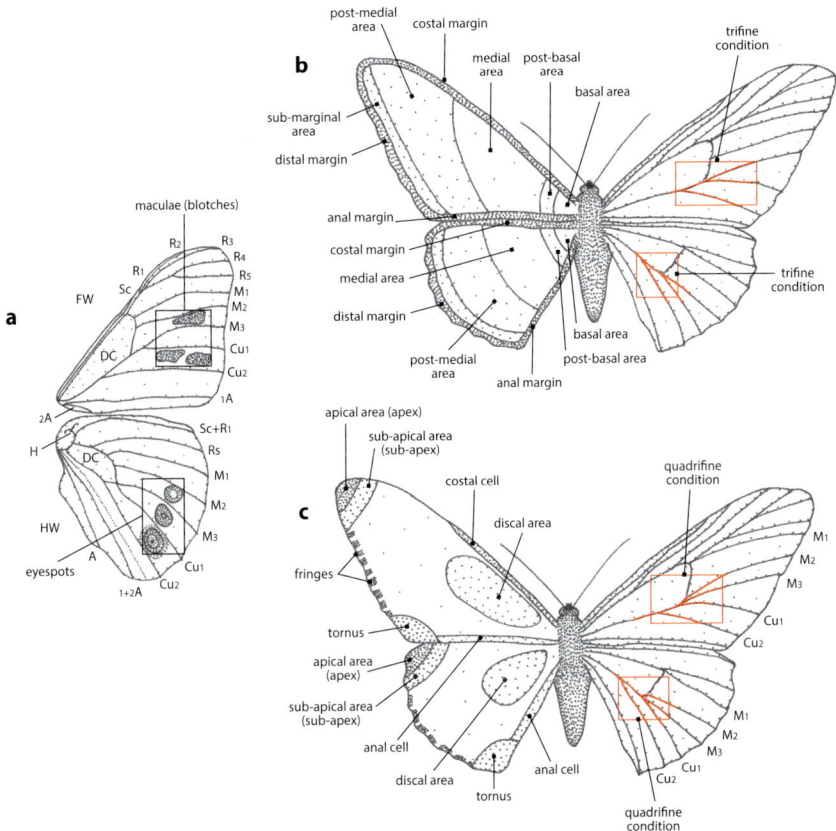

Fig. 16. Wings, showing venation, areas of coloration, and other patterns. (a) Representative schematic of the wing veins and two pattern types. FW: forewing, HW: hind wing, A: anal veins, Cu: cubital veins, DC: discal cell, H: humeral vein, M: medial veins, R: radial veins, Sc: sub-costal vein. (b) Basic wing pattern and structure terminology (left); trifine venation condition (right). (c) Additional wing pattern and structure terminology (left); quadrifine venation condition (right).

takes place when the male genitalia couple with the female genitalia (Figs. 17 and 18). As many species of butterflies and moths are very similar in appearance, male genitalia play a role like a key in a lock: the complex of spines, hairs, and form fits only with the genitalia of females of the same species. Since males must stimulate the female to finish the copulatory process, male genitalia is much more complex than female genitalia, which causes faster evolution of its characters. Therefore, male genitalia, and to a lesser degree female genitalia, are an important tool in Lepidoptera taxonomy.

The sex of butterflies and moths can be determined by the external reproductive structures at the tip of the abdomen. These are, in the case of males, a pair of ventral claspers, or valvae, and in the case of females, a complex organ housing the anus, the egg pore, and the copulatory pore (DeVries 1987) (Fig. 18). The reproductive structures may be difficult to distinguish, but in general terms, the male abdomen is longer and thinner, and the female abdomen is shorter and thicker.

Reproductive Behavior

Generally, lepidopteran courtship takes place in specific places previously selected by males. Many butterflies are very territorial, and males establish their territories in places like forest light gaps or the canopy. A male selects an emergent leaf and rests there, chasing and expelling any other male that gets too close and then returning to its perch. Sometimes males use the same perch for several days until being bitten by another male or dying, behavior seen in *Adelpha* and *Archaeoprepona*. Others, including *Morpho*, *Heraclides*, and *Siproeta*, select open areas with a lot of light and fly in specific circuits during the morning. These butterflies will have encounters and aerial dogfights with other males, and the one that wins gets to keep the territory. Some other butterflies, such as species of *Greta*, establish leks (DeVries 1978), or congregations of males in specific places inside the forest understory. There, in groups, the males display their colors and release pheromones, allowing females to find them and then choose a mate from among them.

For moths, most courtship is mediated by pheromones, since colors and behavior are not as visible at night. For example, the female of *Rothschildia lebeau*, on the same night she emerges from her cocoon, releases pheromones that are perceived by males; practically the first male that finds her, mates with her. Pheromones are also important in butterflies, but they act in shorter distances, since at first sight, females respond to males' visual cues.

As in other animals, many butterfly species present extreme sexual dimorphism, with male coloration very different from female coloration. *Heraclides torquatus* and *Morpho cypris* are good examples of sexually dimorphic species (Fig. 19), but the sexual difference is driven by different forces; in the first, a highly colorful male attracts females' attention, while in the second, the female is protected by resembling another, inedible butterfly (*Parides*). Some sexual dimorphism is not visible to the human eye; butterflies have a light wavelength perception capacity similar to that of humans but also can perceive ultraviolet colors. Therefore, ultraviolet hues are part of butterflies' color patterns and have an important role in male–male interactions and male–female courtship and recognition. For instance, butterflies of a Pieridae subfamily (Coliadinae) appear to the human eye to have bright pigmentary yellow colors and show little sexual dimorphism, but in ultraviolet light, sexual dimorphism becomes visible (Fig. 20).

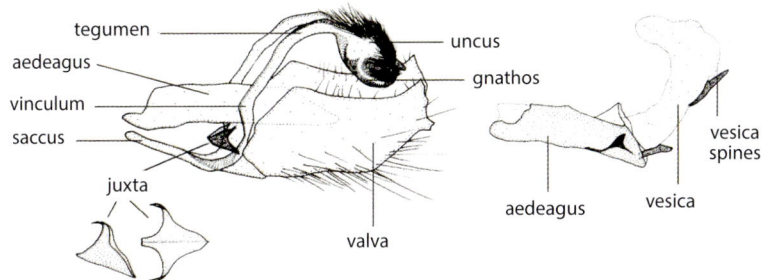

Fig. 17. Male genitalia of a skipper (*Buzyges rolla*, Hesperiidae). Murillo-Hiller et al. 2019.

a, ♂
b, ♀
c, ♂
d, ♀

Fig. 18. Primary sexual structures of *Parides iphidamas* (Papilionidae): a, b. Foreleg morphology (ventral view). The secondary sexual character, of *Prepona laertes* (Nymphalidae): c, d.

Fig. 19. Sexual dimorphism in *Heraclides torquatus* (Papilionidae): a, b; and *Morpho cypris* (Nymphalidae): c, d.

Fig. 20. Sexual dimorphism in selected butterflies (Pieridae) under normal human vision (yellow images) and under ultraviolet vision (purple images). *Aphrissa statira* (a, b, c, d); *Zerene cesonia* (e, f, g, h); *Phoebis philea* (i, j, k, l). JL & LRHM

Defense

Butterflies and moths are primary consumers, eating directly from plants. Their larvae can be thought of as plant predators. Plants have developed a highly specialized defensive tool kit against caterpillars as part of a coevolution process among plants, butterflies and moths, and the predators of butterflies and moths.

In each stage of development, lepidopterans need different defensive strategies since they live in very different ways and places. Predation, by birds and other animals, pushes butterflies and moths to evolve defense mechanisms, the best known of which is camouflage, or crypsis. In a broad sense, camouflage is in effect when an organism tries to avoid being recognized in the place where it usually is. With crypsis, form and pattern fit the resting place (Cott 1940). But this defensive strategy is much more complex than blending into the background. The type of camouflage clothing commonly worn by soldiers employs a strategy known as disruptive coloration. It works by making an enemy believe that what is seen is something different from what it really is, and this prevents, or at least delays, recognition.

There are different kinds of disruptive coloration (Fig. 21). A disruptive marginal pattern uses lines or waves to obscure the borders of different individuals, so that each one is not seen as an individual. With differential blending, light or dark coloration breaks up the animal's shape, depending on where it is resting. Maximum disruptive contrast uses strongly contrasting colors to suggest a different form, contradicting the animal's true form. In coincident disruptive patterns, the coloration breaks up the continuous surface of the body. Most pupae are protected only by crypsis, although some can violently shake and produce noise to startle predators.

Crypsis is probably the most common defense mechanism among Lepidoptera, but another system, aposematism, is also well represented. Since many plants have developed chemical compounds meant to keep larvae from feeding on their leaves, natural selection has favored the survival of those larvae more tolerant to them; in the long term, these larvae, through bioaccumulation, gain the benefits of those toxic compounds. The coevolution of a lepidopteran species with its host that allows the larva to efficiently sequester its host plant's secondary compounds, or metabolites, brings it the benefit of becoming unpalatable itself. At the same time, colors have become a great tool for communicating a species' bad taste to predators, and the result is very bright, or aposematic, coloration in many butterflies and moths—difficult for a predator to forget after it undergoes a bad gastronomic experience. Because predators generally do not teach their offspring what to eat and what not to eat, each predator will have to learn on its own, so at least one insect is going to need to die in order to save others. Aposematic butterfly and moth species generally have a slow and floppy flight, making it is easier to recognize them. Cryptic species generally are fast fliers, to make predators lose track of them as they seek to hide. Those fast fliers even have their imitators; in what is called "escaping mimicry," a palatable species resembles a butterfly that predators avoid not for being unpalatable but for being so hard to catch it is not worth the effort.

In some cases, palatable species have evolved bright colors that resemble those of unpalatable aposematic species, gaining automatic protection from predators. This defensive strategy is called Batesian mimicry. The scheme works as long as the unpalatable species is more common than the palatable one, so the probability of a bad experience is higher. Müllerian mimicry, in contrast, is when an unpalatable aposematic

Fig. 21. Disruptive coloration and crypsis in butterflies and moths: (a) disruptive marginal pattern, *Heliconius doris* larvae (Nymphalidae); (b) differential blending, *Lirimiris truncata* (Notodontidae); (c) maximum disruptive contrast, *Oressinoma typhla* (Nymphalidae); (d) coincident disruptive pattern, *Arawacus togarna* (Lycaenidae); (e) crypsis, *Zaretys ellops* (Nymphalidae).

species is imitated by a second and a third and a fourth (and more) unpalatable species. With their similar color pattern, all the species become protected by a single bad predator experience. If two unpalatable species look alike, the probability of an individual of a species being attacked by a predator is reduced to 50% for that species; if there are three, 33%, and so on. That is why it is not unusual to find many species that are very similar in a given locality.

Eyespots are deflective marks that help protect some butterflies and moths from predators. They may deflect the attack of an enemy to a less important part of the body, allowing the lepidopteran to survive. In some species, eyespots can be exposed suddenly and unexpectedly, frightening or shocking a predator. The first type of deflective eyespots can be seen in the morpho and owl butterflies, while the second type is represented by the moth *Automeris banus* (Fig. 22).

Eyespots play an important role in caterpillar protection in some species, such as *Xylophanes tersa*. Additionally, larvae of many butterflies of the families Riodinidae and Lycaenidae are protected by establishing mutualistic relationships with ants;

the larvae offer a nutritive sugar solution that is expelled by special eversible tentacle-like organs on its body, and ants, in exchange, will protect the larvae against predators and parasitoids. In Riodinidae, if help is needed, the larvae employ a stridulatory noise-production system, located between its thorax and head, to call the ants. Furthermore, many caterpillars have venomous, stinging hairs on their bodies. Caterpillar envenomation is known as "erucism"; in Costa Rica, where unfortunate encounters with *Megalopyge* caterpillars often end with visits to the hospital, the word commonly used is *ortigar*.

Other strategies are found in the family Papilionidae (swallowtails); when alarmed, larvae employ the osmeterium, an eversible gland located in the dorsal T_1 segment that releases a bad odor cause by isobutyric acid, which expels many invertebrate predators. Swallowtail larvae of the genus *Pterourus* present a marking on the thorax that resembles the face of a lizard or a snake; the everted osmeterium is forked, enhancing the impression of a reptilian head (Fig. 23). A different eversible gland, found in *Morpho* and *Caligo* butterflies on the ventral surface of the T_1 segment, releases predator-deterring volatile odors when the larva is alarmed (DeVries 1987).

Fig. 22. Cryptic (left) and threatening (right) postures of *Automeris banus* (Saturniidae).

Fig. 23. Intimidation posture, with osmeterium gland everted (bright red), of fifth-instar larva (dark form) of *Pterourus garamas syedra* (Papilionidae).

BUTTERFLY AND MOTH CONSERVATION

In general terms, butterflies and moths are resilient organisms; this is attributable to features of their biology, such as generations composed of thousands, if not millions, of individuals and fast generation replacement. In ecology, they are known as *r*-strategist species, which means a female will invest a lot of energy in producing a large number of eggs, but none in taking care of them other than choosing the safest place to abandon them. In *r*-strategist species, it is expected that most of the individual offspring will die before reaching the reproductive stage.

In order to maintain a healthy population, lepidopterans need two or three things: enough availability of the larval host plant, a big supply of nectar flowers (or rotting fruits) for the adults, and the habitat they evolved in to enable their defense mechanisms to be efficient against their predators. Among these three things, the last is the most complicated to ensure. Some species are more resilient that others; for instance, a species that evolved on an open savanna will probably also adapt well to a cropland or a city park. But a species that evolved in the forest understory is likely to perish under the same conditions.

Because of their dependence on very specific host plant species and habitats, butterflies and moths have gained rapid popularity as bioindicators of habitat conservation (Scoble 1992, Koh and Sodhi 2004). A forest habitat might look unchanged in form and characteristics, but butterfly and moth species could be locally extinct as a consequence of altered plant species composition by a natural succession process or by human exploitation. For instance, at least fourteen Lepidoptera species have become extinct in Barro Colorado Island (Panama) in the past 50 years, and the most probable reason is the loss of their host plant; reintroducing the plants could spur the recovery of local extinct populations (Basset et al. 2015). Habitat degradation, fragmentation, and climate change are also threats to insect populations, and unfortunately not all habitats can be restored. There are some sad cases of completely extinct butterflies, such as *Glaucopsyche xerces* from San Francisco, California, and *Salamis augustina vinsoni* from Mauritius island. There are, in many places, butterfly and moth species facing the risk of extinction, but because of a lack of solid studies, they are not protected. The future of butterflies and moths is subject to many trends, and if conditions continue in their current state around the globe, most butterflies and moths will become extinct in the next 100 years (Vane-Wright 2015).

To prevent butterfly and moth extinctions, we should also protect every native plant, regardless of our understanding of its function in the ecosystem, and accelerate lepidopteran life-history studies (Basset et al. 2015). Eradicating introduced plants is also a major priority, since they compete with, and even replace, native plants. These measures must go together with protecting existing natural vegetation patches—no matter whether they are early successional or mature forest, they are all important to a number of butterfly or moth species. Reconnecting isolated patches of habitat by planting key species is an important strategy. In Costa Rica, plants such as *Lantana camara*, *Stachytarpheta mutabilis*, *Impatiens walleriana*, *Ageratum conyzoides*, *Asclepias curassavica*, and *Aristolochia anguicida* are needed to recover natural butterfly populations (Murillo-Hiller 2016).

HOW TO SEE BUTTERFLIES AND MOTHS

A rural area or a well-conserved habitat such as a national park is an ideal place to look for Costa Rican butterflies. Sunshine, as bright as can be, is necessary. Excellent places to observe high butterfly diversity include a midsize river course, an open area in a forest, a beach along the coast, and a dirt road through a forest. Places with flowers, whether parks, hotel gardens, or natural growth in rural areas, are excellent places. Binoculars are a helpful tool for detailed observation of butterflies that are far away.

To observe moths, on the other hand, seek lights at night. Pay attention to the phase of the moon: the new moon, and the four days before and after it, are the best nights. Set up your own light, shining it on a white sheet, in the middle of the forest in an open area; seek the moths on the walls beneath outdoor lights in rural areas at hotels or cabins; or look beneath streetlamps along roads—as long as they are white lights (yellow lights are not very attractive for moths). Do not forget to bring a good flashlight with you.

ABOUT THIS BOOK

Butterflies and moths are an important part of our environment, not only because of their role in the ecosystem but also because of their beauty and charm. The author hopes that making more information available will increase awareness and encourage people to become more familiar with these enchanting insects.

The species accounts of butterflies are presented by family. Those of the moths are divided into two groups: Lower Ditrysia Moths, before the butterfly accounts; and Higher Ditrysia Moths, after the butterflies. This sequence represents the order in which butterflies evolved from lower moths, and then a second group of higher moths evolved from a common ancestor with butterflies.

Species Accounts. The treatment for each species begins with scientific name, family and subfamily, and English common name (if the species has one). The forewing length (**FWL**) is the measurement in millimeters (mm) of the forewing from its base at the thorax to its apex (wing tip). **Description** provides identifying features of the adult butterfly or moth, including color and pattern of dorsal (upper-wing) and ventral (underwing) surfaces of forewings and hind wings. The abbreviations used for the wing surfaces are: **DFW** (dorsal forewing), **DHW** (dorsal hind wing), **VFW** (ventral forewing), and **VHW** (ventral hind wing). These descriptions should be used in concert with the explanatory text and illustrations in the "Morphology" section of the Introduction. **Similar species** gives clues to help distinguish the subject species from similar moths or butterflies. **Habitat** describes the places where the species may be seen in Costa Rica, and is followed by the species' complete (worldwide) **Distribution**. **Seasonality** gives the time of year when the species is active as an adult butterfly or moth. Dry season is December to April, and rainy season is May to November. **Natural history** highlights information on the behavior and diet of the adult moth or butterfly, its flight, territorial, and defensive actions, and other information about its life history. Some species accounts provide data from LAREBUB, the laboratory of butterfly breeding research at the University of Costa Rica, which in some cases, has provided new information about Neotropical species. **Early stages** describes the eggs and where they are laid; the larvae through the various instars; and pupae and their placement. **Host plants** lists the plants on which the caterpillars feed, providing each plant's species or genus name as well as its family name. An appendix provides a complete list of the plants treated in this book with common names in English and Spanish.

Distribution maps. A map showing the butterfly or moth's potential distribution in Costa Rica is included with each species description. The yellow shading indicates the species' range within Costa Rica. Beneath the map is the range of elevation (with "0" indicating sea level) in meters (m) at which the species occurs within the habitat specified. The maps have been constructed from multiple sources, including historical collection records, recent publications, and reported range extensions. Given that many species are geographically isolated, knowing which species is expected in different parts of the country can be key to making an accurate identification. That said, almost all range maps are based on a certain amount of guesswork—and ranges change over time for any given species—so expect an occasional surprise.

Photographs. Each species account includes color photographs of the butterfly or moth species; these may include larvae, pupae, and/or male and female adults. Photographs show each species in nature, and many show additional images of mounted specimens.

Groupings. The species accounts section is divided into the following eight groupings of butterflies and moths. The species counts given for each grouping indicate the numbers of species selected for this book.

- Lower Ditrysia: 9 species of moths. The ancestors of this group gave rise to the modern butterflies.
- Hesperiidae: 11 species of butterflies (skippers). Many species of this family look like moths but generally are day fliers (some nocturnal).
- Papilionidae: 9 species of butterflies. These are true butterflies in the strict sense: they fly only during the day and are all colorful and small-bodied.
- Pieridae: 15 species of butterflies. The word *butterfly* probably originated with this group, since many of its species, known as yellows and whites, are the color of butter.
- Lycaenidae: 7 species of butterflies. Although this group of small blue butterflies is very diverse, most of the species are very similar in size and shape.
- Riodinidae: 10 species of butterflies. The vast majority of the species in this family are native to the Neotropics. Known as metalmarks, they are small and very colorful.
- Nymphalidae: 50 species of butterflies. This family contains most of the better-known butterflies, including famous species such as the Monarch and the morphos. Many are big, colorful, common, and widespread.
- Higher Ditrysia: 61 species of moths. This group of moths originated from the same common ancestor as the butterflies.

Species Accounts

Lower Ditrysia Moths

The Lower Ditrysia Moths group comprises many lineages of moths that adapted to live in different environments. The moths in this group are very variable in size, shape, and morphology, but many are microlepidopterans. These tiny, drably colored moths are poorly known and understood; most of the information available is about species that are home or crop pests. Only two species represented in this book are microlepidopterans: *Phereoeca allutella* and *Plutella xylostella*. Other families represented in this section are small to medium-size insects and therefore are more often seen by people. Many are better known as caterpillars, because of striking colors or appearance or because many are armed with urticant setae (stinging hairs) and can cause severe skin injuries when touched.

The first two species treated in this section belong, respectively, to the families Tineidae and Psychidae, in which the larvae build and carry with them a very strong cocoon (shelter), providing them protection from predators. In adults, the proboscis is stunted, and therefore all nutriments are gathered in the caterpillar stage. Tineidae and Psychidae are distributed worldwide and comprise 3000 and 600 species, respectively.

Plutellidae, also found worldwide, is a family of small moths that has around 300 described species. Zygaenidae contains many species active in daytime; this behavior has a different evolutionary origin than diurnal behavior in butterflies, making it an example of ethological convergence—the independent appearance of a trait in unrelated groups. The larvae of Zygaenidae and Limacodidae are known as slug caterpillars because their movement is sluglike, produced by ventral wave-like contractions. Most slug caterpillars bear poisonous spikes for protection. Zygaenidae and Limacodidae are present on all continents, and each family has about 1000 described species worldwide; in Costa Rica there are 9 and 117 species, respectively.

Megalopygidae, a member of the same superfamily (Zygaenoidea) as Zygaenidae and Limacodidae, is found only in the New World, with 240 species total and 42 species in Costa Rica. The larvae are covered in venomous setae that cause extremely painful reactions when they come into contact with human skin.

Castniidae is present on all continents, but they are most diverse is in the Neotropics. Like zygaenids, castniids are day-flying moths; most species are completely diurnal, though a few fly only at dawn or dusk. Members of Castniidae are often confused with skippers (family Hesperiidae) because their antennae also end in a small hook, but castniids can be distinguished from skippers by the presence of a frenulum in the ventral hind wing. There are about 150 described species, of which 12 are present in Costa Rica.

Phereoeca allutella (Tineidae: Tineinae)
Household Casebearer

0–2500 m

FWL: 5-7 mm. Description: *Female:* DFW is grayish-white with three or four dark brown dots; DHW is gray. *Male:* DFW is brown and thin with disperse dark spots of varying sizes; DHW is gray. In both sexes, wings have long scales on distal margins. **Similar species:** Female *Phereoeca uterella* has transverse bands instead of dots on DFW; males may not be possible to distinguish by external coloration (pers. obs.). **Habitat:** Houses, storage rooms, fabrics, any buildings where humans or animals live. On both Caribbean and Pacific slopes; absent from forests and any habitats without humans. **Distribution:** Hawaii, Costa Rica, Panama, Canary Islands, Madeira, Sierra Leone, Seychelles, Sri Lanka, India, Java, Samoa. **Seasonality:** Present all year. **Natural history:** This is an anthropogenic species, living in and around human habitations. As adults these moths do not feed or drink at all; nutriments are stored from larval stages. Females attract males to mate with a signaling organ in tip of abdomen. The larvae are attacked by parasitoid wasps (*Apanteles*, Braconidae). The true distribution of this species is uncertain, but it is likely present in all Central and South American countries since it is easily transported by travelers and merchants. The same may be true of *P. uterella*, whose presence in Costa Rica has been demonstrated by molecular evidence, and both species probably share the same distribution. **Early stages:** Eggs are pale blue, with a soft outer membrane, and are laid on walls. Right after hatching, caterpillars build a larval shelter of found particles of the right size, binding them together with silk. Caterpillars pass through seven instars; when mature, the shelter has a rhomboid shape, and the larva carries it around while eating any available organic matter, such as insect carcasses, hairs, spiderwebs, fungi, or algae. They do not eat clothing or other fabrics or manufactured items. Larva is small and whitish with brown thoracic plates. Pupation occurs inside the shelter. **Host plants:** None.

Phereoeca allutella larva carrying its shelter.

Phereoeca allutella pinned.

Adult male *Phereoeca allutella*.

Oiketicus kirbyi (Psychidae: Oiketicinae)
Giant Bagworm

0–2000 m

FWL: 22 mm (male). **Description:** *Female:* Body is vermiform (or maggotlike), lacking wings, legs, antennae, eyes, or proboscis; covered in scales, it is cream-colored with a longitudinal dark brown dorsal band. *Male:* Body is robust; DFW is narrow and dark brown with two tones, lighter in basal area and anal margin, and darker distally, with a short white transverse line in post-medial area close to costal margin; DHW is dull brown. **Similar species:** None. **Habitat:** Very common on both Pacific and Caribbean slopes in all habitats except primary forest. Found in gardens, backyards, and parks in any part of the country but almost always seen in caterpillar stage, as bagworm. Adults are seldom seen, probably because they are short-lived and do not do much more than find a mate and mate. **Distribution:** Mexico to Brazil, Caribbean islands. **Seasonality:** Present all year. **Natural history:** Giant Bagworms are familiar to most Costa Ricans as larvae, which are very common and can feed on more than fifty species of plants, from over forty different families, many of which are ornamental or crop plants. The bagworms can grow large, up to 15 cm long. They walk around their host plant carrying their bag (shelter) with them, and if threatened they will hide inside the bag and stay there for several minutes. When ready, the caterpillar closes the bag's upper entrance and pupates inside. The females of the litter emerge first and mate with unrelated males, forcing their later-emerging brothers to go find other females; this phenomenon, called protogyny, diminishes inbreeding. Adults do not feed; males live for 2–3 days, during which they search for females and mate by introducing their telescopic abdomen tip inside a female's bag. Females live a few days longer than males, releasing pheromones so mates can find them inside their bag. This species has become a serious pest of Costa Rican crops such as oil palm, banana, peach palm (*pejibaye*), coconut,

dorsal

Oiketicus kirbyi pinned.

Oiketicus kirbyi larva feeding.

Adult female *Oiketicus kirbyi* removed from shelter.

citrus, and others. Among the Giant Bagworm's natural enemies are parasitoid wasps (*Conura oiketi-cusi*, Chalcididae; *Digonogastra diversus*, Braconidae; *Tetrastichus pseudoceticola*, Eulophidae) and a fly (*Sarcophaga lambens*, Sarcophagidae). **Early stages:** Eggs are small, cream-colored, and rounded, and are laid in tidy masses of up to 6000 eggs inside the bag the female lives in. When the eggs hatch, the larvae leave the shelter and, held aloft by a silk string, let the wind transport them to any plant. There, the tiny caterpillars first cut small pieces of leaves to build their own bag; as they grow, they continuously enlarge the bag, also employing carefully cut and arranged sticks of wood. Mature larva is dull gray in segments A_1–A_{10}, which always stay inside the bag; thorax segments are light gray with black longitudinal bars and markings; head capsule is smooth, light gray with many black bands, spots, and marks, and bears a few long white and black setae. Female pupa is vermiform and black; male pupa is long and black; the pupal case is almost completely outside the bag when adult moth emerges, exiting from the opposite end the larva used for walking and eating. **Host plants:** Many plants (polyphagous).

Plutella xylostella (Plutellidae: Plutellinae)
Diamondback Moth

0–3000 m

FWL: 6 mm. **Description:** A small, elongate moth with cream-colored diamond-shaped pattern on FWs when resting; dorsal side of head cream-colored. DFW is thin and brown with anal margin cream-colored. DHW is brown with very acute apex and a long fringe of scales in margins; tornus almost absent. **Similar species:** None. Distinguished from other small moths by cream-colored diamond pattern on folded FWs. **Habitat:** Both Pacific and Caribbean slopes; commonly found on broccoli, cabbage, cauliflower, and radish crops in rural areas; also found where these products are stored, such as supermarket storage rooms and houses. Absent from forest habitats. **Distribution:** Worldwide. **Seasonality:** Present all year. **Natural history:** This is a very important pest of crops of the family Brassicaceae. It can even resist the cold of supermarket and home refrigeration, and caterpillars on vegetables may complete their life cycle, the small moths emerging to fly in unusual places such as kitchens or inside cars. Adults are not very good fliers, but strong winds can carry them, sometimes as far as 400 km in a single night. The life cycle is very

Plutella xylostella larva on broccoli.

Plutella xylostella pupa.

Adult *Plutella xylostella*.

Plutella xylostella pinned.

accelerated, and up to fifteen generations may occur per year. These features, together with great resistance to agrochemicals, has made the species an important problem for farmers. Adults feed on nectar of flowers of their own host plants at dusk, and use the rest of the night to mate, lay eggs, and disperse. Larvae are attacked by parasitoid wasps (*Apanteles ruficornis*, Braconidae; *Diadegma fenestralis*, *D. insulares*, and *Diadromus subtilicornis*, Ichneumonidae; *Ceratosmicra* and *Spilochalcis*, Chalcididae). **Early stages:** Eggs are pale green, oval, and flattened, and are laid singly or in groups in leaves or inflorescences of host plants. Mature larva is solitary; it is green with small dark green spots surrounded by pale white rings all over body and a few, almost invisible, rigid setae; head capsule is green. It makes a light silk cocoon, usually on host plant, in which it pupates; pupa is yellowish-green, long, and cylindrical. **Host plants:** *Brassica oleracea*, *Raphanus raphanistrum sativus*, and many other related plants (Brassicaceae).

Acoloithus totusniger (Zygaenidae: Procridinae)

0–1300 m

FWL: 10–14 mm. **Description:** FWs and HWs black. **Similar species:** *A. totusniger* can be distinguished from *Gardinia magnifica*, *Opharus procroides*, and *Inopsis scylla* (all members of Erebidae) by venation: its DFW has two parallel anal veins, and the radial and medial veins are separated and parallel, all arising from discal cell. **Habitat:** Secondary and primary forest; rare individuals are seen flying by day, in places such as University of Costa Rica campus in San Pedro and Caribbean slope of Guanacaste volcanoes. **Distribution:** Guatemala and Costa Rica. **Seasonality:** Present during rainy season. **Natural history:** Both sexes fly 1–10 m high on sunny days among vegetation. Males often perch higher up on exposed leaves. Feeding habits are unknown, but other zygaenid species consume flower nectar. **Early stages:** Caterpillars rest on upper side of host plant leaves or are seen walking along stalks. Larva has whitish-yellow longitudinal dorsal band, orange longitudinal lateral bands, and is crossed by black rings, one in the middle of each segment and another in segment joints, which in turn have six warts covered by short black setae tufts; longer white setae arise from ventrolateral area in segments T_1 and A_{10}; head is concealed inside body, making it hard to determine which end is anterior. Pupa forms inside a white silk cocoon covered by a white waxy substance, hidden in holes or between dry leaves. **Host plant:** *Cissus verticillata* (Vitaceae).

Acoloithus totusniger fifth instar larva.

Acoloithus totusniger cocoon.

Adult *Acoloithus totusniger*.

Acoloithus totusniger pinned.

Phobetron hipparchia (Limacodidae: Limacodinae)
Monkey Slug Caterpillar

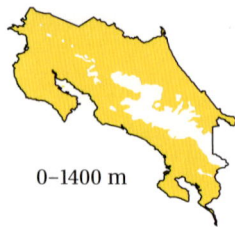

0–1400 m

FWL: 10–14 mm. **Description:** *Female:* DFW is light brown, with sub-marginal orange band on distal margin, large orange blotches in post-medial area forming an irregular band surrounded by brown from costal margin to tornus, and two more orange markings in post-basal area; DHW is pinkish-orange. *Male:* DFW is dark brown with few grayish markings and several orange blotches: one basal, one in tornus, one in sub-apical area, and two or three surrounding a conspicuous translucent patch in post-medial area; DHW is dark gray with large translucent area from medial area to distal margin. Male has narrower wings than female. **Similar species:** Adults are distinguished from similar species by the multiple orange spots and blotches and their arrangement on DFW. Encounters with the caterpillar are more common; *Phobetron hipparchia* larva has thick and curved lateral projections, distinguishing it from *P. cypris* larva, with straight and thin projections. **Habitat:** Occurs on both Pacific and Caribbean slopes in all habitats. Caterpillar is easier to find than adult because of its strange appearance and wide range of host plants; it may be seen in gardens, backyards, and parks in Central Valley, in most

Phobetron hipparchia larva.

Phobetron hipparchia cocoon. JMR, MNCR

Mating *Phobetron hipparchia.* CH

Phobetron hipparchia
pinned. Dorsal view.

Phobetron hipparchia
pinned. Dorsal view.

Costa Rican towns, and also in rural areas. Adults are sometimes seen perched on leaves in daytime or inside houses and buildings, attracted by lights. **Distribution:** USA to Argentina. **Seasonality:** Present all year. **Natural history:** Encounters with the caterpillar of this species usually generate confusion and intrigue among spectators. Its strange appearance is for protection, as its form and pattern disrupt a predator's image of a target prey item, resembling instead a random piece of organic matter. If it is discovered, the spiderlike appearance might alarm and deter an enemy. Venomous setae all over its body can injure any curious observer. The adult also uses crypsis for protection; resting on the upper side of a leaf, it resembles the droppings of a bat that has been eating figs. Despite their excellent defense mechanisms, larvae are attacked by parasitoid wasps (*Barycerus*, Ichneumonidae; *Lathrapanteles*, Braconidae) and flies (*Systrophus*, Bombyliidae; *Austrophorocera*, Tachinidae). This species is considered a minor pest of various Costa Rican crops. **Early stages:** Eggs are shiny, orange, flat, with an irregular shape, and are laid on any part of a plant. The caterpillars disperse and grow slowly. Larvae are variable in coloration, from very dark gray to orange-reddish, green, or yellow; they have six fleshy appendages at sides of body, smaller at the ends and longer in the middle. As with all slug caterpillars, the ventral area and head are concealed by the setae and appendages. A rounded pupa is formed inside a spherical cocoon built with silk and the old larval appendages. **Host plants:** *Dypsis lutescens* (Arecaceae), *Mangifera indica* (Anacardiaceae), *Persea americana* (Lauraceae), *Coffea arabica* (Rubiaceae), *Ochroma pyramidale* (Malvaceae), *Citrus* (Rutaceae), and many others.

Acharia apicalis (Limacodidae: Limacodinae)
Apicalis Saddleback

400–1400 m

FWL: 13–18 mm. **Description:** DFW is different tones of dark brown, with white spot on costal margin in post-medial position, and sometimes a smaller, post-basal one just posterior to discal cell. DHW is all light brown. VWs are all brown. **Similar species:** *Acharia hyperoche* has a small, undivided white spot in post-medial costal margin of DFW, while in *A. apicalis* the spot is larger and often divided into two or more small spots. *A. sarans* does not have a white spot on DFW. Larva of *A. hyperoche* does not have a dorsal white spot on A$_1$, and larva of *A. sarans* has two spots instead of one. **Habitat:** Occurs in all habitats but more commonly seen when associated with humans as pest on crops and backyard plants. In Central Valley it is very common in gardens and often approaches house lights. Most encounters are with the caterpillar, thanks to its bright colors and abundant status. **Distribution:** Mexico to Costa Rica. **Seasonality:** Present all year. **Natural history:** This species has become a considerable pest of many crops since its larvae can feed from dozens of different plants. The caterpillar's spines sting when touched. Adult does not feed but drinks water with a rudimentary proboscis. It is a strong and powerful flier. When resting, it holds the body with the anterior pair of legs and twists the posterior ventrally at a 90° angle; together with the brown coloration, it creates the impression of a dry, twisted small

Acharia apicalis larva.

Acharia apicalis cocoon.

dorsal

Adult *Acharia apicalis*.

Acharia apicalis pinned.

leaf. Larvae are attacked by wasps (*Apanteles*, Braconidae; *Conura* and *Brachimeria*, Chalcididae; and *Casinaria*, Ichneumonidae) and a fly (*Lespesia aletiae*, Tachinidae). **Early stages:** Eggs are pale yellow, flattened, and laid in groups on leaf undersides. Young larvae rest in groups, but mature larva becomes solitary and rests on undersides of leaves. Anterior segments of larva are dark green with a white dorsal spot; on segment A_1 are two fleshy dorsolateral warts from which many spines arise; smaller warts are on T_2 and T_3. Abdomen is light green dorsally and laterally, with a white border and a central dorsal dark green circular patch surrounded by a white ring; at lateral position on each segment is a small fleshy wart, also covered in small venomous spines. Posterior abdominal segments are dark green with a white dorsal spot on A_8 and two on A_9 and fleshy dorsolateral warts with spines on A_8 and A_9. The larva looks like it has a saddle on its back, hence its popular name. The caterpillar's spines release an irritating toxin when the tip breaks, causing a very painful burning sensation. **Host plants:** Citruses (Rutaceae), palms (Arecaceae), bananas (Musaceae), and many other plants.

Megalopyge lanata (Megalopygidae: Megalopyginae)

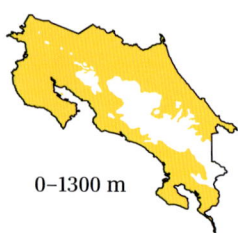

0–1300 m

FWL: 20–31 mm. **Description:** A robust-bodied moth with small wings in comparison to its body. At rest it has a triangular shape, and only FWs are visible. DFW is whitish with many variably sized and shaped brown and gray blotches. DHW is whitish-gray with dark gray veins. The thorax is patterned with gray and white blotches and a few pink markings, and the abdomen is transversely pink-striped. **Similar species:** *Megalopyge lanata* and *M. albicollis* are both variably dark-colored. DFW of *M. albicollis* has a longitudinal black band in costal margin (absent in *M. lanata*); in medial area of DFW, *M. lanata* has a brown patch from vein Cu_1 to vein 2A, while in *M. albicollis* the patch is reduced to veins Cu_1–Cu_2. **Habitat:** Very common on both Pacific and Caribbean coastlines, where its host plants are widespread. More frequently noticed in caterpillar stage because of its conspicuous appearance. In Cahuita, Tortuguero, and Corcovado National Parks, adults are often attracted to lights at night. **Distribution:** Mexico to Brazil, Caribbean islands. **Seasonality:** Present mostly during rainy season. **Natural history:** Not much is known about the habits of this common species. Adults do not feed; the rudimentary proboscis is probably used only to drink water. The caterpillar is better known, mainly for its capacity to produce tremendous reactions on human skin after physical contact with the rigid setae, arranged in two groups on each body segment. Each seta is associated with a gland at its base that injects a toxic cocktail of acids and other substances; stings can cause necrosis, fever, paralysis, and systemic reactions. **Early stages:** Eggs are whitish, rounded, and covered by many scales from the adult female's abdomen; they are laid in small groups on a surface close to host plants. Mature larva is white in dorsal area of each segment except for T_1, which is brown; its sides are brown, and there is a black

Megalopyge lanata larva.

Megalopyge lanata cocoon.

Adult *Megalopyge lanata*. HBö

Megalopyge lanata
pinned. Dorsal view.

Megalopyge lanata
pinned. Dorsal view.

ring between segments; each segment has four red scoli, two dorsolateral and two lateral, from which arise tufts of long, soft, black and reddish setae as well as small white ones. Pupa is reddish-brown and forms inside a grayish silk cocoon covered in stinging setae. Pupae are often arranged in groups on trunk and branches of host plants. **Host plants:** *Terminalia catappa* (Combretaceae), *Byrsonima crassifolia* (Malpighiaceae), *Swietenia macrophylla* (Meliaceae), *Roupala montana* (Proteaceae), *Rhizophora mangle* (Rhizophoraceae).

Telchin atymnius drucei (Castniidae: Castiniinae)
Orange-patched Sun-moth

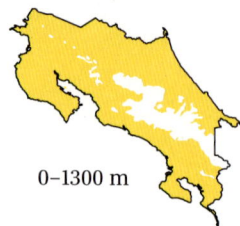

0–1300 m

FWL: 35–48 mm. **Description:** DFW is brown with a transverse pale white line from costal margin to tornus; female also has a series of about five white spots in post-medial area close to costal margin. DHW is mostly orange, with dark orange-brown basal area and a white patch that begins in M_3 in medial area and ends in tornus. VFW has the transverse white line more strongly marked on brown background color. VHW is pale brownish-white. **Similar species:** None in Costa Rica. **Habitat:** Occurs in secondary forest and open areas such as riversides and croplands on both Pacific and Caribbean slopes. Generally absent from urban areas unless forest patches are present. Localities include Alajuela, Ciudad Colón, San Pedro, and Heredia in Central Valley; and San Carlos, Siquirres, and Sarapiquí on Caribbean slope. **Distribution:** Honduras to Brazil. **Seasonality:** Present all year. Museum specimens collected at end

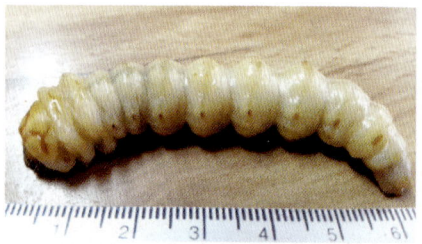

Telchin atymnius drucei larva, taken from inside a stem. JDS

Telchin atymnius drucei cocoon, found on the ground. JDS

Adult *Telchin atymnius drucei.*

Telchin atymnius drucei pinned.

of dry season are very deteriorated, while those from June and later part of rainy season are in good condition; it is possible that the species is univoltine (undergoing an annual cycle), and the adult life span may be almost a year. **Natural history:** Both sexes feed from a wide variety of flowers during the hot hours of the day. Males are very territorial; they perch in clearings on branch tips or exposed spots and chase away any other individual with powerful circular flights. Females lay eggs during midday; they choose a host plant, land on the stalk, and then descend, walking backward as the abdomen tip touches the stalk, then lay a single egg in the ground, buried about 5 mm deep, less than 1 cm from the stalk (at times eggs are laid in wounds in base of stems). This species has become a pest of sugarcane plantations, where larvae burrow into and damage the stalks. Ants that destroy most of the eggs and early instars are one of the species' most important enemies. **Early stages:** Eggs are gray and elongate. Upon hatching, the larva burrows into the host plant's stem, where it lives throughout its development. It can have up to ten growing instars. In each stage, larva is whitish and semitranslucent, with a reddish-brown plate on the prothorax (segment T_1), and its segments are well marked; it has reduced prolegs, but these still have two bands of crochets; head is orange, with smooth, sclerotized surface. Pupa is cylindrical, reddish-brown, with transverse rings of small spikes in abdomen; it is formed inside the plant rhizome in a chamber built with plant fibers. When adults hatch, the pupa is half-unburied with the anterior part aboveground (pers. obs.); this seems to suggest that the pupa actually moves to the surface while the adult is emerging. **Host plants:** *Heliconia latispatha* (Heliconoaceae), *Musa* (Musaceae), *Calathea* (Marantaceae), *Saccharum officinarum* (Poaceae).

Skippers (Hesperiidae)

Skippers are small to medium-size butterflies named for their very fast, powerful, straight flight, which resembles a hop or skip. They live in most places in the world (except for permanently frozen lands and New Zealand) and are among the more common and abundant lepidopterans of Costa Rican habitats. They feed on the nectar of a wide variety of flowers. Most species are diurnal, although a few are known to be nocturnal, such as *Celaenorrhinus fritzgaertneri* (DeVries, Schull & Greig 1987), or crepuscular (*Dyscophellus*).

Skipper eggs are very variable: some are spherical with a smooth surface, others bear longitudinal ribs or are covered by dense long setae, and some are semispherical or bean-shaped. They are always laid singly. In most species, the larva lives alone in a self-built shelter on the host plant's leaves, though a few are burrowers or build silk tents; these are strategies to reduce predation and control conditions such as temperature and humidity. The larvae do not develop bodily ornamentations or structures since they are hidden inside their shelter most of the time; some have a translucent exoskeleton, revealing part of the respiratory, circulatory, and digestive systems. In many species, the larvae secrete a white waxy powder on which they live; it is even present during the pupal stage, but its function is poorly understood. The pupa is formed inside the shelter; a silk girdle keeps it attached to the leaf. Most pupae are white, and many have prolongations of the proboscis.

Adult skippers are generally dark brown and dull-colored, although many species of the subfamily Pyrrhopyginae are bright metallic blue or marked with red, orange, white, and black. The thorax is thick and robust, and the eyes are small and well separated from each other; these characters together with the hooked antennae are the best way to distinguish hesperiids from all other Lepidoptera families. Hesperiidae comprises nine subfamilies, of which six (Pyrrhopyginae, Pyrginae, Heteropterinae, Eudaminae, Hesperiinae, and Tagiadinae) occur in Costa Rica. Species of Pyrginae and Hesperiinae, the most common and easily found skippers in Costa Rica, are covered in this book. Worldwide the family has around 5000 species (Chacon & Montero 2007); about 490 species occur in Costa Rica, and 2365 species in the Neotropical region (Mielke 2004).

Quadrus lugubris lugubris (Hesperiidae: Pyrginae)
Tanned Blue-Skipper

0–1400 m

FWL: 16–18 mm. **Description:** A small brown skipper with multiple variable semitransparent patches in medial area of FW. Both DWs have distinct brown tones; DHW has darker brown post-basal band. VWs are similar to DWs but paler. **Similar species:** *Zera zera* has substantially fewer semitransparent patches on FW. *Gindanes brebisson* does not have brown post-basal band on DHW. **Habitat:** Primary and secondary forest habitats on both Pacific and Caribbean slopes. **Distribution:** Mexico to Costa Rica. **Seasonality:** Present all year. **Natural history:** This skipper flies very fast and close to the ground. It feeds during the morning from flowers of *Lantana* and from Asteraceae such as *Ageratum*, *Zinnia*, and many other genera. Nothing is known about its habits or ecology. **Early stages:** Eggs are laid on underside of host plant's leaves. The caterpillar builds a shelter by cutting a piece of leaf border and folding it. When mature, larval shelters are pierced with several round holes. Larva is pale green, darker dorsally at mid-abdomen, smooth, and semitranslucent, revealing part of its respiratory system; head capsule is black with faint red markings in front area. Pupa is cylindrical and white, with black markings on ventral side of abdomen. **Host plant:** *Piper psilorhachis* (Piperaceae).

Quadrus lugubris lugubris larva (shelter opened by hand).

Quadrus lugubris lugubris pupa (shelter opened by hand).

Adult *Quadrus lugubris lugubris*.

Eantis pallida (Hesperiidae: Pyrginae)
Pale Sicklewing

500–1900 m

FWL: 27–32 mm. **Description:** A shiny, variegated dark brown butterfly with patches of different tones on DWs and pale yellowish in medial area of VFW. **Similar species:** None. **Habitat:** Open areas such as gardens, parks, and backyards in cities and towns, as well as croplands and cattle ranches on both Pacific and Caribbean slopes. Localities include Monteverde, Central Valley, and San Vito. **Distribution:** USA to Bolivia. **Seasonality:** Present all year but more common during rainy season. **Natural history:** This is a common skipper species in disturbed habitats at mid-elevations. It flies fast during hot mornings, actively visiting flowers of *Lantana, Stachytarpheta, Pentas, Zinnia*, and many other plants. It is also commonly seen drinking from moist and decaying organic matter on the ground and sometimes feeding from bird droppings on leaves. The female lays eggs on small trees. **Early stages:** Eggs are yellowish-green, semispherical, and laid on underside of host plant's leaves. Larva is semitranslucent, with lateral line of small irregular yellow patches (around five per segment) along the body; head capsule is heart-shaped and red, with stemmata and mouthparts black. The larva builds a shelter where it hides when not eating; the pupa is formed inside the shelter. Pupa is green, covered in a white waxy powder, and cylindrical with a short, conspicuous spike on the head. It is attached to the leaf by a silk girdle around segment T$_3$. **Host plants:** *Zanthoxylum melanostictum, Z. caribaeum, Z. americanum, Z. setulosum, Pilocarpus racemosus, Citrus aurantiifolia, C. medica, C. reticulata, C. sinensis, Stauranthus perforatus, Toxosiphon lindenii* (Rutaceae).

Eantis pallida larva (shelter opened by hand).

Eantis pallida pupa (shelter opened by hand).

Adult *Eantis pallida*.

Zera hosta (Hesperiidae: Pyrginae)
Hosta Skipper

900–1800 m

FWL: 15–19 mm. Description: DFW is dark brown in basal area, light brown in medial and post-medial areas, and darker again at distal margins; it has three small translucent spots at costal margin in sub-apical area, and a single translucent spot in post-medial area between veins M_3 and Cu_1. DHW is dark brown in basal area, light brown elsewhere, and dark brown again in distal margins. VFW is brown with tornus area brownish-orange. VHW is whitish-silver with brownish-orange costal margin. Anterior half of abdomen is dark brown, and posterior part is light brown, mirroring HW pattern. **Similar species:** In *Zera hyacinthinus hyacinthinus* and *Z. tetrastigma tetrastigma*, basal area of DFW is light brown, while in *Z. hosta*, DFW is continuously dark brown from basal to medial area. **Habitat:** Open areas near secondary forest patches, such as riversides, light gaps, gardens, and croplands. Occurs on both Pacific and Caribbean slopes but associated with rainy habitats. **Distribution:** Mexico to Colombia. **Seasonality:** Present all year but more common during rainy season. **Natural history:** This species inhabits forested areas, but adults like to feed in direct sunshine from small white and yellow inflorescences during the hottest hours of the day. Males claim territories by perching with the wings wide open on leaves along trails and in forest understory, where they also obtain food from decaying matter and bird droppings close to ground level. This species is a low-impact pest of avocado plantations. Its caterpillars are attacked by a parasitoid fly (*Siphosturmia rafaeli*, Tachinidae). **Early stages:** The caterpillar builds a shelter with the host plant's leaves and makes a perimeter of small holes around the place where its rests. Mature larva is light green and semitranslucent; heart-shaped head capsule is red with black in front and mouth area. Pupa is cylindrical, white, with a small black wart between the eyes in the head area, and is attached to the leaf inside the shelter by a silk girdle around segment T_3. **Host plants:** *Persea americana*, *Ocotea puberula*, *Cinnamomum triplinerve*, *Nectandra umbrosa*, *N. hihua* (Lauraceae).

Zera hosta larva (shelter opened by hand).

Zera hosta pupa (shelter opened by hand).

Adult *Zera hosta*. Dorsal view.

Zera hosta pinned. Ventral view.

Telegonus audax (Hesperiidae: Pyrginae)
Audax Telegonus

400–1200 m

FWL: 25–35 mm. **Description:** DFW is brown with basal and post-basal area iridescent bluish-green; FW has a semitranslucent compound band in medial area from costal margin to vein CU_2 in sub-margin; and a second very small semitranslucent band in sub-apical area from costal margin to vein R_5. DHW is also brown with basal area iridescent bluish-green. VFW is light brown with costal area iridescent blue in basal area; semitranslucent bands are as in DFW. VHW is light brown with white costal margin in basal area. **Similar species:** Can be distinguished only by the molecular sequence from related similar species *Telegonus boreas*, *T. favilla*, *T. augeas*, *T. fruticibus*, *T. inflatio*, *T. obstupefactus*, *T. procrastinator*, *T. synecdoche*, *T. viracocha*, and *T. fulgerator*. **Habitat:** Secondary and primary forest areas, along trails and forest edges, on both Pacific and Caribbean slopes. **Distribution:** Costa Rica. **Seasonality:** Present all year but more common during rainy season. **Natural history:** *Telegonus audax* is one of ten species recently split from the widely distributed species *T. fulgerator*, which had long been considered to have a wide range of host plants and different larval forms; Lepidoptera taxonomists continue to discuss the viability of these ten species. Not much is known about the newly established *T. audax*. Adults visit flowers such as *Stachytarpheta*, *Lantana*, and a wide variety of Asteraceae species; they also visit bird droppings, which the skipper dilutes with tiny urine droplets in order to obtain

Telegonus audax larva (shelter opened by hand).

Telegonus audax pupa (shelter opened by hand).

Adult *Telegonus audax*.

Telegonus audax
pinned. Ventral view.

the dry solids. **Early stages:** Eggs are laid on underside of host plant's leaves. The caterpillar builds a shelter when young by folding a small piece of leaf, but when mature it usually sticks two leaves together with silk. Larva is black with yellow and white transverse rings on each segment, and very few short white setae dispersed along the body; head capsule is black with four faint, thin yellow longitudinal lines, and is covered in long, dense, soft white setae. Pupation occurs inside leaf shelter. Pupa is dark blue, covered in dusty white wax, and cylindrical, and bears two small black protuberances at base of wing pads. **Host plant:** *Celtis iguanaea* (Cannabaceae).

Telegonus alardus latia (Hesperiidae: Pyrginae)
Frosted Flasher

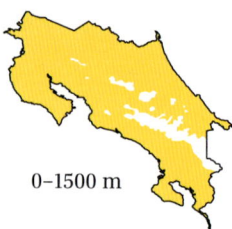

0–1500 m

FWL: 27–33 mm. **Description:** A triangle-shaped butterfly. DWs are dark brown with metallic greenish-blue basal area and white distal margins. VWs are brown with distal margins washed white. **Similar species:** *Telegonus alector hopfferi, T. cretatus cretatus,* and *T. creteus crana* have similar dorsal pattern but lack white wash to distal margins of VWs, having instead a single white mark in medial and post-medial areas of VFW anal margin. **Habitat:** Secondary forest and forest edges of both Pacific and Caribbean slopes at mid-elevations; seen flying along semi-open areas and trails. Localities include Santa Ana, San José, Cartago, Coto Brus, San Carlos, Sarapiquí, and Monteverde. **Distribution:** Mexico to Colombia. **Seasonality:** Present all year but more common during rainy season. **Natural history:** Adults fly in morning sunshine, actively feeding on flowers such as *Lantana, Stachytarpheta, Pentas,* and several Asteraceae genera, which they visit for a few seconds before flying quickly to another flower. Males set up territories inside forest along trails and in light gaps, which they protect for several days in a row. Females lay eggs in host plant saplings from mid-morning to midday, alternating with feeding periods. Males feed from bird droppings; this behavior is thought to be a way to obtain nitrogen, which is then passed to females through copulation. Larvae are attacked by parasitoid wasps (Ichneumonidae; Braconidae). **Early stages:** The caterpillar builds a leaf shelter on which it rests when not eating. Mature larva is light green with many yellow dots dispersed throughout thorax and abdomen; head is heart-shaped and black, with two conspicuous red patches anterior to stemmata. Pupa, a rounded cylinder, is black and covered in a white waxy powder; it forms inside the caterpillar shelter and is fixed to the leaf by a silk girdle around segment T$_3$. **Host plants:** *Erythrina poeppigiana, E. costaricensis, E. gibbosa, E. lanceolata, E. berteroana* (Fabaceae).

Telegonus alardus latia larva (shelter opened by hand). *Telegonus alardus latia* pupa (shelter opened by hand).

Adult *Telegonus alardus latia*.

Telegonus alardus latia
pinned. Ventral view.

Atarnes sallei (Hesperiidae: Pyrginae)
Orange-spotted Skipper

0–1500 m

FWL: 17–19 mm. **Description:** DFW is brown with white in medial area from costal to anal margin, followed by a set of thin white lines in post-medial area; three small translucent spots between radial veins in costal margin right after distal end of discal cell; and conspicuous orange spot in anal margin of medial area. DHW is also brown with a medial white patch from costal to anal margin, followed by white lines extending to distal margin. VW patterns similar to DW patterns. Abdomen is brown with white rings. **Similar species:** None. **Habitat:** Occurs on the Pacific slope in open areas and light gaps close to forest patches; it is associated with second-growth habitats such as abandoned croplands. In the Central Valley it is found along riversides, and in rural areas it visits gardens and backyards. **Distribution:** Mexico to Colombia. **Seasonality:** Present all year but more common during rainy season. **Natural history:** Both sexes fly at low altitude above the ground and feed from a wide variety of flowers; they prefer small white inflorescences, such as those of *Alternanthera pubiflora, Bidens pilosa*, and *Melanthera nivea*, but also feed from yellow-flowered species of Asteraceae such as *Melampodium divaricatum*. This species always rests with the wings wide open. It is a low-impact pest of anona and guanabana plantations. Larvae are attacked by parasitoid flies (*Chrysotachina, Hyphantrophaga*

Atarnes sallei larva (shelter opened by hand).   *Atarnes sallei* pupa (shelter opened by hand).

Adult *Atarnes sallei*. JA

virilis, Tachinidae). **Early stages:** The caterpillar builds a shelter and makes a perimeter of small holes around it. Larva is green, slightly hairy, covered in many small whitish dots, and has a faint white lateral line along body; head capsule is green with some black lines. Pupa is tubular and white, with thorax covered in white setae; front of head and eye area have many small black dots. It is attached to the leaf by a silk girdle around segment T$_3$. **Host plants:** *Annona holosericea, A. cherimola, A. purpurea, A. reticulata, A. rensoniana, Sapranthus palanga* (Annonaceae).

Urbanus proteus proteus (Hesperiidae: Pyrginae)
Long-tailed Skipper

0–1400 m

FWL: 22–25 mm. **Description:** DFW is reddish-brown with turquoise iridescence and several semitransparent patches: two in costal cell, one in distal end of discal cell, one on veins Cu_1–Cu_2, a smaller one on Cu_2–2A, three small ones in sub-apical area from R_2 to R_5, and two or three smaller ones in post-medial area between M_1 and M_3. DHW is metallic blue, green, red, and brown with white distal margins, and a long tail (vein 2A). VFW has same pattern of semitransparent patches as DFW, but background color is brown, not metallic, and a darker brown shade surrounds post-medial semitransparent patches. VHW is brown with two dark spots between Sc+R_1 and Rs in medial area; another dark spot with a white distal border inside discal cell; one more reaching 2A, forming a broken band; a dark post-medial band at distal margin; and a white distal margin. **Similar species:** *Urbanus esta, U. viterboana, U. belli, U. pronta, U. pronus, U. esmeraldus, U. evona, U. viridis,* and *U. prodicus* are extremely similar to *U. p. proteus.* Examination of male genitalia can aid in identification (Fig. 24, p. 54). Externally, in

Urbanus proteus proteus larva (shelter opened by hand).

Urbanus proteus proteus pupa (shelter opened by hand).

Urbanus proteus proteus. Dorsal view.

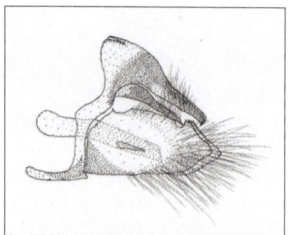

Fig. 24. Male genitalia of *Urbanus proteus proteus*.

male specimens, look for the costal fold; if it is present, the dark basal spots on the VHW must be separated from each other, the ventral part of the antenna clubs must be pale yellow, and the spots between Sc+R$_1$ and Rs must be equal in size. A female will have the same combination of characteristics except for the costal fold. These characteristics in combination generally separate *U. p. proteus* from the others; for further details see the key presented by Steinhauser (1981). **Habitat:** Open areas and second-growth forest habitats, including cattle ranches, croplands, and gardens, almost everywhere in its area of distribution on both Pacific and Caribbean slopes. It is common in urban areas and is often seen during hot mornings in backyards and parks in the Central Valley. **Distribution:** USA to Argentina. **Seasonality:** Present all year. **Natural history:** This species is considered a pest of bean plantations throughout the Americas. Adults feed in open areas from flowers such as *Stachytarpheta*, *Pentas*, *Lantana*, and *Hamelia patens*; and Asteraceae species including *Zinnia*, *Bidens pilosa*, and many others. That plasticity, together with the capacity of the larvae to eat from a wide variety of wild bean vines that pioneer open areas, has allowed the species to perfectly adapt to *Phaseolus vulgaris* bean crops. Adults live up to 25 days. Larvae are attacked by parasitoid wasps (*Apanteles leucostigmus*, Braconidae; *Grotiusomya nigricans*, Eulophidae); eggs are parasitized by wasps (*Trichogramma*, Trichogrammatidae). **Early stages:** Eggs are whitish, semispherical, with faint ribs, and are laid in new shoots or on underside of leaves. The caterpillar builds a shelter to live inside. Mature larva is light green with a longitudinal black dorsal line, a pair of conspicuous longitudinal yellow dorsolateral lines, and many small black dots mottling the body; head capsule is reddish-brown on top and sides, and black in front with orange patch at sides of jaws. Pupa is dark brown, covered in a white waxy powder, and shaped like a short, curved cylinder. **Host plants:** *Centrosema macrocarpum*, *C. pubescens*, *C. sagittatum*, *Desmodium distortum*, *D. glabrum*, *D. incanum*, *D. infractum*, *D. nicaraguense*, *Galactia striata*, *Mucuna*, *Phaseolus lunatus*, *P. vulgaris*, *Vigna vexillata* (Fabaceae).

Adult *Urbanus proteus proteus* pinned. Dorsal view.

Quinta cannae (Hesperiidae: Hesperiinae)
Cana Skipper

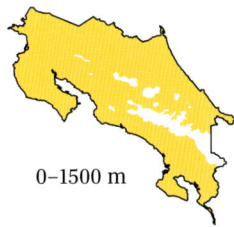

0–1500 m

FWL: 13–16 mm. **Description:** DFW is brown with two semitransparent patches in post-medial area, one in M_3–Cu_1, the other in Cu_1–Cu_2; and one or two very small white spots in costal sub-margin in post-medial position. DHW is brown with light brown distal margin. VFW is similar to DFW but grayish-brown in post-medial area, especially Cu_2–2A. VHW is brown with lighter brown markings in basal area, an irregular longitudinal band in medial area, and a grayish-brown sub-marginal and marginal area. **Similar species:** It is difficult to distinguish *Quinta cannae* from *Cynea cynea* and other similar skippers without knowledge of the variations among the different species and a careful analysis of each spot, band, and color. In the case of *Q. cannae*, the VHW characters are especially useful. **Habitat:** Common in gardens and open areas like croplands and cattle ranches, on both Pacific and Caribbean slopes. Localities include the Central Valley, Guanacaste, Upala, Siquirres, Garabito, and Osa Peninsula. **Distribution:** Mexico to Argentina. **Seasonality:** Present all year. **Natural history:** This is a very common inhabitant of

Quinta cannae larva (shelter opened by hand).

Quinta cannae pupa (shelter opened by hand).

Adult *Quinta cannae*.

Quinta cannae pinned. Dorsal view.   Quinta cannae pinned. Ventral view.

gardens and urban areas throughout the Central Valley, where its host plant *Canna indica* is used extensively as an ornamental, providing suitable habitat. The species is active during hot mornings, flying swiftly from flower to flower, feeding on *Stachytarpheta*, *Lantana*, *Pentas*, *Duranta erecta*, and many Asteraceae species. Larvae are attacked by parasitoid wasps (*Alphomelon xestopyga* and other species, Braconidae; *Creagrura nigripes*, Ichneumonidae) and flies (*Winthemia*, Tachinidae). **Early stages:** Eggs are pale green, semispherical, with a smooth surface, and are laid on underside of leaves. During all instars, the larva builds a leaf shelter, where it lives alone. Early instars are smooth, translucent green, with a whitish lateral line and no setae. Mature larva is yellowish-pinkish with many darker dots, and short setae at the posterior abdominal segments; head capsule is brown with a longitudinal white band on each side. Pupation occurs inside leaf shelter. Pupa is pinkish, cylindrical, and covered in a white waxy powder, with proboscis as long as entire pupa. **Host plants:** *Canna indica* (Cannaceae), *Heliconia* (Heliconiaceae), *Calathea* (Marantaceae).

Talides hispa (Hesperiidae: Hesperiinae)
Hispa Skipper

1000–1500 m

FWL: 25-29 mm. **Description:** *Female:* DFW is brown with three sub-apical clear spots at costal margin; one at distal end of discal cell; and three in a row in post-medial cell, at veins M_3–Cu_1, Cu_1–Cu_2, and Cu_2–2A, the largest one in the middle. DHW is brown with orange distal margin and a single circular clear spot at M_3–Cu_1. VFW has same clear spots as DFW except the one in Cu_2–2A is replaced by a diffuse cream patch; distal margin is grayish from Cu_2 to apex. VHW has whitish areas, varying in contrast, around the circular clear spot. *Male:* DFW is brown with a conspicuous red mark in medial area from discal cell to 2A; clear spots are same as in female, but the one in Cu_2–2A is absent. DHW is same as in female. VWs are similar to female's, but diffuse cream patch in Cu_2–2A in VFW is much reduced. **Similar species:** Extremely similar to *Talides cantra* and *T. sinois*, and variable, which makes identification difficult without examination of male genitalia (Fig. 25). Some external characters, while not definitive, may help: Whitish shade on distal margin of VFW clearly ends close to margin in *T. hispa* but is diffused with rest of the wing in *T. cantra*. Also, *T. cantra* has a small vestigial clear spot on DFW at Cu_2–2A, which is totally absent in *T. hispa*. Head capsule of fifth-instar larva of *T. sinois* has black in front and mouth area inside an orange background color, while head capsule of *T. hispa* is all orange. Finally, *T. hispa* has longer FWL (≥25 mm) than *T. sinois* and *T. cantra*. **Habitat:** Adults are difficult to see in nature, but larvae may be found on host plants in understory of primary and secondary forest. Localities include Leonelo Oviedo Ecological Reserve at the University of Costa Rica campus in San José and probably other Central Valley riparian forest sites. **Distribution:** Costa Rica to Brazil.

Seasonality: Present all year. **Natural history:** This species is still poorly known. Adults may fly in very early morning, late evening, or at night (pers. obs.), although photographs exist of adults in daylight, including some feeding from *Costus* flowers. *Talides* taxonomy is unresolved; research is ongoing (J. Burns and D. Janzen) and will probably result in splitting of species. *T. hispa* as presented in this work comes from a single population at the University of Costa Rica campus, and it seems to be a mix: adults looks like *T. sinois* but are the size of *T. hispa*; larvae look like *T. sergestus* (the adults of which are very different). This is a clear example

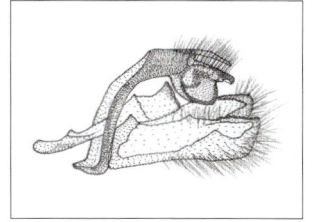

Fig. 25. Male genitalia of *Talides hispa.*

of how much research is needed on the taxonomy, biology, and ecology of tropical skippers. **Early stages:** Eggs are whitish, semispherical, with smooth surface, and are laid on underside of leaves. The caterpillar builds a shelter along the vein on the leaf underside. Mature larva is light green and translucent, revealing internal systems and organs; head capsule is heart-shaped, orange, with granulated surface. Head and body are covered in a white waxy powder. Pupa is white, cylindrical, with a conspicuous forward-pointing peak at the head; proboscis is longer than body. **Host plant:** *Heliconia pogonantha* (Heliconiaceae).

Talides hispa larva (shelter opened by hand).

Talides hispa pupa (shelter opened by hand).

Adult *Talides hispa.*

Talides hispa, dorsal view.

Talides hispa, ventral view.

Talides hispa, dorsal view.

Talides hispa, ventral view.

Carystus cynaxa (Hesperiidae: Hesperiinae)
Black-veined Ruby-eye

0–1600 m

FWL: 20–24 mm. **Description:** DFW is dark blue with a transverse, irregular semitranslucent band composed of three patches from sub-marginal area of costal margin nearly to tornus but not reaching the margin. DHW is all greenish-brown. VFW is bluish-black with same band as DFW, but olive-green costal margin and sub-apical and apical areas. VHW is brownish-yellow with black veins and gray anal margin. Eyes are noticeably red. **Similar species:** None. **Habitat:** Associated with rainforest and seen inside secondary-forest patches along trails, in light gaps, and on riversides, on both Pacific and Caribbean slopes. Localities include San Pedro and San José in the Central Valley, and Rincón de la Vieja, Irazú Volcano, and Corcovado National Parks. **Distribution:** Mexico to Costa Rica. **Seasonality:** Present all year but more common during rainy season. **Natural history:** This species has been poorly studied. It seems to have very localized populations, although its host plant is very common. Adults are fast fliers and not often seen in nature; they feed from flowers such as *Impatiens* and *Stachytarpheta* with a very long proboscis seemingly adapted to other kinds of flowers. A species of rainy areas, it prefers dark understory habitats where it feeds from bird droppings and seeks out flowers close to ground level. **Early stages:** Eggs are yellowish, semispherical, and laid on undersides of leaves. The caterpillar builds a shelter longitudinally aligned with the leaf. Larva is white and covered in a dense layer of waxy powder; head capsule is white with a diffuse dark longitudinal band in the center. Pupa is pale green and cylindrical with a conspicuous forward-pointing spike on head, and proboscis a little longer than pupa itself. It is attached to the leaf inside the shelter by a silk girdle around segment T_3. **Host plant:** *Chamaedorea costaricana* (Arecaceae).

Carystus cynaxa larva.

Carystus cynaxa pupa (shelter opened by hand).

Adult *Carystus cynaxa*. KN

Carystus cynaxa pinned. Dorsal view.

Thracides phidon (Hesperiidae: Hesperiinae)
Jewel-studded Skipper

0–1500 m

FWL: 25–29 mm. **Description:** DFW is dark brown with metallic bluish-green basal area, and three semitranslucent patches in medial area. DHW is also dark brown with metallic bluish-green basal area. VFW is brown with basal area of costal cell white; a small white patch inside discal cell followed by white patches in areas of veins M_3–Cu_1, Cu_1–Cu_2, and Cu_2–2A; and a yellowish-brown patch at costal margin in sub-apical area. VHW has a yellowish-white basal area and four small yellowish-white patches in medial area at M_1–Cu_2. Thorax and anterior half of abdomen are metallic bluish-green. **Similar species:** Although many skippers look very similar, this species can be recognized by its combination of characters. *Perichares deceptus deceptus* is slightly similar, but DFW has white patch on Cu_2–2A that is smaller than the one in Cu_1–Cu_2, while in *T. phidon* the Cu_2–2A patch is larger. **Habitat:** Secondary and primary forest, forest

Thracides phidon larva (shelter opened by hand).

Thracides phidon pupa (shelter opened by hand).

Adult *Thracides phidon*.

edges, and riversides with some natural vegetation, on both Pacific and Caribbean slopes. Localities include Tortuguero, Corcovado, and Rincón de la Vieja National Parks. It is a sporadic visitor to Central Valley gardens and riversides. **Distribution:** Mexico to Brazil. **Seasonality:** Present all year. **Natural history:** This quick-moving species is not often seen, possibly spending most of its time flying in the canopy. It is easier to find larvae on host plants than to see adults in the wild. Although it is not very common in Costa Rica, it is a potential pest of ornamental *Heliconia* plant crops, as in Brazil, where it attacks four *Heliconia* species. Adults feed from flowers of *Heliconia*, *Impatiens*, and *Stachytarpheta*. Larvae are attacked by a parasitoid fly (*Spathidexia marioburgosi*, Tachinidae). **Early stages:** Larva is white; head capsule is white with two black blotches in front, one in upper mouth area, and another around the stemmata. Larva is densely covered in waxy powder, as is its shelter, built in underside of host plant's leaves. Pupa is cylindrical, with a conspicuous forward-pointing spike at the head; proboscis is a little longer than pupa. **Host plants:** *Heliconia irrasa*, *H. latispatha*, *H. longa*, *H. longiflora*, *H. mathiasiae*, *H. metallica*, *H. pogonantha*, *H. tortuosa*, *H. umbrophila*, *H. wagneriana*, *H. vaginalis* (Heliconiaceae), *Musa acuminata* (Musaceae).

Thracides phidon
pinned. Dorsal view.

Thracides phidon
pinned. Ventral view.

Swallowtails, Kite-Swallowtails, Swordtails, and Cattlehearts (Papilionidae)

The butterflies of the family Papilionidae are common inhabitants of virtually every habitat. They live on all continents, from the frozen tundra to the tropics. Their large size and colorful patterns make them conspicuous and difficult to ignore. They are active during sunny periods and can be abundant where their host plants and nectar plants are present. Adults of all species feed on flower nectar, but recently emerged males can also be seen, either alone or in large conglomerations, drinking from wet ground along rivers or dirt roads, a behavior called "puddling."

The eggs are rounded and smooth-surfaced, but in some genera (*Battus, Parides*) are covered in a nutrient-rich waxy paste that provides newborn larvae with extra sustenance to enhance growth. Egg color can be orange, reddish, green, pale yellow, or brown; most species lay single eggs, but some lay clusters of up to forty (*Heraclides anchisiades*). The larva bears short spines or setae in first instar but loses them and becomes smooth-surfaced. Some also have bright-colored fleshy spikes (*Battus, Parides*). The head capsule is always rounded and smooth. Unique to the family, and present in all species, is the osmeterium, a bright yellow to red Y-shaped prothoracic gland on segment T1. It has a defensive purpose, releasing strong-smelling volatile acids that deter ants and parasitoids. In contrast to most other butterflies, pupae are generally formed with the head pointing up. Before pupation, the larva spins a silk girdle around its body and attaches it to the substrate. All papilionid pupae are cryptic, resembling young leaves or sticks of wood.

Species in this family are medium-size to large. In Costa Rica, the most common colors are black, yellow, red, white, green, and (to a lesser degree) blue. It is easy to recognize papilionids, because they generally do not stop fluttering their wings while feeding. The adults are distinguished from all other butterflies by the presence of the 2A vein on the forewing. There are about 600 species worldwide; 36 species occur in Costa Rica, at elevations from sea level to 2500 m.

Battus polydamas polydamas (Papilionidae: Papilioninae)
Polydamas Swallowtail

0–1500 m

FWL: 40–55 mm. **Description:** Medium-size butterfly. DWs are mostly iridescent greenish-black with a sub-marginal row of yellow marks, more prominent in DHW. VWs are dark; VHW has line of red sub-marginal marks and sometimes has white scales. Thorax and abdomen are dotted with red spots, and a red line runs along each side of abdomen. **Similar species:** *Papilio polyxenes stabilis* (melanistic form; p. 75) and *Pterourus coroebus laetitia* (p. 71) lack red markings on thorax and abdomen. **Habitat:** Open and semi-open areas of both slopes, from dry Guanacaste to the rainy Pacific and Caribbean. Easy to see feeding on flowers in disturbed areas, such as gardens, parks, suburban areas, towns, and roadsides; less common in primary forest. **Distribution:** USA to Argentina, except Chile. **Seasonality:** Present all year but most abundant during rainy season. **Natural history:** A fast-flying butterfly, it feeds on a wide variety of flowers such as *Lantana camara* and *Stachytarpheta mutabilis*. In forest, males patrol territories and feed from tree flowers in the canopy. Females lay eggs in sunshine from mid-morning to early

dark form

Fifth instar larva of *Battus polydamas polydamas*.

brown form

Fifth instar of *Battus polydamas polydamas*.

Polychromatism in *Battus polydamas polydamas* pupa.

Adult *Battus polydamas polydamas*.

afternoon. *Battus polydamas polydamas* is considered unpalatable to predators because of its ability to sequester chemicals from larval host plants. Larvae are attacked by solitary wasps (*Macrojoppa*, Ichneumonidae), the single adult of which emerges from the pupa by making a large hole in the wing area; and a fly (*Patelloa xanthura*, Tachinidae). **Early stages:** Eggs are orange, round, and covered in a nutrient-rich waxy substance; they are laid in small clusters of three to ten on new shoots of host plants and sometimes in adjacent substrate. Young larvae feed in small groups and rest on underside of young leaves. Mature larvae are more solitary and rest on the host plant's stalks. There are two forms of larvae: one is streaked black and brown with orange fleshy warts on each segment; the other is velvet black with some orange fleshy warts and others black. Osmeterium is orange. Pupation can take place on the host plant, but caterpillars often move to other places as far as 10–12 m away. Pupa is polychromatic, from dark brown to light green. **Host plants:** *Aristolochia leuconeura*, *A. elegans*, *A. anguicida*, *A. maxima*, and other species (Aristolochiaceae).

Parides iphidamas iphidamas (Papilionidae: Papilioninae)
Iphidamas Cattleheart

0–1200 m

FWL: 30–45 mm. **Description:** Medium-size butterfly, mostly black, with thorax and abdomen covered in red spots. Both sexes are very variable and hard to distinguish from similar species. *Female:* DFW has a white band from costal vein to medial area composed of lobes of variable size and shape; DHW has a red patch in medial area always composed of four or more red lobes. *Male:* DFW is black with a central green patch, variable in shape and size, accompanied by a variable cream-yellow spot; DHW has a patch ranging from iridescent red to fuchsia that is always composed of four or more lobes. **Similar species:** *Parides panares*, *P. erithalion*, *P. eurymedes*, *Heraclides anchisiades idaeus* (p. 70), *H. isidorus rhodostictus*, *H. torquatus tolmides* male, and *Mimoides ilus branchus*. Male *P. i. iphidamas* always has four or more red iridescent lobes on DHW; female's DFW has white spot on M$_2$ that is larger than spot on M$_3$ and is cut obliquely at the base (this needs further study to be confirmed). **Habitat:** Common in farmlands and secondary and old-growth forest habitats all over the country. Found in forest light gaps, along rivers and creeks, in open

Fifth instar larva of *Parides iphidamas iphidamas*.

Pupa of *Parides iphidamas iphidamas*.

Dorsal view of pinned male and female
Parides iphidamas iphidamas.

Parides iphidamas iphidamas courtship, male flying.

areas between forest patches, along trails, and less often in gardens of rural areas. **Distribution:** Mexico to Panama. **Seasonality:** Present all year in rainforest habitats but virtually absent in northern dry forest in dry season. **Natural history:** Adults generally feed energetically from a wide variety of flowers but prefer *Lantana camara*, *Stachytarpheta mutabilis*, *S. calderonii*, and *Impatiens walleriana*. Both sexes feed on nectar in early morning (7:30–9:00 a.m.) in shaded areas and then take a break, after which females start looking for oviposition sites. Males return to visiting flowers amid chasing other males out of their territories or trying to conquer receptive females. During courtship, the male rubs the female's head, eyes, antennae, and proboscis with white scent scales (androconia), probably to stimulate the female to mate. The female lays single eggs on underside of host plant's leaves, in leaves of other plants close to the host plant, or sometimes on rocks or tree trunks where the host is growing. The latter could be a technique to avoid egg parasitoids. Larvae are attacked by a parasitoid wasp (*Meteorus papiliovorus*, Braconidae) and fly (*Patelloa xanthura*, Tachinidae). This butterfly is considered unpalatable to predators because it sequesters chemicals from larval host plants. In the LAREBUB, male individuals lived up to 54 days; female pupae lasted 24.9 days (n = 19); male pupae in dry season lasted 28.8 days (n = 10) and in rainy season 24.1 days (n = 8). **Early stages:** Eggs are reddish and covered in a waxy nutritive substance. Young larva is reddish with white warts. Mature larva is velvet-black with white, black, and red fleshy warts, and sides of segments A_3–A_5 crossed diagonally by a broken white line. Osmeterium is yellowish. Larvae are cannibalistic, and a larger larva will not hesitate to eat a smaller one or a pupa. Pupa is light green; it undergoes no diapause, at least in rainforest habitats. **Host plants:** *Aristolochia leuconeura*, *A. anguicida*, *A. tonduzii*, and other species (Aristolochiaceae).

Parides photinus (Papilionidae: Papilioninae)
Pink-spotted Cattleheart

0–1200 m

FWL: 40–50 mm. **Description:** Medium-size butterfly. DFW and VFW are black; male has slight bluish iridescence on DFW. DHW is black with two rows of red spots, the basal spots round and the distal spots crescent-shaped; male has distinguishable fuchsia-blue iridescence on DHW, lacking in female. Thorax and abdomen are black with red dots. **Similar species:** *Parides montezuma* has only one row of red spots on HWs, all crescent-shaped. **Habitat:** Present only on the Pacific slope; commonly seen in gardens and tree plantations in rural areas that are linked to forest patches. Localities include Ciudad Colón, Puriscal,

Santa Ana, Orotina, Esparza, Miramar, Acosta, Abangares, Liberia, Sámara, La Cruz, Carara National Park, Jacó, and Tárcoles. **Distribution:** Mexico to Costa Rica. **Seasonality:** Most common during rainy season; rare adult individuals may be seen in dense forest patches or along rivers during dry season. Larvae may be found all year. **Natural history:** This species reaches the southern limit of its distribution around the Costa Rican central Pacific coast. Both sexes fly in the morning (8:00–9:00 a.m.) searching for flowers such as *Caesalpinia pulcherrima*, visiting the same flowers day after day in a circuit. After feeding, they stop their activity until around 10:00 a.m., when females start flying again, searching for host plants and ovipositing single eggs on the underside of leaves. Males also start flying again at that time, alternating feeding with patrolling flights along open areas such as trails or forest edges, where they chase other males and females. **Early stages:** Eggs are orange, laid singly on underside of leaves of host plant or on surrounding plants or objects; oviposition on nonhost vegetation may help prevent egg-parasitic wasps from finding them. The larvae rest on leaf undersides, eating from time to time, whether day or night; they may rest alone or in small groups of two or three. Until fourth instar, larva is shiny and black with white and red warts. Fifth-instar larva has lighter-colored thorax and abdomen, contrasting with head capsule and dorsal part of prothoracic plate, which are black; each proleg (except on A_{10}) bears a red wart on its side. Pupa is light green, resembling a young leaf. **Host plants:** *Aristolochia arborea, A. anguicida* (Aristolochiaceae).

Fifth instar larva of *Parides photinus*.

Dorsal view of pinned *Parides photinus*.

Pupa of *Parides photinus*.

Parides photinus mating, male on top.

Protographium epidaus epidaus (Papilionidae: Papilioninae)
Mexican Kite-Swallowtail

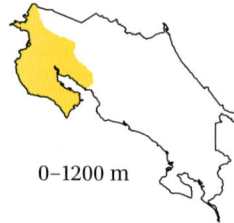

0–1200 m

FWL: 40–45 mm. **Description:** Medium-size, predominantly translucent-white butterfly with wings crossed longitudinally by black lines, four on FWs, two on HWs. HW bears a very long projection of vein M_3. **Similar species:** *Protesilaus protesilaus* has discal cell on VHW divided by a black and red longitudinal band producing a small white area distally inside discal cell, which never occurs in *P. epidaus*. In *Protographium agesilaus*, the black and red longitudinal line on VHW has the red coloration proximal to the body; in *P. epidaus* the red is on the other side of the black line. **Habitat:** Well-conserved forest, but also secondary forest and cattle pastures with riverside forest, on the northern Pacific slope from Limonal and Abangares north to Nicaragua. Males puddle in sand along rivers and dirt roads. Common in Santa Rosa National Park, Cañas Dulces, Guachipelín, Liberia, Bagaces, and Cañas. **Distribution:** Mexico to Costa Rica. **Seasonality:** Highly seasonal. Adults begin to appear in May with the first rains and are in great abundance in June and July; individuals can be seen commonly until November, when they all but disappear. A report of an abundance of adults in late March 1985 is an isolated case that must have been caused by a particular climatic event, such as an off-season rain that activated dormant pupae. **Natural history:** When just emerged from pupation, males puddle along rivers in groups or singly, and sometimes are seen puddling along beaches and reefs that are exposed at low tide. Both sexes can be seen in open pastures and forests, drinking nectar from flowers such as *Lantana camara*, *Inga*, and others. Females stay away from rivers and spend most of their time in pastures and forest areas searching for host plants. Larvae are parasitized by wasps (*Trogus picadoae*, Ichneumonidae; *Meteorus papiliovorus*, Braconidae). **Early stages:** Eggs are white, spherical, with smooth surface, and are laid on upper side or underside of young leaves. Larva rests on leaf upper side along central vein

Fifth instar larva of *Protographium epidaus epidaus*.

Pupa of *Protographium epidaus epidaus*.

Adult *Protographium epidaus epidaus*.

♂

Protographium epidaus epidaus pinned. Dorsal view.

with head pointing toward plant stem. First, second, and third instars are black with black setae and a white transverse band in A_4–A_5 segments. In fourth instar, larva becomes polychromatic; body may be entirely light green with a longitudinal dorsolateral white band on each side, or it may have black dorsal area with white patches on segments A_4–A_5 and lateral areas light green. Intermediate forms also appear, with white and yellow markings on all segments within black dorsal area. Pupa resembles a wooden stick with a small dorsal distal projection. This species spends the dry season (5 months) in pupal stage, and sometimes pupa remains in diapause until the next year. **Host plants:** *Annona reticulata*, *A. pruinosa*, *A. purpurea*, *A. holosericea* (Annonaceae).

Heraclides thoas autocles (Papilionidae: Papilioninae)
Thoas Swallowtail

0–1200 m

FWL: 55–67 mm. **Description:** A large black butterfly with a yellow band, variable in width, from apex of FW to anal margin of HW, and a sub-marginal row of yellow spots on DHW and half of DFW. HW has long spatulate tail (vein M_3). **Similar species:** This species is very easy to confuse with *Heraclides rumiko*; the following characteristics should help ensure identification. *H. thoas autocles* always has four sub-marginal yellow spots from tornus up to halfway point on DFW, and the spots are aligned; *H. rumiko* sometimes has four small spots but not aligned. *H. thoas autocles* usually (but not always) has a yellow spot in discal cell of DFW, and usually (but not always) has thin black longitudinal lines inside discal cell of VFW; *H. rumiko* always lacks both of these features. *H. astyalus* lacks the four large, yellow sub-marginal spots from tornus to halfway point on DFW. **Habitat:** This is a common species in all habitats in its elevation range except mangrove. It can be seen in forest light gaps or along medium-size to large

Fifth instar larva of *Heraclides thoas autocles*.

Pupa of *Heraclides thoas autocles*.

Adult *Heraclides thoas autocles*.

Heraclides thoas autocles pinned. Dorsal view.

rivers inside primary forest, and prefers open or less dense areas where there is some light. In disturbed habitats, croplands, and gardens it is easily seen feeding on flowers such as *Caesalpina pulcherrima*, *Stachytarpheta mutabilis*, *S. calderonii*, *Lantana camara*, *Zinnia elegans*, *Pentas lanceolata*, and many other species. **Distribution:** USA to Panama. **Seasonality:** Present all year but much more common in rainy season. **Natural history:** Males fly along rivers, puddling from time to time, and females search for oviposition sites no more that 3 m above the ground. Caterpillars feed on one of the most diverse and abundant genera of Costa Rican plants, *Piper*. The larva rests on upper side of leaves, mimicking a bird dropping; when alarmed, it abruptly raises the head and thorax and releases the osmeterium, an offensive behavior. Adult males can be seen hill-topping in hottest hours of day. Janzen & Hallwachs (2019) reported seven parasitoid flies (*Lespesia*, Tachinidae) emerging from a single larva, and also a single parasitoid wasp (*Pedinopelte latipennis*, Ichneumonidae) emerging from one side of a pupa. **Early stages:** Eggs are reddish-brown, with smooth surface, and are laid singly on upper side of leaves. First instar larva is reddish, covered in tiny setae; second to fifth instars resemble a bird dropping with many small warts and no visible setae. Osmeterium is red-orange, and thorax is enlarged and resembles a reptilian head. Pupa is very cryptic, resembling a wooden stick. In the LAREBUB, pupae lasted 33 days (n = 4), but some pupae undergo diapause for almost 100 days during dry season. **Host plants:** *Piper auritum*, *P. aduncum*, *P. marginatum*, and many other *Piper* species (Piperaceae).

Heraclides anchisiades idaeus (Papilionidae: Papilioninae)
Ruby-spotted Swallowtail

0–1400 m

FWL: 50–60 mm. **Description:** Medium-size, predominantly black butterfly. DFW and VFW are black with a diffuse white band from medial area at discal-cell level to costal border. DHW is all black with a red sub-medial patch from anal cell to around vein M_1. VHW is black with red or pink sub-medial spots and sometimes some small red spots in sub-marginal area close to costal border. These color patterns are variable: the white line on FWs can be absent in some males and very large in some females, and the red patch on DHW is generally larger in females than in males. **Similar species:** Distinguished from female *Heraclides torquatus tolmides* by lack of a yellow line on sides of abdomen; and from *H. isidorus rhodostictus*, *Parides panares*, *P. erithalion*, *P. eurymedes*, and *P. iphidamas iphidamas* (p. 65) by white line on DFW that is diffused. *Mimoides ilus branchus* can be distinguished by conspicuous red spots on VHW basal area. **Habitat:** This butterfly may be seen in open areas of secondary forest and in disturbed habitats such as croplands and urban areas, including gardens and backyards in Central Valley suburban

A group of fifth instar larvae of *Heraclides anchisiades idaeus* (note a fourth instar, which is lighter green).

Pupa of *Heraclides anchisiades idaeus*.

Adult *Heraclides anchisiades idaeus*.

Heraclides anchisiades idaeus pinned. Dorsal view.

neighborhoods. **Distribution:** Texas (USA) to Colombia. **Seasonality:** Present all year. **Natural history:** This common species visits flowers such as *Lantana camara*, *Stachytarpheta*, *Zinnia elegans*, and *Caesalpinia pulcherrima*, among others, during hot, sunny mornings. It is considered palatable to predators such as birds, but it mimics the highly unpalatable *Parides* (an example of Batesian mimicry). However, it is more common and widely distributed than *Parides*, so when sharing the same habitat, the agile, fast-flying *H. anchisiades* may educate predators that it is too much effort to catch it, resulting in mutualistic "escaping mimicry." Before the introduction of oranges, lemons, and other citruses to the Americas around 1500 CE, these butterflies used five or six native tree species as host plants, but with habitat destruction and other changes in land use, cultivated citrus trees have become important host plants, and larvae may be an occasional minor pest in plantations. The species is attacked by parasitoid wasps (*Meteorus papiliovorus*, Braconidae; *Pedinopelte*, Ichneumonidae), which enter the larva and emerge from the host when it is a pupa. **Early stages:** Eggs are yellow and laid in groups of up to fifty on underside of leaves. After emerging, young larvae split into smaller groups of around twenty individuals and rest together on upper sides of neighboring leaves; sometimes one group rests on the underside and another on the upper side. They rest during the day and move to other leaves to eat at night. First- to fourth-instar larvae are light green with a few thin white lines. Larvae rest in the leaves until fourth instar, when they move together to tree branches and trunks, although sometimes fifth-instar larvae remain resting on leaves. When resting on trunks, fifth-instar larvae may gather in groups numbering several dozen, resembling tree bark in color and texture. To pupate, the larvae generally spread out, walking as far as 20 m from the host plant. Pupa looks like a wooden stick. **Host plants:** *Esenbeckia berlandieri*, *Amyris pinnata*, *Pilocarpus racemosus*, *Casimiroa edulis*, *Zanthoxylum*, *Citrus* (Rutaceae).

Pterourus coroebus laetitia (Papilionidae: Papilioninae)
Victorine Swallowtail

800–1800 m

FWL: 50–60 mm. **Description:** A large black butterfly with a yellow band in medial area from costal vein on DFW to anal margin of DHW, and a row of small yellow sub-marginal spots on both DWs. Female occurs in two forms: one similar to male; and another with DFW mostly black with few sub-marginal yellow spots close to anal margin, and DHW with a large sub-medial green patch and a row of yellow crescents in sub-marginal area. **Similar species:** *Battus polydamas polydamas* (p. 64) has a less robust body with red marks on thorax and abdomen. *Pterourus birchallii bryki* has yellow spots on costal and/or anal borders of VHW. *P. coroebus vulneratus* does not have a yellow bar across discal cell of DFW. In the regions of Tilarán and Guanacaste, hybrids of the two *P. coroebus* subspecies may be seen with intermediate characteristics. **Habitat:** Fields, open areas, and forest canopy, always in association with well-preserved forest patches, in mountains. Rare to see in the field, but localized populations

Fifth instar larva of *Pterourus coroebus laetitia*.

Pupa of *Pterourus coroebus laetitia*.

Adult *Pterourus coroebus laetitia*.

Pterourus coroebus laetitia
pinned. Ventral view.

green
form

Pterourus coroebus laetitia
pinned. Dorsal view.

occur. May be seen in mountains surrounding Central Valley, in places such as San Isidro and San Rafael of Heredia, Cerros de Escazú, Cerros de Ochomogo, and the Carpintera region; and also Cerro de la Muerte, Las Alturas, and Monteverde, and the slopes of Poás, Irazú, and Turrialba volcanoes on both Caribbean and Pacific sides. **Distribution:** Costa Rica and Panama. **Seasonality:** Present from April to November; most common from June to September. **Natural history:** This montane species flies only in bright sunshine; it sometimes comes out of the forest to feed on garden flowers or, in the case of males, to drink water from puddles on dirt roads. Both sexes visit flowers of *Lantana camara*, *Stachytarpheta calderonii*, *S. mutabilis*, *Euphorbia pulcherrima*, *Inga*, *Eupatorium*, and *Mikania*. Larvae are sometimes found on upper side of leaves of avocado trees in gardens or plantations close to forest patches. **Early stages:** Eggs are honey-colored, round, with smooth surface, and are laid on mature or young trees. First-instar larva is dark brown and orange with fleshy warts bearing short black setae. Second- to fourth-instar larvae resemble bird droppings. Fifth-instar larva looks like a green lizard or snake head anteriorly and has a brown X on the back. Pupa polychromatic, presenting two forms: one resembles a green stick with lateral brown bands, a dorsal brown band from cremaster to head, and a brown dorsal projection on thorax; the other (pictured) is all brown. **Host plants:** *Aiouea montana*, *Persea caerulea*, *P. americana* (Lauraceae).

Pterourus garamas syedra (Papilionidae, Papilioninae)
Magnificent Swallowtail

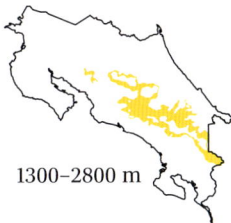

1300–2800 m

FWL: 58–70 mm. **Description:** DFW is black with two prominent yellow bands, a wide medial one and a thin post-medial one. DHW is also black with a wide yellow band from costal to anal margin; this band is spiky along its distal margin, which is followed by a row of large blue spots. HW has three tails, extensions of veins M_3 (spatulate), Cu_1, and Cu_2. VFW is same as DFW but brown instead of black. VHW is also brown, with the same yellow band but followed distally by a row of thin bluish lines and reddish-brown sub-marginal bars. **Similar species:** None. **Habitat:** Well-preserved montane forest habitats on both Pacific and Caribbean slopes, from Tilarán to Talamanca. It can be seen gliding along steep rocky hills, forest edges, and light gaps, and sometimes visits flowers in gardens or pastures surrounded by forest. Localities include Monteverde, Barva Volcano, Coronado, Cerro de la Muerte, and Las Alturas. **Distribution:** Costa Rica and Panama. **Seasonality:** Present all year. **Natural history:** This is one of the most impressive butterflies of Costa Rica, attracting the attention of observers with its huge size and bright colors. Rare individuals may be observed flying at sub-canopy level, but if there are suitable flowers, this species feeds as low as 40 cm above the ground. It flies only when it is sunny and warm, from mid-morning to early afternoon. Males actively patrol territories about the size of a soccer field, where they fly up with strong wingbeats and then glide for several meters along the light gaps. Both sexes feed from flowers

Fifth instar larva of *Pterourus garamas syedra*.

Pupa of *Pterourus garamas syedra*.

Adult *Pterourus garamas syedra*.

such as *Fuchsia arborea*, and males puddle along rivers. Both sexes also feed from flowers of *Impatiens walleriana* at ground level (pers. obs.). Females search at 2–4 m altitude for oviposition places on low branches of their host plants or young trees along forest edges. **Early stages:** Eggs are green, round, with smooth surface, and are laid on new shoots of host plants. Young caterpillars are dark green and white, with a dorsolateral row of small blue dots, and resemble a bird dropping; they rest on upper side of leaves, in the center. Mature larva is green dorsally and ranges from light brown to dark green laterally. Its swollen thorax (which gives the impression of a snake head) has, in segment T_3, a brown and pink transverse dorsal band with irregular margins and a yellowish dot at each side, resembling eyes; and four blue dots on T_3 followed posteriorly by a profuse pattern of dozens of pinkish small dots that extend to A_1 and end with a white line with four blue dots. The abdomen has a dorsal gray band (A_4–A_7) that forms an X figure with its center on dorsal A_5–A_6; the X has four blue dorsolateral dots. The X figure and the T_3 band are variable; they can be white or completely brown or dark green (see Fig. 23, p. 27). Below each spiracle is a blue dot. Pupa is brown with green, resembling a broken wooden stick, with white in dorsal area of segment T_1, suggesting lichen growth on the wood; a small dorsal protuberance on segment T_2; and two rows of small dorsolateral warts on A_4–A_6. **Host plants:** *Magnolia sororum* (Magnoliaceae) and at least two unidentified trees species in Lauraceae.

Papilio polyxenes stabilis (Papilionidae: Papilioninae)
Central American Black Swallowtail

1200–1800 m
(sporadically as
low as 250 m)

FWL: 35–43 mm. **Description:** Medium-size butterfly with a small beaklike tail on HW (M_3 vein). Two forms: One is black with a wide yellow band from FW apex to HW anal margin and a row of yellow sub-marginal spots on DFW and DHW. The other (melanistic) form is all black but with yellow sub-marginal spots. **Similar species:** It is distinguished from *Battus polydamas polydamas* (p. 64), *Pterourus menatius victorinus*, and *Heraclides astyalus* by its abdomen, which has three lateral rows of yellow spots (no lines). **Habitat:** Open areas in mountains such as pastures, fields, gardens, crops, prairies, roads, and rocky hills. Can be seen gliding and landing on low vegetation in any montane habitat. **Distribution:** Costa Rica and Panama. **Seasonality:** Present all year. **Natural history:** Eggs are laid on underside of leaves or on stems. Generally only one or two larvae are found on a given plant, since the plants are small and short-lived, dying back after a few months. Larvae are attacked by wasps and bugs (Hemiptera), which indicates they do not sequester secondary compounds from plants or produce any defensive chemicals besides the one in the osmeterium. Pupae present polychromatism, ranging from dark brown to light green; some pupae undergo several months of diapause. Adults live only 3–4 weeks, feeding on flowers such as *Emilia sonchifolia*, *Lantana camara*, and *Melanthera nivea*. Males establish territories in high, exposed places–a behavior known as hill-topping. Females sometimes mate more than once. In general terms, the melanistic form is less common than the yellow form,

Fifth instar larva of *Papilio polyxenes stabilis*.

Pupa of *Papilio polyxenes stabilis*.

although in some localities the black form is represented in a higher proportion. In the LAREBUB, pupae lasted 19.5 days (n = 16) during rainy season and 14.7 days (n = 4) during dry season. **Early stages:** Eggs are laid singly; they are yellow and become black before they hatch. The first four instars are covered in black, white, and red seta-bearing warts arranged in transverse bands. The fifth instar is transversely banded with yellow lines and black lines with a row of orange spots inside them; but color may vary from almost all black and orange to yellow and orange. **Host plants:** *Cyclospermum leptophyllum, Spananthe paniculata, Arracacia xanthorrhiza, Foeniculum vulgare* (Apiaceae).

Adult "melanic form" of *Papilio polyxenes stabilis.*

Adult *Papilio polyxenes stabilis.*

Whites, Yellows, Sulphurs, and Longwing Mimics (Pieridae)

It is said that the word *butterfly* has its origin in the common small, butter-yellow pierids of European pastures. Members of Pieridae are small to medium-size and generally white or yellow with black markings. Their bright hues often make them the most visible butterflies of any habitat—although some species are cryptic or mimic other species. There are seventy species in Costa Rica and around a thousand worldwide. They are distributed all around the world, from temperate regions to tropics, and are active during the hottest hours of the day, flying in direct sunshine. Many species are abundant inhabitants of open areas, gardens, and parks, where they can be seen feeding from flower nectar. In Costa Rica, pierids may be seen at elevations as high as 3800 m above sea level. Pierid males drink from wet sand or mud when recently emerged; they also form huge conglomerations that can in some cases reach thousands of individuals, sometimes mixed with butterflies of the families Papilionidae, Hesperiidae, Nymphalidae, and Riodinidae, and the day-flying moths of genus *Urania*.

In Costa Rica, Pieridae comprises three subfamilies: Pierinae, Coliadinae, and Dismorphinae. The first two subfamilies have the typical pierid characteristics, as described above. Dismorphinae diverges from most pierid patterns; many species evolved as mimics of the glasswings (Nymphalidae: Danainae) and the longwings (Nymphalidae: Heliconiinae), and therefore behave like them, while others present unique color patterns. All pierid species can be identified by forked tarsal claws (small claws at tips of legs) when observed under magnification.

Pierid eggs are small (relative to those of other families) and bottle-shaped, sometimes with a spiked crown and longitudinal ribs. They are laid singly (*Phoebis, Glutophrissa*) or in groups (*Catasticta, Pereute*), on the underside or upper side of mature leaves or on young leaves and new shoots. The larvae are smooth or hairy and generally very cryptic; the head capsule is always smooth-surfaced. The pupa forms with the head pointing up and is held by a silk girdle, as in Papilionidae; it is always very cryptic, resembling young green leaves (*Dismorphia, Hesperocharis*), wood or detached bark (*Pereute*), or bird droppings (*Melete, Ascia*). Adults feed on nectar from flowers that they actively visit again and again during hot days. They like a wide variety of flower species. *Phoebis* is a particularly insistent visitor of *Hibiscus* and *Malvaviscus* flowers, which few butterflies visit in Costa Rica. Except for the Dismorphinae, pierids are fast and erratic fliers, which makes them hard to observe unless they are perched.

Remarkably, given that this is an extensively collected and well-known group, a new pierid species for Costa Rica, *Cunizza hirlanda* (Chacón et al. 2018), was discovered recently.

Dismorphia amphione praxinoe (Pieridae: Dismorphiinae)
Tiger Mimic-White

0–1300 m

FWL: 35–40 mm. **Description:** An orange-and-black-striped butterfly with a few yellow dots on DFW. Wing shape long and narrow, characteristic of the longwings (Nymphalidae: Heliconiinae). *Female:* DFW has three transverse orange lines from basal to medial area, in costal margin, medial area, and anal margin; two rows of yellow dots in medial and sub-apical areas; and a spike-like tip in distal margin. DHW is black with an orange transverse band from basal to medial area and a brownish-orange band in anal margin. *Male:* Similar to female but FW is thinner than HW, and DHW has a very large, bright silver-white patch in costal area. **Similar species:** Easily distinguished by wing shape from butterflies of other families, such as *Mechanitis* (Nymphalidae: Danainae; p. 131), *Eueides isabella eva*

Fifth instar larva of *Dismorphia amphione praxinoe.*   Pupa of *Dismorphia amphione praxinoe.*

Adult female *Dismorphia amphione praxinoe.* LRMH

♂   ♀

Dismorphia amphione praxinoe pinned. Dorsal view.

Dismorphia amphione praxinoe pinned. Dorsal view.

(Nymphalidae: Heliconiinae; p. 142), and many others: *Dismorphia* has wider HWs and more acute apex to FW. It can be distinguished and from other Costa Rican members of Dismorphiinae by orange-striped color pattern on DHW. **Habitat:** This is a forest-dependent species, found on both Caribbean and Pacific slopes except in dry forest. May be seen along forest trails or on flowers close to forest patches, or flying along rivers and roads with forest on either side. **Distribution:** Mexico to Colombia. **Seasonality:** Present only during rainy season. **Natural history:** *D. a. praxinoe* has a slow and floppy flight, and it lands on leaves with the wings closed. Its flight behavior differs from the fast and powerful flight of the pierids of the other two subfamilies. In flight, wing shape, and coloration, this species mimics the distasteful *Mechanitis* and *Hypothyris* (Nymphalidae: Danainae) and some members of Heliconiinae, such as *Eueides*. It also flies close to the ground in dark areas, as its models do. Both sexes feed from flowers of *Cosmos bipinnatus, Zinnia elegans, Comaclinium montana,* and *Lantana camara.* Larvae are parasitized by a wasp (*Glyptapanteles,* Braconidae). **Early stages:** Eggs are white, bottle-shaped, and laid upon mature leaves of host plants. Larva is green with a granular texture and covered in short setae; head capsule is green. It rests on underside of leaf along veins it has exposed by eating, or on the border of the eaten part of the leaf. Pupa is green and elongate, resembling a young leaf; it has a long beak on head and not too prominent, somewhat concave wing pads. **Host plants:** *Inga sapindoides, I. densiflora, I. chocoensis, I. punctata, I. vera, I. oerstediana* (Fabaceae).

Phoebis philea philea (Pieridae: Coliadinae)
Orange-barred Sulphur

0–1500 m

FWL: 40–45 mm. **Description:** *Female:* Wing coloration is very variable, from light yellow to deep orange, with many brown markings on distal border and sub-marginal area of DFW and brown markings on distal border of DHW. *Male:* Wings are bright yellow with a prominent orange bar on DFW from costal vein through discal cell to medial area, and orange in distal border of DHW. **Similar species:** Female is similar to but larger than females of *Phoebis sennae, P. argante argante* (p. 80), and *P. agarithe.* No species is similar to male with its characteristic orange bar on DFW. **Habitat:** May be seen visiting flowers in gardens around San José and rest of Central Valley; along beaches or around swimming pools; in backyards and parks in towns in rural areas; in disturbed open areas; and in the canopy in forests on both Caribbean and Pacific slopes. **Distribution:** USA to Brazil. **Seasonality:** Present all year but most common during rainy season. **Natural history:** Males are fast and erratic fliers that fight other males to protect territories by crashing in the air and injuring each other with sharp, serrate scales on costal border of FW. In an encounter, the two males chase each other in a spiral up to 30 m high and then abruptly plummet to the ground. A male-female encounter is similar but usually ends in copulation.

Fifth instar larva of *Phoebis philea philea.*

Pupa of *Phoebis philea philea.*

Adult *Phoebis philea philea*.

Adult *Phoebis philea philea*.

Fresh males are often seen puddling along dirt roads or riverbanks at midday but never in the large numbers seen with other pierid species. Males have lived as long as 30 days in the greenhouse of the LAREBUB. Both sexes visit flowers such as *Stachytarpheta, Lantana, Impatiens, Malvaviscus arboreus, Turnera, Combretum farinosum, Delonix regia, Caesalpina pulcherrima*, and many others during the hottest hours of the day. **Early stages:** Eggs are whitish, small relative to adult butterfly size, and are laid on edges of younger leaflets or in flower structures, in greater quantity than many other butterflies. Young larvae are yellowish-green with a light yellow lateral line. In third instar, larval body has disperse short black spikes surrounded by white, and a yellow lateral line. Mature larva spends more time on leaves and becomes dark green with a wide brown lateral line and transverse dark green rings; it is also covered in many short black spikes. Right before pupation, it becomes yellow with transverse dark green rings. Pupa is green with a prominent head beak and convex ventral area, resembling a young leaf. **Host plants:** *Senna atomaria, S. hayesiana, S. hirsuta, S. papilosa, S. spectabilis, S. septemtrionalis, S. bicapsularis, S. alata, S. pallida, S. reticulata, S. occidentalis, Cassia grandis, C. emarginata, Caesalpinia pulcherrima* (Fabaceae).

Phoebis argante argante (Pieridae: Coliadinae)
Dark Apricot Sulphur

0–1700 m

FWL: 32–37 mm. **Description:** Medium-size butterfly. *Female:* DWs are yellow to white; DFW has black or brown distal margin and apex, and spots in medial area; DHW has dark spots in distal margin. *Male:* DWs are all yellow-orange with thin dark spots on distal margin of DFW that are partially or completely spaced apart. **Similar species:** Female *P. argante argante* is similar to female *P. philea philea* (p. 79), but the latter is much larger. Female *P. argante argante* has thicker black border on DFW apex than rest of border, while in *P. sennae* female, black border on DFW is even from apex to distal margin. Male *P. sennae* is lemon-green, not orange-yellow as in male *P. argante argante*. Male *P. argante argante* is distinguishable from *P. hersilia* by marginal black pattern on DFW: *P. argante argante* has a thinner and disconnected series of black spots, while in *P. hersilia* most of the spots are thicker and connected. Female *P. hersilia* has dark area on DFW twice as thick as the area on *P. argante argante* female. *P. agarithe* has a straight dark line on VFW from apex to inner margin, distinguishing it from both sexes of *P. argante argante*, which have a broken line. **Habitat:** Common in disturbed habitats such as gardens, parks, croplands, and cattle pastures. Seen visiting red flowers or puddling along rivers and roadcuts. In Guanacaste and Puntarenas it can be very abundant along rivers and beaches. Also found in Central Valley and Caribbean coast. During migration can be seen flying in Costa Rica's highest mountains. **Distribution:** Mexico to Brazil. **Seasonality:** Present all year in all habitats but can be extremely abundant during first half of rainy season. **Natural history:** This species is active during the hottest time of day. Both sexes actively visit flowers such as *Stachytarpheta*, *Lantana camara*, *Malvaviscus arboreus*, and *Caesalpina pulcherrima*. Can be seen in forest sub-canopy when males chase females up to 20 m high. Females oviposit on new shoots of host plants. North Pacific populations of *P. argante argante* are thought to migrate during dry season across the central mountains to the Caribbean slope; there are unconfirmed observations of individuals heading southward along the coastline to wetter environments. Chalcid wasps (Chalcididae) and a fly (*Blepharipa albicauda*, Tachinidae) have been recorded as pupal and larval parasitoids, respectively. **Early stages:** Eggs are yellowish-white and laid singly in new leaves of host plants. Larva is green with many transverse streaks, a light blue lateral line, darker blue dots covering dorsal area, and a light green ventral area; head capsule is green with many granulations. **Host plants:** *Inga vera, I. ruiziana, I. longispica, Zygia longifolia, Cassia biflora, Senna papilosa, S. pallida, Pentaclethra macroloba, Cojoba arborea* (Fabaceae).

Fifth instar larva of *Phoebis argante argante*.

Pupa of *Phoebis argante argante*.

Adult female *Phoebis argante argante*.

Phoebis argante argante pinned.
Dorsal view.

Phoebis argante argante pinned.
Ventral view.

Phoebis argante argante pinned.
Ventral view.

Anteos clorinde (Pieridae: Coliadinae)
White Angled-Sulphur

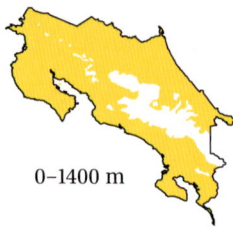

0–1400 m

FWL: 40–49 mm. Description: Medium-size butterfly. Both sexes: DWs are white, VWs are light green, and each wing has a black dot at distal end of discal cell; FW is angled to a point at apex; HW has spiked tail in distal margin. *Female:* DFW has yellow-orange bar that is diffuse or absent. *Male:* DFW has very bright yellow-orange bar from costal vein to discal cell and sometimes beyond. **Similar species:** *Anteos maerula* can be distinguished from male *A. clorinde* by yellow-green color of DWs. Females are distinguished by dark spot in DHW: in *A. clorinde* it is a black dot surrounded by orange; in *A maerula* the black dot is C-shaped or absent. **Habitat:** Seen in open areas, gardens, along rivers, and on beaches on both slopes but more common on Pacific slope. **Distribution:** USA to Argentina. **Seasonality:** Present all year but rare during dry season and very abundant during rainy season. **Natural history:** This butterfly always flies in direct sunshine and on hot days. Males puddle on beach sand or riverbanks alone or in large conglomerations. Female oviposits single eggs on young leaves of host plants, from 10:00 a.m. to 2:00 p.m., usually high in the canopy, and larvae never come down to ground level. Depending on place and reach of sunlight, female will lay eggs on lower branches, always preferring mature trees to saplings. She lays many single eggs on the same plant, returning to it day after day. Both sexes are fast and powerful fliers and visit flowers of *Lantana camara*, *Pentas lanceolata*, *Stachytarpheta*, *Hibiscus*, *Malvaviscus arboreus*, and *Impatiens walleriana*. **Early stages:** Eggs are yellowish-white, bottle-like, and laid singly. Young larva is green with many yellow and black dots all over body. Mature larva is polymorphic: one morph is yellow and feeds on flowers; another is all green; a third (pictured) is green with a cream lateral line and small black spots with blue dots above the lateral line; and a fourth has

Fifth instar larva of *Anteos clorinde*.

Pupa of *Anteos clorinde*.

♀

Adult *Anteos clorinde*.

Anteos clorinde
pinned. Dorsal view.

a yellow lateral line with a thick parallel black line above it and many transverse black lines all over the body. Head capsule is always green with many short hairs. Pupa is yellowish-green with variable black stains on wing pads, thorax, and abdomen; head beak is dark, and a dark green line runs laterally from head to abdomen tip. **Host plants:** *Cassia emarginata, Senna papilosa, S. spectabilis, S. atomaria, S. obtusifolia, S. pallida* (Fabaceae).

Abaeis salome jamapa (Pieridae: Coliadinae)
Salome Yellow

700–2700 m

FWL: 22–27 mm. **Description:** Both sexes: Small, lemon-yellow butterfly with a small spike-like tail on distal margin of HW, and a few disperse light brown markings on VWs. *Female:* Only DFW has black distal margin. *Male:* DFW and DHW have black distal margins. **Similar species:** Easily confused with *Abaeis xantochlora* (p. 84); both sexes of *A. salome jamapa* can be distinguished by prominent spike-like tails and lemon-yellow background color; *A. salome jamapa* male has wider black distal margin of DHW than *A. xantochlora*. Female *A. arbela boisduvaliana* has less acute tail on HW, and apex of DHW, which is variable, may have a black patch or lines, lacking in female *A. salome jamapa*. **Habitat:** Seen in gardens, parks, and open areas associated with riparian forest or forest edges. Common in Central Valley urban and suburban areas and on Pacific and Caribbean volcanic slopes. **Distribution:** USA to Panama. **Seasonality:** Common during rainy season but absent on Pacific slope during dry season. **Natural history:** This highly seasonal species is a common inhabitant of San José gardens and parks. Both sexes feed from flowers such as *Stachytarpheta, Lantana camara, Turnera ulmifolia,* and *Cosmos sulphureus.* Females fly during midday, looking for oviposition places, usually in vegetation close to the ground, as this species does not fly very high. **Early stages:** Eggs are white, bottle-like, and laid singly on underside of leaves. First-instar larva is whitish and covered in small black setae. Second instar is green with disperse bluish-white warts, from each of which arises a single black seta; head capsule is green with setae as on body. Third, fourth, and fifth instars are green with a yellowish lateral line and many short setae covering the body. In all instars, an evident droplet exudes from the tip of each seta, the function of which is unknown. First and second instars skeletonize leaves; third instar rests on underside of leaves. Fourth and fifth instars rest on either stalks or tops of leaves, raising up more than half of body, holding on by the last three prolegs. Pupa is green, with very concave wing pads, medium-size head beak, and a black stain ventrally where wing pads meet abdomen. **Host plants:** *Diphysa americana, Senna septemtrionalis, S. pallida* (Fabaceae).

Fifth instar larva of *Abaeis salome jamapa*.

Pupa of *Abaeis salome jamapa*.

Adult *Abaeis salome jamapa*.

Abaeis salome jamapa pinned.
Dorsal view. ♂

Abaeis salome jamapa pinned.
Dorsal view. ♀

Abaeis xantochlora xantochlora (Pieridae: Coliadinae)
Tropical Yellow

0–2500 m

FWL: 20–25 mm. **Description:** Both sexes: Small yellow butterfly with a small spike-like tail on distal margin of HW; VWs have few, disperse light brown markings. *Female:* Only DFW has black distal margin. *Male:* DFW and DHW have black distal margins. **Similar species:** Easily confused with *Abaeis salome jamapa* (p. 83): both sexes can be distinguished by smaller spike-like tails in *A. x. xantochlora*; male *A. salome jamapa* has wider black distal margin of DHW than *A. x. xantochlora*. Female *A. arbela boisduvaliana* distinguished by apex of DHW, which is variable and might present a black patch or lines, lacking in *A. x. xantochlora*; also, black margin of DFW reaches tornus in *A. x. xantochlora* female but in *A. arbela* does not. **Habitat:** Common in secondary-growth forest and open areas, seen visiting flowers along dirt roads and gardens. In the Central Valley, it visits flowers in parks and flies along rivers. **Distribution:** Mexico to Colombia and Venezuela. **Seasonality:** Present all year but most common in rainy season. **Natural history:** This common species reproduces all year long, even in dry season in the central Pacific region. Females look for young or mature trees in understory of forest patches on which to oviposit. Both sexes feed from flowers such as *Stachytarpheta*, *Lantana camara*, *Turnera ulmifolia*, and *Cosmos sulphureus*. Larvae are parasitized by a fly (*Patelloa xanthura*, Tachinidae). **Early stages:** Eggs are yellowish-white and laid in large clusters of up to 100 on underside of leaves. Larvae are almost transparent when they hatch but become light green once they start eating. First and second instars skeletonize leaves; broods are gregarious, feeding synchronously on the leaf underside, producing characteristic damage. Mature larvae rest on undersides of leaves but move to leafstalks when larger. Mature larva is bluish with a green lateral line and light brown head. As with *A. salome jamapa*, all instars have setae from which tiny droplets exude. The broods' pupae are placed together in the eaten host plant's branches; they are yellow with irregular black stains and have convex wing pads and a beaklike head, resembling small dried leaves hanging from the stalks. **Host plants:** *Senna papilosa*, *S. septemtrionalis* (Fabaceae).

Fifth instar larva of *Abaeis xantochlora xantochlora*.   Pupa of *Abaeis xantochlora xantochlora*.

Adult males of *Abaeis xantochlora xantochlora* puddling.

Abaeis xantochlora xantochlora pinned.
Dorsal view.

Abaeis xantochlora xantochlora pinned.
Dorsal view.

Whites, Yellows, Sulphurs, and Longwing Mimics | 85

Abaeis albula marginella (Pieridae: Coliadinae)
White Yellow

0–1600 m

FWL: 15–22 mm. **Description:** Small white butterfly. DFW has a wide black margin from apex to distal border. DHW is all white but variable, sometimes with black distal margin. VWs are yellowish-white to white with a few brown flecks. **Similar species:** *Eurema agave agave* and *E. agave millerorum* have gray-black costal border on DFW. *Leptophobia aripa elodia* (facing page) has a wavy brown distal border on DFW. **Habitat:** This species is less tolerant of open areas than other *Abaeis* species; it prefers secondary forest and is seen along trails and light gaps. In Central Valley, it is common in forest patches and along rivers, where it finds protection in riparian vegetation. In rural areas it can be seen along forest borders, roads, or rivers. **Distribution:** Nicaragua to Ecuador. **Seasonality:** Present all year on Caribbean slope but only during rainy season on Pacific slope. **Natural history:** Adults feed on nectar from small flowers of Asteraceae species such as *Melanthera nivea* and *Emilia fosbergii*, but also from larger flowers such as *Lantana camara* and *Stachytarpheta frantzii*. This species feeds from flowers close to the ground, always flies close to the ground, and lays its eggs on young sapling host plants no more than 1 m tall. **Early stages:** Eggs are less than 1 mm long, white, and bottle-like, and are laid singly on underside of leaves. The caterpillars feed alone on undersides of leaves in all instars. Young larva is light green. Mature larva is dark green dorsally and bright green ventrally, separated by a whitish lateral line; head capsule is green and granulated. In all stages, larval body is covered in dense short setae that exude droplets. **Host plants:** *Senna papilosa*, *S. septemtrionalis* (Fabaceae).

Fifth instar larva of *Abaeis albula marginella*.

Pupa of *Abaeis albula marginella*.

Adult *Abaeis albula marginella*.

Abaeis albula marginella pinned.
Dorsal view.

Leptophobia aripa elodia (Pieridae: Pierinae)
Mountain White

500–2000 m

FWL: 23–28 mm. **Description:** Small butterfly. DFW is opaque white with thick, wavy black apical and distal borders. DHW is all opaque white. VHW is glossy white with a small black dot at distal end of discal cell, and a yellow basal area. **Similar species:** Distinguishable from *Abaeis albula marginella* (opposite page), *Eurema agave agave*, *E. agave millerorum* (Pieridae), and all other Costa Rican small white butterflies by distinctive wavy black pattern to apical and distal border of DFW. **Habitat:** Preferred habitats are open areas, forest borders, and croplands. This butterfly is common at mid-elevations in semi-rural habitats and open areas, and in Central Valley is easy to see in parks and in vegetation along rivers. In Cervantes, Pacayas, Puriscal, and many other mid-elevation towns, it is a common inhabitant of backyards and gardens. **Distribution:** Mexico to Panama. **Seasonality:** Present all year. **Natural history:** This butterfly flies close to the ground in fast zigzag movements. When startled, it can fly high and perch with closed wings, becoming very hard to see. Adults feed from flowers such as *Impatiens*, *Lantana*, and *Tropaeolum* but also from *Zinnia elegans*, *Comaclinium montana*, *Melanthera nivea*, and *Emilia fosbergii*. This species has become a pest of radish and cabbage crops and ornamental flowers such as *Tropaeolum majus*; larvae feed on the leaves and can destroy small crops in a short time. Many parasitoids have been

Fifth instar larva of *Leptophobia aripa elodia*.

Pupa of *Leptophobia aripa elodia*.

Adult *Leptophobia aripa elodia*.

reported to attack the larvae, including flies (*Chetogena scutellaris, Hyphantrophaga virilis, Lespesia aletiae, Winthemia,* Tachinidae; *Sarcodexia sternodontis, Helicobia morionella,* Sarcophagidae) and wasps (*Apechthis zapoteca, Lymeon,* Ichneumonidae; *Brachymeria ovata, Brachymeria mnestor, Conura immaculata,* Chalcididae). **Early stages:** Eggs are small, yellow, and bottle-like, and are laid in groups on upper side of leaves. Young larvae are pale green with dark green heads and feed gregariously. When mature, they disperse to feed alone. Mature larva is green with a yellow lateral line and transverse black and yellow-green lines across the body; head capsule is green. In all stages, larval body is covered in abundant short setae. Pupa is green, covered in disperse black dots, with a longitudinal dorsal keel and a black lateral spike at segment A_4; head has a small, forward-pointing green spike with a yellow tip. **Host plants:** *Tropaeolum majus, T. moritzianum* (Tropaeolaceae), *Nasturtium officinale, Brassica rapa* (Brassicaceae), *Capparidastrum discolor* (Capparaceae).

Ascia monuste monuste (Pieridae: Pierinae)
Great Southern White

0–1700 m

FWL: 30-35 mm. **Description:** *Female:* DWs are mostly white, with a brown border that is thicker on DFW and extends to DHW, forming small triangles; VWs are variable, from dirty white on VHW and VFW apex to light yellow with brown markings. *Male:* DWs are almost all white; DFW has thin, wavy brown border in apical and distal areas; VWs are all white. **Similar species:** When alive, both sexes of *A. m. monuste* have light blue antennal tip. Female is distinguishable from all other species by conspicuous brown wavy border on DFW and DHW distal border. Both sexes are distinguished from *Glutophrissa drusilla tenuis* (facing page) by lack of yellow on base of VFW, and from *Itaballia demophile centralis* by shape of brown apex and border of DFW. **Habitat:** A common inhabitant of open areas such as parks, gardens, towns, and croplands in most of Costa Rica. This butterfly is especially common in major towns such as San José, Heredia, Cartago, and Alajuela because its host plants grow as weeds in gardens and backyards; so it is not uncommon to find pupae attached to walls and fences of houses and buildings and adults flying among people and traffic. Nonetheless, it is also possible to find it in preserved secondary forest along rivers and forest edges. **Distribution:** USA to Venezuela. **Seasonality:** Present all year; in Guanacaste, it is more abundant during rainy season, but in the rest of the country it persists all year. **Natural history:** The species is considered a pest, causing more damage to small crops and orchards than to large plantations. Adults feed from many flowers such as *Tournefortia hirsutissima, Lantana camara, Hamelia patens,* and numerous Asteraceae species. Larvae are parasitized by flies (*Lespesia archippvora, L. parviteres, Zenilla blanda,* Tachinidae); pupae are parasitized by a wasp (*Brachymeria incerta,* Chalcididae). **Early stages:** Eggs are bright yellow, bottle-like, and are laid, if the adult female is not disturbed, in groups of about forty on both underside and upper side of leaves. When young, larvae feed and rest in groups, resting on underside of leaves. When mature,

Fifth instar larva of *Ascia monuste monuste.*

Pupa of *Ascia monuste monuste.*

Adult *Ascia monuste monuste.*

most individuals disperse throughout the plant and may rest on stems. Mature larva is grayish with five yellow longitudinal lines, and entire body is covered in disperse black warts of different sizes and soft setae; head capsule is beige, granulated with black warts, and covered in setae. Pupa is white with a small yellowish dorsal keel on abdomen, a dorsal hump on thorax, a black thorn on each side of segment A$_3$, a forward-pointing spikelike projection from head, white wing pads with a brown border, and brown patches as well as small black dots dispersed throughout the body. **Host plants.** *Cleome viscosa, C. gynandra, C. spinosa* (Cleomaceae), *Crateva tapia, Quadrella cynophalloflora, Morisonia americana* (Capparaceae), *Nasturtium officinale, Lepidium virginicum, Brassica rapa* (Brassicaceae), *Tropaeolum* (Tropaeolaceae).

Glutophrissa drusilla tenuis (Pieridae: Pierinae)
Florida White

0–1300 m

FWL: 30–35 mm. **Description:** *Female*: Two forms. One is all white except for black apex on DFW; the other is white with a thick brown distal margin on both DWs and a yellowish gloss on DHW. *Male:* All white. Both sexes have a yellow patch on base of VFW. **Similar species:** Distinguished from *Ascia monuste monuste* (opposite page), *Ganyra phaloe limona*, and *G. josephina josephina* by yellow patch on base of VFW. **Habitat:** Seen flying rapidly along beaches, rivers, and gardens, on both slopes. Sporadic individuals feed on garden flowers in Santa Ana, La Guácima, and La Garita. Other localities include Guápiles, San Carlos, Limón, Garabito, Santa Rosa, and Osa. **Distribution:** Mexico to Peru. **Seasonality:** Present all year on the Caribbean slope; most common during rainy season on the Pacific. **Natural history:** This is not a rare species, but it is intolerant of dense urbanization. Both sexes can be seen feeding from flowers such as *Lantana camara, Stachytarpheta*, and many others in gardens of rural areas. Males puddle along rivers with dense forest cover, and females look for oviposition sites on young saplings of host plants in forest understory. This behavior indicates that this is a forest species more than an open-area species, as are the host plants, which are forest species rarely planted by people. **Early stages:** Females look for the base of young host plants, landing on the mature part and then walking up the trunk until they find the young green stalks (P. Venegas, pers. comm.). If there are no eggs there already, they will lay single, white, bottle-like eggs. Young larva is all green; it rests along central vein of youngest leaves, where it is completely camouflaged. Mature larva is green with a granulated texture composed by many small black and yellow warts; body and head are covered in short, dense setae, and the tail is slightly two-forked; head is also green with small black warts. Pupa is light green with disperse small yellow and black spots; a light orange and black longitudinal dorsal keel, which is interrupted in central part of pupa by a wide, flat winglike expansion, ending on each side in black peaks; and a prominent beaklike projection from head. **Host plants:** *Dichapetalum morenoi* (Dichapetalaceae), *Drypetes standleyi* (Putranjivaceae), *Crateva tapia* (Capparaceae).

Fifth instar larva of *Glutophrissa drusilla tenuis*.

Pupa of *Glutophrissa drusilla tenuis*.

Dark form adult of *Glutophrissa drusilla tenuis*.

Glutophrissa drusilla tenuis pinned.
Dorsal view.

White form *Glutophrissa drusilla tenuis*
pinned. Dorsal view.

Hesperocharis crocea crocea (Pieridae: Pierinae)
Orange White

500–1400 m

FWL: 30–34 mm. **Description:** DWs and VWs are entirely orange-yellow; VHW has a few purple-brown blotches on costal border and sub-medial area. **Similar species:** Distinguished from *Pyrisitia westwoodii* (Coliadinae) by wing shape, with more acute apex to FW and HW than *Pyrisitia*; and by the small purple-brown blotches on VHW. **Habitat:** Occurs in rural areas in middle elevations, in towns in the Central Valley, and also around crops and rivers. In San José, Alajuela, Heredia, and Cartago, individuals fly in parks and gardens searching for host plants. **Distribution:** Mexico to Panama. **Seasonality:** Present during rainy season. **Natural history:** Individuals are seldom seen along forest edges or open areas, flying fast as they visit flowers such as *Acnistus arborescens*, *Stachytarpheta mutabilis*, *Hamelia patens*, *Lantana camara*, and many others. Populations have multiple generations per year, but it is not known whether the pupa undergoes diapause during the dry season. Larvae are parasitized by a fly (*Hyphantrophaga virilis*, Tachinidae). **Early stages:** Eggs are yellow-orange, bottle-shaped, and laid in large clusters of up to sixty-eight eggs. Larvae feed gregariously on leaves and flowers. All instars are entirely covered in short spine-like setae, which exude liquid droplets in all instars except fifth. Mature larva is dull orange and covered in many black protuberances with white and black setae at the tip; head capsule is red or black with many black warts. Pupa resembles a typical *Phoebis* pupa (pp. 79-81), with somewhat concave wing pads and a small beaklike protuberance from the head; a white lateral line runs from cremaster to wing base, and there are some brown-gray spots on sides of abdominal area. For a detailed description see Braby & Nishida (2007). **Host plant:** *Struthanthus orbicularis* (Loranthaceae).

Fifth instar larva of *Hesperocharis crocea crocea*. KN

Pupa of *Hesperocharis crocea crocea*. KN

Adult *Hesperocharis crocea crocea*. KN

Hesperocharis crocea crocea
pinned. Dorsal view.

Melete lycimnia isandra (Pieridae: Pierinae)
Creamy White

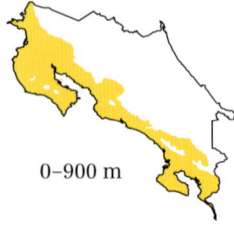

0–900 m

FWL: 28–33 mm. **Description:** Both sexes: DWs are white, yellow, or a shade in between; DFW has variably thick brown apex; VFW has brown bar at distal end of discal cell. *Female:* VWs are white, yellow, or orange. *Male:* No matter DW color, apical, sub-apical, and costal areas of VFW are yellow, the rest white, and VHW is yellow. **Similar species:** Yellow females can be confused with *Hesperocharis crocea crocea* (p. 91), which is smaller and has brown dots on VHW. White females can be confused with *Glutophrissa drusilla tenuis* (p. 89), which has yellow patch on base of VFW. Males can be recognized by combination of yellow and white on VWs. Both sexes can be distinguished from *Melete polyhymnia florinda*, which has brown apex to DFW that in females encloses diffuse yellow markings, and a much more strongly marked and thicker brown bar on VFW at distal end of discal cell; in females of *M. p. florinda*, this brown bar on VFW is visible on DFW, a character rarely seen or absent in *M. l. isandra*. **Habitat:** This species seems to require some degree of forest conservation to exist and may be seen in rural areas or in gardens close to forest patches. Found only on Pacific slope; localities include Santa Rosa, Carara, Corcovado, and

Fifth instar larva of *Melete lycimnia isandra*. KN

Pupa of *Melete lycimnia isandra*. KN

Adult *Melete lycimnia isandra*.

Melete lycimnia isandra pinned.
Dorsal view.

Melete lycimnia isandra pinned.
Dorsal view.

other national parks. **Distribution:** Texas (USA) to Costa Rica. **Seasonality:** Adults start flying in early rainy season, and breeding takes place in late rainy season. Absent in north Pacific lowlands during dry season. **Natural history:** This species has a fluttering flight that can be fast and erratic when it is alarmed. Males patrol territories during hot hours. Both sexes visit flowers such as *Oyedaea verbesinoides*, *Stachytarpheta*, *Lantana*, and *Zinnia elegans*. It is not known what happens to adults during dry season in northern dry forest; it has been suggested that they migrate to wetter habitats to survive and return in early rainy season. It is also not known why in some years the population explodes into hundreds of individuals, but in others it is very rare to see a single one. Males puddle along dirt roads and riverbanks. **Early stages:** Eggs are yellow, short, bottle-like with a crown, and laid in clusters of about sixty. First-instar larva is light green with colorless setae that exude fluid droplets from their tip; head capsule is black. Mature larva is brown with many small cream-colored plates dispersed along the body, from which arise short, white hairy setae; head capsule is dark brown, also covered by white setae. Pupa is variable in color, from mostly white to covered in many brown and black markings; it has a dorsal abdominal row of short, two-forked thorns (one per segment, although absent from some segments); a black lateral spine on segment A_4, followed anteriorly by a shorter white spine; and a rounded dorsal keel on T_1 and T_2; head bears a short, curved, two-forked projection. For a detailed description see Braby & Nishida (2010). **Host plant:** *Phoradendron quadrangulare* (Santalaceae).

Catasticta cerberus (Pieridae: Pierinae)
Costa Rican Dartwhite

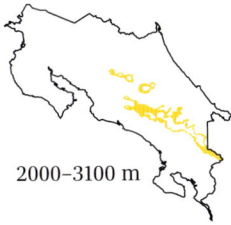

2000–3100 m

FWL: 25–29 mm. **Description:** Small butterfly. DWs are black with a longitudinal white band crossing both wings, followed distally by a row of sub-marginal white dots. VWs have variegated pattern of black-and-white zigzag bands crossed by bright yellow lines. **Similar species:** Distinguished from other similar *Catasticta* species by wider white band on DWs and a slight depression in FW costal margin. Ventrally, it is the only *Catasticta* with long, straight yellow lines. **Habitat:** Found in high-elevation forest on Cerro de la Muerte, Talamanca mountains, and tops of Irazú and Barva volcanoes. **Distribution:** Costa Rica and Panama. **Seasonality:** One generation occurs during dry season in February–April, and another, smaller generation occurs during rainy season in August–October. **Natural history:** This rare species has a fast and erratic flight and is active in morning sunshine; males set and patrol territories in the forest canopy and in *Chusquea* habitats. Both sexes feed from flowers such as *Senecio*, *Solanum*, *Eupatorium*, and *Fuchsia paniculata*. **Early stages:** Eggs are yellow, with a compressed bottle-like shape and a crownlike tip, and are laid in small clusters of about fifteen. First-instar larva is yellow and covered by long whitish setae, with a black head capsule. Mature larva is green and covered in fuzzy white setae emerging from cream-colored warts; head capsule is shiny black with white setae. Pupa resembles a bird dropping, with an irregular combination of white, reddish-brown, and black areas; a short, wavy dorsal ridge in abdominal area; two smooth keels on segments T_2 and T_1; and a short, beaklike forward projection from head. For a detailed description see Braby & Nishida (2010). **Host plant:** *Dendrophthora costaricensis* (Santalaceae).

Larvae of *Catasticta cerberus*. KN

Pupa of *Catasticta cerberus*. KN

Adult *Catasticta cerberus*. KN

Catasticta cerberus pinned.
Dorsal view.

Catasticta cerberus pinned.
Ventral view.

Catasticta teutila flavomaculata (Pieridae: Pierinae)
Pure-banded Dartwhite

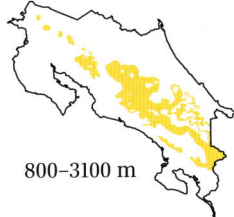

800–3100 m

FWL: 28–32 mm. **Description:** *Female:* DFW is black with a wide orange band from costal to anal margins in medial area; DHW has same pattern as DFW but also has small white marginal triangles between veins. VFW is brown with same orange band as on DFW, a faint row of small white-orange spots parallel to distal margin, and sharp orange-white triangles at distal margin between veins; VHW is gray with many orange spots all over wing, a faint white band in post-basal area from costal to anal margin, a zigzag whitish line in sub-marginal position parallel to distal margin, and sharp white and orange triangles in distal margin between veins. *Male:* DFW is black with a bluish-white medial band from costal to anal margins, and a row of sub-marginal whitish spots at distal margin; DHW has same pattern as DFW but also has small white marginal triangles between veins. Male's DHW pattern is similar to female's, but the band is whitish instead of orange. VWs are very similar to female's. **Similar species:** Female *Catasticta flisa melanisa* lacks orange band that is present on female *C. teutila flavomaculata*. Males are differentiated by VHW patterns: male *C. flisa* has very few and small yellow spots and zigzag patterns, while these are large and copious in *C. teutila flavomaculata*. **Habitat:** Open areas such as pastures, roadsides, riversides, and forest edges associated with forest patches, in mountains and hilly habitats. Locations include Monteverde, Cerro de la Muerte, Cerros de Escazú, Barva Volcano, Irazú Volcano, La Amistad International Park, and Chirripó National Park. **Distribution:** Costa Rica and Panama. **Seasonality:** Present all year but more common between February and April (Braby & Nishida 2010). **Natural history:** This is a common species that flies fast and erratically, often close to the ground but also seen perching and visiting flowers higher in the forest sub-canopy. Males often visit mud on roads or riversides. Adult activity is restricted to the hottest hours of the day, in sunshine, when both sexes visit flowers of Asteraceae plants, such as *Senecio*, *Stevia lucida*, and *Comaclinium*

Larvae of *Catasticta teutila flavomaculata*. KN

Pupa of *Catasticta teutila flavomaculata*. KN

Adult *Catasticta teutila flavomaculata*.

Catasticta teutila flavomaculata pinned. Dorsal view.

Catasticta teutila flavomaculata pinned.
Dorsal view.

Catasticta teutila flavomaculata pinned.
Ventral view.

montana, and also *Maianthemum* (Asparagaceae), *Symphonia* (Clusiaceae), and *Inga* (Fabaceae). The Squirrel Cuckoo (*Piaya cayana)* is an important predator of larvae (J. Corrales, pers. comm.). **Early stages:** Eggs are yellow-orange, barrel-shaped, and laid in clusters of up to ninety on upper side or underside of young leaves. Mature larvae rest, feed, and molt synchronously but disperse in smaller groups. Mature larva is greenish-yellow with a longitudinal black dorsal line, and is covered in many small yellowish dots and disperse soft white and yellow setae; segments T_1 and A_{10} have a black dorsal plate; head capsule is black and covered in white setae. Pupa resembles a bird dropping, with mixed areas of white, reddish-brown, and black; it has dorsal keels on T_1, T_2, and abdominal segments, and a short, curved, forward-projecting horn on head. For a detailed description see Braby & Nishida (2010). **Host plants:** *Dendrophthora costaricensis, Phoradendron tonduzii* (Santalaceae).

Pereute charops (Pieridae: Pierinae)
Darkened White

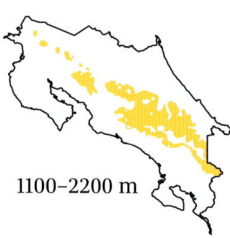

1100–2200 m

FWL: 35–42 mm. **Description:** *Female:* Slightly larger than male; DWs are dark brown; DFW has a wide orange band from costal margin to distal margin; VWs are similar, but VHW also has a bright yellow line from basal area to middle of costal border and a smaller, very short yellow line on base of VHW along costal margin. *Male:* DFW is black with a disperse bluish-gray dusty pattern; VWs are similar to female's, but band on VFW is yellower and thinner. **Similar species:** Male *Pereute cheops* has a yellow band on DFW from costal to distal margins; female *P. cheops* has orange bands on VHW, not yellow as in *P. charops.*

Habitat: Montane habitats; around the Central Valley, it is common in San José de la Montaña, Monte de la Cruz, Cerro Ochomogo, La Carpintera, Coronado, San Antonio de Escazú, Tapantí National Park; it is also common in Monteverde. **Distribution:** Mexico to Peru. **Seasonality:** Present all year but more abundant during dry season. **Natural history:** This butterfly can be observed gliding in forest edges and gardens and perching in forest sub-canopy. It feeds from flowers such as *Pentas lanceolata, Duranta, Euphorbia pulcherrima, Clibadium surinamense, Eupatorium,* and *Fuchsia arborescens.* J. Corrales (pers. comm.) has observed that Squirrel Cuckoos (*Piaya cayana)* have learned where to look for mature larvae and will periodically inspect those places to find them in July–September. Larvae and pupae are parasitized by flies (*Jurinia, Trichophora,* Tachinidae) and wasps (*Brachimeria, Conura,* Chalcididae). **Early stages:** Eggs are bright yellow, bottle-like, and laid in large clusters of about 180 eggs on underside of host plant's leaves (J. Corrales, pers. comm.). After hatching, larvae are yellow with black heads; some older larvae eat younger ones, considerably reducing number in cluster. Larvae rest in place until they finish eating the leaf they hatched on, then they spread in smaller groups to neighboring leaves. Third-instar larvae rest during the day in clusters at base of hemiparasitic host plant; at night they move synchronously to the leaves to eat. Mature larvae are dark brown, covered in long, not too dense, soft yellow setae, which arise from small yellow warts; head capsule is black with yellow setae. They rest during the day at base of trunk of host tree, and move during the night, in a processionary formation, to the host plant to feed (normally up in the canopy). Pupa is shiny black, with a spiky dorsal keel on abdomen, then a depression followed by a smooth, rounded dorsal keel on first thorax segment; head bears a short, two-forked peak pointing dorsally. As in all its early stages, pupae are highly gregarious, grouped at base of trunk of host tree, and all adults emerge synchronously. Details about the biology and early stages of *P. charops* have been published by Braby & Nishida (2010). **Host plants:** *Phoradendron undulatum*, *P. tonduzii* (Santalaceae). Commonly called *matapalos*, these are hemiparasitic plants that grow on host trees; their roots invade the host's vascular system to obtain water and minerals, but *matapalos* have chlorophyll and photosynthesize.

Group of larvae of *Pereute charops*. JCo

Adult female *Pereute charops*. JCo

Adult *Pereute charops* emerging from their pupae. JCo

Pereute charops pinned.
Dorsal view.

Pereute charops pinned.
Dorsal view.

Blues and Hairstreaks (Lycaenidae)

Lycaenidae is a very diverse family of small, delicate butterflies distributed worldwide in habitats with vegetation during at least some portion of the year. The world's smallest butterfly, *Micropsyche ariana*, from Afghanistan, with a wingspan of about 7 mm, belongs to this family. There are about 5200 species of lycaenids worldwide, 1200 in the Americas, and some 250 in Costa Rica. Lycaenidae has seven subfamilies; four (Miletinae, Lycaeninae, Theclinae, Polyommatinae) occur in the Americas, the latter two in Costa Rica.

Lycaenidae has been poorly studied in the Neotropics; there are no complete field guides, and the taxonomy is constantly changing. The most important diversity work of recent years is that of Robbins & Lamas (2004). The lycaenids are difficult to study because their wings are very fragile and easily destroyed if handled by an inexperienced entomologist. Their populations are often highly localized and therefore not easy to find, and they are fast and erratic fliers, hard to see in the wild and to catch.

Most adults of Theclinae (hairstreaks) are recognized by the peculiar behavior of moving their hind wings gently up and down when perched, causing the movement of small tails (extensions of veins Cu_1 and Cu_2), which mimic the movement of antennae and mouthparts. This is probably a defense mechanism to distract predators to a nonvital part of the body. Adults of Polyommatinae (blues), often seen puddling on mud and wet sand during hot days, are slightly smaller; some also have hind-wing projections but much smaller ones. Adults of both subfamilies always perch with the wings closed.

Most lycaenid species lay their eggs on host plants, where the caterpillars feed on leaves and flower parts (although some African, Asian, and European species also prey on insect nymphs and ant larvae). Myrmecophily, or relationships with ants, occurs with some lycaenid larvae, which have organs on the abdomen (segments A_7 and A_8) that provide food for ants.

Adults of Lycaenidae are distinguished from adults of all other butterfly families by the absence of the humeral (H) vein in the hind wing (Murillo-Hiller 2008). They feed on flower nectar.

Strymon megarus (Lycaenidae: Theclinae)
Megarus Scrub-Hairstreak

0–1300 m

FWL: 14–19 mm. **Description:** HWs have tail-like extensions of veins Cu_1 and Cu_2. *Female:* DFW is all brown. DHW is brown with an orange sub-marginal patch from Cu_1 to Cu_2 and sometimes, anterior and posterior to it, a sub-marginal row of small blue spots, variable in size and number; another small orange patch on tornus; and a thin sub-marginal white line on distal margin. VFW is grayish-white with post-medial row of orange dots from costal margin to Cu_2, but broken at M_3 level; and reddish distal margin. VHW is same color as VFW, with red spots at $Sc+R_1–R_5$ in basal area; another inside discal cell; a broken row of red spots from $Sc+R_1$ to anal margin in medial area; and an orange sub-marginal spot from Cu_1 to Cu_2, followed by a small black and red spot on tornus. *Male:* DFW is dark greenish-brown with some transverse diffuse metallic-blue lines from basal area to post-medial area along Cu_2 and 2A veins; and a prominent circular patch of scent scales (androconia) at end of discal cell. DHW is dark greenish-brown with blue transverse lines from basal area to distal margin, along M_3 to 2A; and a red crescent-shaped sub-marginal spot from Cu_1 to Cu_2. VWs are similar to female's. **Similar species:** *Strymon serapio* lacks red spots in basal area of VHW. Male *S. ziba* lacks blue on DFW. Females of *S. ziba* and *S. megarus* are not distinguishable by external characters, only by structure of genitalia. **Habitat:** Secondary forest and forest edges on both Pacific and Caribbean slopes; seen flying 1–4 m above the ground. Especially common in pineapple crops during flowering. **Distribution:** Mexico to Brazil. **Seasonality:** Found all year but more common during rainy season. **Natural history:** Males establish territories on hilltops and forest edges during afternoons in secondary forest and croplands where *Heliconia* and bromelias are abundant; adults like to feed on nectar of *Heliconia* flowers. This species is an important pest of pineapple

Larva of *Strymon megarus*. HBI

Pupa of *Strymon megarus*. HBI

Adult male *Strymon megarus*. EVC

Strymon megarus pinned. Dorsal view.

Strymon megarus pinned.
Dorsal view.

Strymon megarus pinned.
Ventral view.

crops and can damage up to 50% of production if not properly managed. Females lay eggs in pineapple bracts two weeks after flowers bloom. Adults live for seven days. **Early stages:** Mature larvae burrow into pineapples; while feeding, they produce a sticky fluid that causes decomposition of the pineapple flesh. Larva is pink, covered in short black setae, and somewhat dorsoventrally compressed, with body segmentation distinctly marked; head capsule is pink with black mouth area. Pupa is round and dark brown with some dark marks on dorsolateral areas of abdomen and thorax; it is usually hidden between pineapple bracts. **Host plants:** *Ananas comosus, Bromelia pinguin* (Bromeliaceae).

Ostrinotes keila (Lycaenidae: Theclinae)
Keila Hairstreak

0–2000 m

FWL: 14–17 mm. **Description:** HWs have tail-like extensions of veins Cu_1 and Cu_2. *Female:* DFW is blue in basal half and dark brown or black in distal half; DHW is almost all blue, with brown in costal margin and a thin brown area at distal margin. *Male:* DWs are similar to female's but with less blue on DFW and more blue on DHW. Both sexes: VWs are gray with a thin, white and brown medial line from VFW costal margin to VHW anal margin, its pattern broken up in VHW, and a second, sub-marginal faint grayish line almost parallel to that one; VWs have no red lines or markings except a small amount of orange in eyespot in VHW sub-marginal area. **Similar species:** Many hairstreak genera and species are very similar, and a careful observation, mainly of VWs, is necessary to ensure identification. In *O. keila*, the medial and sub-marginal longitudinal white lines in VHW are almost equally distanced from each other along their trajectory; in other *Ostrinotes* species, including *O. empusa*, and in *Gargina gargophia*, the two lines become closer, almost touching each other at the level of the orange eyespot. On DFW, *G. gargophia* male has a rounded patch of scent scales (androconia) almost touching the costal margin, while in *O. keila* the patch has a straight proximal margin and is separated from the costal margin by a space almost as large as the patch itself. **Habitat:** On both Pacific and Caribbean slopes in primary and secondary forest, riparian forest, and small forest patches in Central Valley and rural areas. Generally seen resting on low vegetation in forest edges or open areas. **Distribution:** Mexico to Colombia. **Seasonality:** Present all year. **Natural history:** Not much is known about this species; because of its fast and erratic flight and small size, it is ignored by most butterfly observers. It is active during the hot hours of the morning. **Early stages:** Mature larva is several tones of brownish-purple, covered in a layer of dense, short, soft setae, and has a very unusual shape, with two longitudinal dorsal ridges in the anterior half and four transverse ridges in the posterior half, all surrounded by a perimeter resembling a tire-like structure. Pupa is brown, rounded, and also covered in small, soft setae. **Host plant:** *Cupania glabra* (Sapindaceae).

Larva of *Ostrinotes keila*.

Pupa of *Ostrinotes keila*.

Adult female *Ostrinotes keila*.

Ostrinotes keila pinned.
Dorsal view.

Ostrinotes keila pinned.
Ventral view.

Cyanophrys herodotus (Lycaenidae: Theclinae)
Tropical Greenstreak

0–1800 m

FWL: 13–19 mm. **Description:** *Female:* DFW is dull blue with apex and distal margin dark brown; DHW is same blue color, with a thin brown distal margin. *Male:* DFW is metallic bright blue with narrow black costal and distal margins, and apex also black; DHW is same color, with distal margin dark brown. Both sexes: VFW is bright green, except for grayish-silver anal margin and brown distal margin. VHW is also bright green, with small post-medial white and brown markings, varying in number from only a tiny one to six or seven reaching the costal margin; tornus is dark brown or black, and distal margin is brown; sometimes there is a short tail extending from vein Cu_2, but it can be absent. **Similar species:** Many other Theclinae species are similar, especially *Cyanophrys amyntor*. Male *C. herodotus* has two short clusters of scent scales (androconia) on DFW, while male *C. amyntor* has one very long cluster, visible only under magnification; and on DFW, between veins R_3 and M_1, *C. herodotus* lacks upper discal-cell vein, which is present in *C. amyntor*. Females can be distinguished only by examination of genitalia. **Habitat:** On both Pacific and Caribbean slopes in disturbed habitats and open areas such as gardens, backyards, parks, and croplands. In Central Valley, common in San Pedro, Moravia, Alajuela, and Heredia. **Distribution:** Mexico to Argentina. **Seasonality:** Present all year. **Natural history:** Females can be seen laying eggs from mid-morning to midday when there is sunshine. They fly swiftly around host plants, land, and spend some seconds walking to find the right place to oviposit, generally on a flower, sometimes on underside of young leaves. Males establish territories on forest edges and hilltops in mid-afternoon. **Early stages:** Mature larva lives in and eats flowers of host plants. It is green with prominent purple dorsal humps on each body segment and purple lateral areas, and is covered in many small white dots and disperse short, black and white setae; head capsule is light green. Pupa is round, dark brown, and covered in many short, dark setae; it has a small yellow dorsolateral mark on segment T_1. **Host plants:** *Stachytarpheta mutabilis*, *S. frantzii*, *Lantana camara* (Verbenaceae).

Dark form larva of *Cyanophrys herodotus*.

Light form larva of *Cyanophrys herodotus*.

Pupa of *Cyanophrys herodotus*.

Adult *Cyanophrys herodotus* laying an egg on host plant.

Cyanophrys herodotus
pinned. Dorsal view.

♂

♀

Cyanophrys herodotus pinned.
Dorsal view.

♂

Cyanophrys herodotus pinned.
Ventral view.

Arawacus sito (Lycaenidae: Theclinae)
Fine-lined Hairstreak

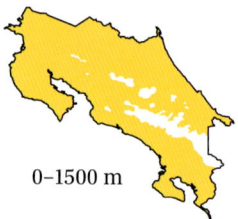

0–1500 m

FWL: 16–19 mm. **Description:** HWs have tails extending from veins Cu_1 and Cu_2. *Female:* DFW is white (sometimes with some blue) with distal third brown and some brown scales in basal area; DHW is white with some brown scales in basal area, a brown sub-marginal line on distal margin, and a small orange patch on tornus. *Male:* DWs are blue where female's DWs are white; VFW has an orange patch anterior to patch of scent scales (androconia). Both sexes: VFW is white with thin, parallel longitudinal dark brown lines from costal margin to discal cell in post-basal and medial areas and to 2A in medial and post-medial areas; distal margin is brown. VHW is also white, with thin, parallel longitudinal dark brown lines from costal margin to tornus, which is orange; distal margin is brown, followed by white scales. **Similar species:** *Arawacus leucogyna* has thinner brown longitudinal lines in VFW than *A. sito*; the two lines in medial and post-medial areas begin in costal area and end at vein 1+2A in *A. sito*, but in *A. leucogyna* end at Cu_2. **Habitat:** Occurs on both Pacific and Caribbean slopes in association with secondary- and primary-forest edges; found in dry lowlands in Guanacaste and rainy lowlands of Caribbean slope. Central Valley localities include Ciudad Colón, San Isidro de Heredia, and Alajuela. **Distribution:** Mexico to Costa Rica. **Seasonality:** Present during rainy season. **Natural history:** Although it is not common, this species is widely distributed and lives in many habitats. It perches on leaves 1–2 m above the ground. Males establish territories in light gaps surrounded by forest (where host plants grow) at low altitudes. They perch on top of leaves and chase away other small butterflies with impressive velocity. Females fly at midday looking for oviposition sites, which are always 20–30 cm above ground level, on host plant saplings. Both sexes feed on flowers of various Asteraceae, including *Bidens pilosa*. The black stripes on the VWs obscure the head while directing attention to the orange tornus area and the tails, resembling false antennae in a nonvital part of the body. **Early stages:** Mature larva is green, elliptic, and covered in dense whitish setae; head is concealed beneath thorax. It rests on central vein on upper side of host plant's leaves. Pupa is rounded, with abdomen thicker than thorax and head; abdomen is light green with a thin, brown longitudinal line on dorsum; thorax is dark green with longitudinal bands on wing pads; dorsal part of thorax and head are brown. Pupa is covered in many disperse long, soft white setae and attached to surface by a silk girdle on segment A_1. **Host plants:** *Solanum rudepannum, S. umbellatum, S. torvum* (Solanaceae).

Larva of *Arawacus sito*.

Pupa of *Arawacus sito*.

Adult female *Arawacus sito*.

Arawacus sito pinned. Dorsal view.

Arawacus sito pinned. Dorsal view.

Arawacus sito pinned. Ventral view.

Pseudolycaena damo (Lycaenidae: Theclinae)
Sky-blue Hairstreak

0–1400 m

FWL: 45–55 mm. **Description:** *Female:* DWs are almost completely metallic pale blue-green; dorsal thorax and abdomen are also bluish-green. DHW has two small sub-marginal black spots from Cu_1 to 2A; Cu_1 extends as a short black tail, and Cu_2 as a longer one. VWs are bluish-white with many disperse and variably sized black markings, some of which in VHW form two somewhat broken lines: one medial and the other sub-marginal; a conspicuous black dot is on sub-marginal area from Cu_1 to Cu_2. *Male:* DWs are similar to female's but with much thinner black margins; VWs also similar to female's; FW shape is variable: some individuals present a very acute apex, others do not. **Similar species:** *Pseudolycaena marsyas* and *P. damo* can be distinguished by characteristics of the genitalia, but in general terms, *P. marsyas* has much thicker black costal and distal margins on DFW, covering almost a quarter of the wing area. However, the presence of *P. marsyas* in Costa Rica is still doubtful and awaits confirmation. **Habitat:** Occurs on both Pacific and Caribbean slopes, in virtually any habitat within its range, from primary rainforest to dry forest to urban areas. Rare individuals may be seen, flying quickly and erratically, in open areas, along rivers and trails, and in forest light gaps. They sometimes visit flowers in gardens, backyards, parks, and croplands. **Distribution:** Mexico to Ecuador and Venezuela. **Seasonality:** Present all year. **Natural history:** This species is a skittish and fast flier, which makes it hard to observe. Isolated individuals are usually seen flying nervously from one flower to another or around a

Larva of *Pseudolycaena damo*.

Pupa of *Pseudolycaena damo*.

♂

Pseudolycaena damo pinned. Dorsal view.

Adult *Pseudolycaena damo*.

host plant while laying eggs. Its wide distribution is likely attributable to the large number of different species and families it uses as host plants. Adults sometimes feed on fallen fruits but mostly feed on small white flowers such as *Cordia alliodora*, *Croton draco*, *Acnistus arborescens*, *Lantana camara*, *Alternanthera pubiflora*, and Asteraceae species. **Early stages:** Eggs are turban-shaped, with a very sculptured surface, and are laid singly in new shoots of host plants. Mature larva has sluglike shape, thicker in thorax area, tapering through abdomen; its color is variable, from red to green, but it always has a brown longitudinal dorsal line and is covered in dense, short white setae; head is concealed beneath thorax. Pupa is round and brown, with a few short, white and black setae dispersed along body; and prominent dorsolateral black dots on segment T_2, which resemble eyes somewhat, probably serving to deter predators. Pupa is attached ventrally to substrate by a silk girdle on segment A_1. **Host plants:** *Tetrapterys discolor*, *Byrsonima crassifolia*, *Heteropterys laurifolia* (Malpighiaceae), *Roupala montana* (Proteaceae), *Thouinidium decandrum* (Sapindaceae), *Licania arborea* (Chrysobalanaceae), *Inga vera*, *Pterocarpus officinalis* (Fabaceae), *Mangifera indica* (Anacardiaceae), *Rosa* (Rosaceae).

Eumaeus godartii (Lycaenidae: Theclinae)
White-tipped Cycadian

0–1300 m

FWL: 29–38 mm. **Description:** *Female:* DFW is black with disperse bluish-green scales inside discal cell and medial area; DHW is black with a row of sub-marginal golden patches on distal border followed by a white rim. *Male:* DFW is black with many blue scales dispersed all over wing; DHW is like DFW but with sub-marginal row of small golden patches followed by a white rim. Both sexes: VFW is black with a white distal rim; VHW is black with three rows of golden spots from post-medial to sub-marginal area, white distal margin, and a prominent red patch on anal margin at abdomen level; ventral part of abdomen is bright red. **Similar species:** *Theorema eumenia* does not have a red abdomen or a red patch on VHW anal margin. **Habitat:** Occurs on both slopes, in rainforest habitats only, aways inside the forest understory in well-preserved habitats. Seen along trails and small creeks in Cahuita, Tortuguero,

Larvae of *Eumaeus godartii*.

Pupae of *Eumaeus godartii*; note Langurid beetles on top of them.

Adult *Eumaeus godartii*.

Braulio Carrillo, Arenal Volcano, Carara, and Corcovado National Parks. In Central Valley, occurs in Ciudad Colón and other areas. **Distribution:** Nicaragua to Ecuador. **Seasonality:** Present all year on the Caribbean slope but only during rainy season on the Pacific slope. **Natural history:** This beautiful butterfly is very sensitive to habitat alteration and therefore dependent on well-preserved habitats. Females live in dark forest understory where there are healthy populations of its host plants; individuals take sporadic flights in morning sunshine and in the afternoon search for males. Males establish territories for several days in the same place, around 15 m high, and perch on the upper side of leaves with the head pointing slightly downward. From this vantage, they rapidly chase other individuals out of their territories; they are mostly active from 3:00–5:00 p.m. Male has prominent, retractile androconial hair tufts in its reproductive organ. The adults' very bright coloration and slow flight suggest that *E. godartii* probably sequesters toxic secondary compounds from its host plants. This species has become a potential pest on the Caribbean slope in plantations of ornamental cycads, where it can become very abundant. Larvae and pupae are often accompanied by several individual beetles (Languriidae), which seem to feed on something in the larval and pupal exoskeletons without harming them; these beetles have also been observed feeding on larval and pupal carcasses. **Early stages:** Eggs are cream-colored, dorsoventrally compressed, with granulated surface, and are laid in groups of up to seventy on underside of leaves of host plant. Larvae feed synchronously and rest on underside of leaves. Mature larva is bright red with yellow transverse dorsal bars from segment T_3 to A_6, and is covered in disperse short black setae; head capsule is black and smooth. Pupa is round and red with many disperse black dots and patches of short white setae; pupae are attached in groups to underside of leaves with abundant silk web, and each pupa has a silk girdle keeping it close to the leaf. **Host plants:** *Zamia neurophyllidia, Z. fairchildiana, Z. skinneri* (Zamiaceae).

Eumaeus godartii pinned. Ventral view.

Eumaeus godartii pinned. Dorsal view.

Metalmarks (Riodinidae)

The metalmark family contains around 1600 species worldwide, of which 281 are known to live in Costa Rica, but because few samples have been taken, there are probably many more to be found; *Parcella amarynthina* has recently been caught in the Costa Rican central Pacific area (Murillo-Hiller, 1st rec.). This group reaches its highest diversity in the Neotropics, which has around 1330 species. Diversity greatly decreases away from the equator, with no species occurring as far as south as Chile and just one (*Apodemia mormo*) reaching as far north as southwestern Canada. Tropical Africa and Asia have very few species compared with the Neotropics.

Riodinidae is a very diverse group of small butterflies, which in Costa Rica vary in size from *Theope nycteis* (FWL: 9 mm) to *Eurybia patrona* (FWL: 35 mm). Riodinid behavior and natural history are poorly known; the major modern work is that of DeVries (1997). In general terms, most species seem to have very localized populations, concentrating in areas where they have all necessary resources within a few meters, as in the case of *Myselasia leucon*, sometimes isolated from others by many kilometers. Fewer species have wide ranges and occur in many different habitats; examples are *Eurybia lycisca* and *Rhetus arcius*.

In Costa Rica, riodinid larvae feed from the leaves and flowers of around 37 different host plant families, although there are some species that feed on items other than plant tissue, such as scale insects. More widespread in the family is myrmecophily, or close relationships with ants of more than 28 species, present in species belonging to several tribes of the subfamily Riodininae. Their larvae have specialized eversible organs, one pair on segment T3 and the other pair on A8, that provide a food secretion for the ants. When the larva requires the ants' protection, it makes an ultrasonic call with an organ situated in the prothoracic plate that vibrates against the top of the head, which is full of granulations that produce the sound signal.

According to Seraphim et al. (2018), Riodinidae comprises two subfamilies: Nemeobiinae and Riodininae, both present in Costa Rica. Adults of this family are distinguished from Nymphalidae by the lack of ventral longitudinal keels on the antennae (Murillo-Hiller 2008); and from Lycaenidae by a modification in male riodinids of the anterior pair of legs, which have five segments fused and are not used for walking (Scoble 1992).

Methone noctula (Riodinidae: Nemeobiinae)
White-rayed Metalmark

0–1800 m

FWL: 22–24 mm. **Description:** DWs are black with a lacy white margin; female has faint white, short, parallel apical lines on DFW. VWs are black with conspicuous and profuse parallel white lines that disappear toward basal area, where a prominent red patch is present on both VWs but larger on VFW. **Similar species:** *Uraneis ucubis* (Riodininae) and females of *Necyria duellona* and *N. ingaretha* (Riodininae) have conspicuous sub-marginal white markings on DHW, lacking in *M. noctula*. *Eumaeus godartii* (p. 107) is distinguished by absence of basal red patch on VFW. **Habitat:** Present on both Pacific and Caribbean slopes.

Flies in understory and light gaps in primary and secondary forest in places where extensive forest patches still persist, such as Osa Peninsula, Guanacaste, Sarapiquí, and San Carlos, and in Central Valley in Ciudad Colón, Palmares, and Tres Ríos. **Distribution:** Mexico to Venezuela. **Seasonality:** Present all year. **Natural history:** This butterfly has shown an abrupt population decline in the past 20 years in the San José area, where it used to be abundant; its populations have diminished in other, rural places as well. Females fly in the hot morning hours with a slow, floppy flight. Males perch on undersides of leaves with the wings closed and the head pointing away from the plant, watching for intruders; they actively protect their territories in the early morning (7:00–8:00 a.m.) and again in the afternoon (3:00–4:00 p.m.), in light gaps 5–10 m above the ground. They are never very active fliers, do

Larva of *Methone noctula*.

Pupae of *Methone noctula*.

Adult *Methone noctula*.

Methone noctula pinned.
Dorsal view.

not move too far from where they were born, and therefore have very localized populations. Both sexes feed from small white flowers such as *Eugenia*. **Early stages:** Eggs are laid in groups of up to ninety in the host plant canopy. Larvae are highly gregarious and move in a processional fashion. They feed at night in the canopy and rest during the day at base of host plant's trunk. Mature larva is black with a grayish-white dorsal band on each segment, and is covered with dense, short setae and many long whitish setae; head capsule is smooth and orange. When ready to pupate, caterpillars gather in a large mass on underside of a leaf close to host plant trunk. Pupae are formed together, side by side, on the leaf underside, with a silk girdle across segment A$_1$ attaching each pupa to the leaf; the ventral area is against the leaf, exposing only the dorsal area, which has black markings on dorsal and dorsolateral areas of all thorax and abdomen segments, accompanied by long black setae; wing pads have black stripes corresponding to future wing venation. **Host plants:** *Spondias mombim*, *Tapirira mexicana*, *Anacardium excelsum* (Anacardiaceae).

Pelolasia chrysippe (Riodinidae: Nemeobiinae)
Golden Euselasia

0–1500 m

FWL: 12–15 mm. **Description:** *Female:* DFW is black with a large, rounded, medial to basal yellow patch; DHW is yellow with brown costal and distal margins. *Male:* DFW is black with a large, rounded medial to basal orange patch; DHW is orange with black margins except for costal margin, which is yellow. Both sexes: VWs are yellow with small sub-marginal spots on VHW. **Similar species:** *Pelolasia aurantia* and *Marmessus gyda gydina* have white VWs, not yellow. *Erythia labdacus reducta* has prominent longitudinal red bands on VWs. *Erythia aurantiaca aurantiaca* has a longitudinal medial line on VWs, a distinctive, acute DW apex, and a different shape to black area of DFW. *Pelolasia bettina* is very similar to *P. chrysippe* ventrally, but DWs are all black. **Habitat:** Occurs in rainforest habitats on Pacific and Caribbean slopes. Both sexes frequent open forest light gaps and riversides. Locally common in Turrialba valley, Cahuita National Park, and Las Alturas de Cotón. **Distribution:** Mexico to Colombia. **Seasonality:** Present all year. **Natural history:** This species persists a long time in local populations if habitat does not suffer degradation. It may be seen flying quickly during morning and early afternoon. Males are territorial and expel male intruders from their territories by chasing them at high speed, making circular flights as they chase each other. Early in the morning, when the sun is not directly overhead, males copulate with females (T. Johnson, pers. comm.). Females search for oviposition spots in the afternoon. Adults feed from extrafloral nectaries (places other than flowers where plants release sugar solutions) and nonfloral structures of plants such as *Ficus* and *Inga* in the wild, and from rotting fruits in laboratory settings. Parasitoid wasps attack eggs (*Encarsia* cf. *porteri*, Ahelinidae) and larvae (*Calolydella*, *Campylochaeta*, Tachinidae). *P. chrysippe* has been studied as a possible agent of

Larvae of *Pelolasia chrysippe*.

Pupa of *Pelolasia chrysippe*.

Adult *Pelolasia chrysippe.*

Pelolasia chrysippe pinned. Dorsal view.

Pelolasia chrysippe pinned. Dorsal view.

biological control of the invasive plant *Miconia calvescens* (Melastomataceae) in Australia and oceanic islands such as Hawaii. **Early stages:** Eggs are pale brown with apex area purplish, barrel-shaped with smooth surface, and are laid on undersides of leaves in clusters of twenty to seventy, attached to plant by a short, stalklike connection. Larvae are highly gregarious and show a strong processionary resting and feeding behavior. There are six larval instars; all stages rest on underside of leaves. Mature larvae are greenish-black with whitish, dorsolateral stripe marks and are covered in many (long and short) black and white setae; head capsule is bright orange with long white setae. Pupa is whitish, covered in dark dorsal and lateral markings and spatulate white and brown setae; abdomen is triangular; head and thorax are rectangular and narrower than first segments of abdomen, and head area has a small notch; wing pads have parallel longitudinal black lines; a silk girdle across A_1 attaches pupa to leaf. **Host plants:** *Conostegia rufescens, Miconia calvescens, M. elata, M. appendiculata, M. donaeana, M. impetiolaris, M. longifolia, M. trinervia* (Melastomataceae), *Marila* (Calophyllaceae).

Myselasia sergia (Riodinidae: Nemeobiinae)
Sergia Myselasia

0–1350 m

FWL: 12–15 mm. **Description:** *Female:* DWs are brown with a blurry orange patch from basal to medial area on FW and blurry orange on anal half of HW. *Male:* DWs are similar to female's but darker brown and with less orange. Both sexes: VWs are gray to brown, with a prominent medial longitudinal orange line across both wings, followed by a blurry gray sub-marginal band; there is a large black sub-marginal spot from M_3 to Cu_1, sometimes surrounded by yellow. **Similar species:** *Myselasia hieronymi bianala, Methone eubule, M. eucrates leucorrhoa,* and *Pelolasia amphidecta* all have brown, rather than gray, VW background color. *Myselasia inconspicua* has medial and sub-marginal longitudinal lines on VWs similar in thickness and tone, but in *M. sergia* the sub-marginal line is more diffuse orange than brown, as in *M. inconspicua.* **Habitat:** Common in secondary forest and open areas on both slopes. In Central Valley, it is common in Leonelo Oviedo Ecological Reserve, Sabanilla, Ciudad Colón, and Alajuela. It also occurs in Turrialba, Monteverde, and many other places where natural vegetation remains. **Distribution:** Nicaragua to Panama. **Seasonality:** Present all year. **Natural history:** Males establish territories early in the morning (7:00–8:00 a.m.) and again in the afternoon (3:00–4:00 p.m.), in light gaps or forest edges, 5–10 m high. They perch on leaves and will chase other intruder males in fast circular flights. Females oviposit on host plant leaves 2–5 m above ground level. The species is highly local and persistent in habitats with the right conditions. Larvae are attacked by parasitoid flies (*Houghia,* Tachinidae). **Early stages:** Eggs are laid in large clusters of about forty on underside of host plant's leaves. Mature larva is black with a pair of dorsolateral orange patches interleaved with white spots

Larvae of *Myselasia sergia.*

Pupae of *Myselasia sergia.*

Adult *Myselasia sergia.*

Myselasia sergia pinned. Dorsal view.

on each segment, and a longitudinal double white dorsal line along body, which is covered in many dense, long and short, white and black setae; head capsule is black with a small orange bar on top. Pupa is pale cream with two small dorsolateral dark spots on segment A_1 and dark dorsal bands on T_2, T_3, A_1, A_2, A_5, and A_6; it is covered in many long, black and white setae arising from small warts; a silk girdle across A_1 attaches pupa to leaf. **Host plants:** *Eugenia acapulcensis, E. hypargyrea, E. monticola, E. salamensis, Psidium friedrichsthalianum, P. guajava* (Myrtaceae).

Eurybia lycisca (Riodinidae: Riodininae)
Blue-winged Eurybia

0–1500 m

FWL: 24–28 mm. **Description:** DFW is brown with a conspicuous metallic-blue eyespot surrounded by a yellow ring inside discal cell. DHW is metallic blue except for margins, which are brown. VFW is light brown with blue eyespot as in DFW. VHW is light brown with small sub-marginal black spots. **Similar species:** *Eurybia caerulescens fulgens* is very similar but has two small white spots distal to eyespot on DFW. **Habitat:** Found in primary and secondary habitats, along trails and in forest understory, always in association with thickets of its host plants. Localities include Limón, Sarapiquí, San Carlos, Monteverde, Osa Peninsula, and Carara National Park. In Central Valley it is found in forested patches along rivers and nature reserves. **Distribution:** Mexico to Ecuador. **Seasonality:** Present all year. **Natural history:** This is a common species, widely distributed around the country and in many habitats except northern Pacific dry forest. Adults can be found in any forested place where their host plants exist, but they must be flowering, since early stages of development take place on flowers. Males will set up and defend territories for several days, as long as flowers persist. These butterflies have a rapid and erratic flight, very hard to follow with the eyes, but they soon land on the underside of a leaf with the wings wide open; if an observer approaches, they will fly away. Adults feed from flowers of various plants, such as *Lantana camara, Asclepias curassavica, Psychotria, Passiflora,* and *Stachytarpheta,* but are seen more often on flowers of *Calathea* and *Heliconia.* This species has one of the longest proboscises among butterflies, an adaptation that is not well understood. The eyespot on the VFW is strikingly similar to the real eye, so it is probably a defense mechanism against predators, drawing attacks away from

Larva of *Eurybia lycisca.*

Pupa of *Eurybia lycisca.*

Adult *Eurybia lycisca.*

the head. Caterpillars are attacked by wasps (*Triraphis*, *Rogas*, Braconidae). **Early stages:** Eggs are green, flattened and disk-shaped, with a granulated surface, and are laid in flower structures and on leaves and stalks. Larva is long and flattened, adapted to live and feed between parts of the host plant's inflorescences, on which it feeds exclusively. Young larvae are light brown. Mature larva is yellow with few disperse setae; segment T_1 has a yellow and black dorsal plate; head capsule is brown and smooth. Younger instars constantly move from one inflorescence to another; mature larvae usually stay between the flowers of the same inflorescence, often with most of the body submerged in water pooled in the flower structures. Pupa is brown, tubular, with a long compartment extending posteriorly from cremaster to house the proboscis. Larvae and pupae are attended by ants of the genera *Paratrechina*, *Pheidole*, *Crematogaster*, *Aphaenogaster*, *Solenopsis*, *Wasmannia*, *Megalomyrmex*, *Camponotus*, and *Pachycondyla*. **Host plants:** *Calathea crotalifera*, *C. latifolia*, *C. warsczewisczia*, *C. lutea*, *C. inocephala*, *C. marantifolia*, *Pleiostachya pruinosa* (Marantaceae).

Rhetus arcius castigatus (Riodinidae: Riodininae)
Long-tailed Metalmark

0–1400 m

FWL: 18–19 mm. **Description:** A stunning butterfly. HWs bear very long tail from M_3 vein. DFW is black with two longitudinal white bands, one post-basal and one medial, those of males a little more bluish than those of females. *Male:* DHW is also black, with a continuation of DFW bands, but at the M_3 level they become fused, turning a bright metallic-blue color that extends through the tail; anal area is black with a red patch from anal margin more or less to M_3 vein. VWs are similar, but VHW lacks metallic blue. *Female:* DHW is similar to male's but less bright and with very little blue; VHW is similar to male's. **Similar species:** *Rhetus arcius castigatus* is immediately distinguished from similar species by its very long tails. *R. periander naevianus* and *R. dysonii caligosus* do not have tails as long or as thin. **Habitat:** Primary and secondary forest as well as gardens, parks, and backyards around Central Valley. Although never common, this species has a wide distribution and is present in most habitats. **Distribution:** Guatemala to Costa Rica. **Seasonality:** Present all year. **Natural history:** Both sexes feed during hot mornings from flowers of *Inga*, *Warszewiczia coccinea*, *Croton draco*, *Cordia*, *Alternanthera pubiflora*, and many Asteraceae species. Males feed from rotting fruits of *Morus* (C. Rojas, pers. comm.), and presumably might visit other small, purple fruits that ripen on the plant, such as those of *Lantana*. Usually, solitary individuals are seen, puddling or feeding from flowers. They fly very fast and erratically, but when they land on a flower or in mud, they stay still with wings wide open. DeVries (1997) observed that males set up territories in forest edges in early mornings and actively protect them against intruders. He also questions whether Costa Rican Central Valley individuals are migrants; caterpillars are seen all year in gardens and backyards, so they probably live year-round in urban areas (pers. obs.). **Early stages:**

Larva of *Rhetus arcius castigatus*.

Pupa of *Rhetus arcius castigatus*.

Adult *Rhetus arcius castigatus.*

Mature larva is velvet-black with a prominent dorsolateral white patch on segments A_4-A_5; and in each segment, two rows of dorsal red tufts as well as lateral red tufts, from which tassels of long white hairs project to the sides (hairs are black on T_2); head capsule is black and smooth on top but covered by short white setae in front. Pupa is lime-green with two short lateral rows of black spikes on abdominal segments, a circular dorsal black patch on segment A_5, a black lateral line on thorax that surrounds head, and a small black spike in lateral area of head; a silk girdle across A_1 attaches pupa to leaf. **Host plants:** *Terminalia catappa*, *T. amazonia* (Combretaceae), *Inga vera* (Fabaceae), *Mabea occidentalis* (Euphorbiaceae), *Phoradendron* (Santalaceae).

Melanis pixie sanguinea (Riodinidae: Riodininae)
Pixie

0–1500 m

FWL: 20–25 mm. **Description:** A black butterfly. DFW has orange apical area and two prominent red spots from veins Cu_2 to 2A, one in distal margin and the other in basal area; sometimes distal-margin red spots extend to one on Cu_1-Cu_2 or even M_3-Cu_1. DHW has a row of large red spots on distal margin and another one in post-basal area from Cu_2 to 2A. VWs are same as DWs but with an extra red spot in costal margin of post-basal area. **Similar species:** *Melanis electron melantho* has a prominent transverse yellow band on FW from sub-costal vein to Cu_2. **Habitat:** This species is a sporadic inhabitant of secondary forest and disturbed habitats, where it always flies close to thickets of flowering host plants. Generally found in great abundance, flying 1–5 m high in open areas such as coffee plantations and abandoned fields. More common on Pacific slope, although also present on Caribbean side. Localities include Ciudad Colón, Abangares, Miramar, Liberia, Sarapiquí, and Cahuita National Park. **Distribution:** Honduras to Colombia. **Seasonality:** Present all year. **Natural history:** This species commonly occurs in small colonies, but colonies are increasingly rare, probably because it is very dependent on presence of larval host plants in combination with availability of the right kind of flowers for adults to feed on, and such habitats are declining. Generally, it prefers open secondary forest, mangrove edges, old-growth savanna, cliffs, and tree-accompanied crops such as coffee plantations. Both sexes feed from flowers of *Terminalia catappa* (pers. obs.), *Albizia*, *Pithecellobium*, *Coffea*, *Cordia*, *Citrus*, and *Lantana*. The species has a slow, fluttering flight, which in combination with its bright colors might indicate it is unpalatable. **Early stages:** Eggs are light green and laid in clusters of up to forty on underside of leaves or on bark of host plants. Hatching and movements are synchronous and gregarious. Mature larva is variable, whitish to greenish, with transverse black lines on dorsal area of each body segment alternating with orange lines, abundant long white setae, and a longitudinal dorsolateral row of small black tufts of short setae; head capsule is light green with a couple of small black bars in front and a smooth surface. Pupa is whitish with transverse black dorsal lines on every segment of abdomen, a longitudinal dorsal patch on thorax, which becomes more conspicuous as it matures, and a dorsolateral row of small black spikes; wing pads also have longitudinal black lines. Pupa is attached to surface (generally trunk or branches of host tree) by a silk girdle across segment A$_1$. **Host plants:** *Albizia adinocephala*, *A. niopoides*, *Inga*, *Pithecellobium*, *Samanea saman* (Fabaceae).

Larva of *Melanis pixie sanguinea*.

Pupae of *Melanis pixie sanguinea*.

Tree trunk fully covered by larvae, pupae, and emerging adults of *Melanis pixie sanguinea*.

Melanis pixie sanguinea pinned. Dorsal view.

Lasaia sula sula (Riodinidae: Riodininae)
Blue Lasaia

0–1300 m

FWL: 14–16 mm. **Description:** *Female:* DWs are checkered white and brown. *Male:* DWs are metallic light blue with many disperse small black and brown bars; some black sub-marginal dots on DHW. VWs are similar in both sexes. **Similar species:** *Lasaia agesilas* has no black sub-marginal dots in DHW distal margin; in *L. pseudomeris* the spots are present but smaller than those of *L. sula sula*. Female *L. sula sula* is much whiter than females of the two similar species. **Habitat:** Occurs on both slopes; on Pacific slope it is more common in the north and not recorded much farther south than Carara National Park. Found in large open areas inside forest patches such as riverbanks and landslides; also in secondary habitats or disturbed places like cattle ranches or croplands. Rare individuals may be seen feeding in gardens or drinking water from the soil. Localities include Ciudad Colón, Guanacaste, Siquirres, and San Carlos. **Distribution:** Mexico to Costa Rica. **Seasonality:** Present all year. **Natural history:** Both sexes are active visitors of flowers such as *Asclepias curassavica*, *Lantana*, and many species of Asteraceae, and

Larva of *Lasaia sula sula*. JMR

Pupa of *Lasaia sula sula*. JMR

Recently emerged adult *Lasaia sula sula*. JMR

Lasaia sula sula
pinned. Dorsal view.

always prefer to fly on hot sunny mornings. Males are commonly seen puddling on dirt roads and exposed damp soil. **Early stages:** Eggs are flat and saucerlike, with highly ornamented surface. Mature larva rests alone on host plant's trunk or branches and is extremely cryptic. It is greenish-gray with some disperse brown markings and many small black dots; extending from each segment in lateral area are many long setae, which touch the substrate, hiding its ventral area and obscuring the caterpillar shape; head capsule is green with smooth surface. Pupa resembles a small piece of wood, with same gray and brown colors as bark, and has curves and peaks all over; it is placed on the trunk in a diagonal position, at about a 45° angle to the substrate. **Host plants:** *Albizia niopoides, Vachellia collinsii* (Fabaceae).

Catocyclotis adelina (Riodinidae: Riodininae)
Adelina Metalmark

800–1600 m

FWL: 18–20 mm. **Description:** *Female:* DFW is brown with bluish curved lines along costal margin, and tornus pale yellow; DHW is pale yellow with basal area brown. Dorsal side of thorax is brown; small anterior part of dorsal side of abdomen is brown; posterior part yellow. *Male:* DWs are similar to female's but with orange instead of yellow. VWs are the same in both sexes. **Similar species:** *Calociasma ictericum* has a conspicuous whitish-yellow patch on DFW costal margin, absent in *Catocyclotis adelina*. **Habitat:** Open areas and forest edges of secondary and primary forest on both Caribbean and Pacific slopes. Localities include Sabanilla, Tres Ríos, University of Costa Rica at San Pedro campus, Monteverde, and Guayabo National Monument. **Distribution:** Costa Rica to Colombia. **Seasonality:** Present all year but more common during dry season. **Natural history:** Males have been observed in late afternoon perching on large tree trunks with wings open and head pointing downward, in places where sunlight hits at 4–5 m high; under these conditions, from two to ten males chase each other in rapid flights from one perch to another–this might indicate lek behavior (K. Nishida, pers. comm.). Both sexes rest on underside of leaves, 1–4 m high, along trails and in darker places in forest interior. Females lay their eggs in young saplings or low-altitude branches of host plants. Adults feed from flowers such as *Dendropanax, Acnistus arborescens*, and *Alternanthera pubiflora* (N. Page, pers. comm.) from mid-morning to midday. Larvae are attended by ants such as *Myrmelachista zeledoni* (K. Nishida, pers. comm.), which protect them against enemies such as parasitic flies and wasps, other insects, and other ants. To reward the ants for providing protection, the caterpillar secretes a food solution; when danger approaches, its thorax-head stridulatory system makes sounds to alert the ants. **Early stages:** Eggs are white, dorsoventrally compressed in a disk shape, with highly ornamented surface, and are laid in branches and new shoots. Larvae build shelters by joining two leaves with silk; generally more than one individual lives in each shelter. Mature larva is light green with a longitudinal brown dorsal line, many brownish dots in lateral area, a dorsal black anal plate on segments A_9–A_{10}, and a dorsal black prothoracic plate on T_1 with two parallel longitudinal white lines from which long, white setae arise; body is covered in disperse long white setae; head capsule is shiny black with long white setae. Pupa, also formed inside leaf shelter, is tubular and light green with a conspicuous dorsolateral black dot on T_1 (resembling an eye). Pupae are also attended by ants. **Host plants:** *Inga vera* (Fabaceae), *Cupania glabra* (Sapindaceae), *Cinnamomum triplinerve* (Lauraceae), *Trichilia havanensis* (Meliaceae).

Larva of *Catocyclotis adelina* being attended by ants. KN

Pupa of *Catocyclotis adelina* being attended by ants. KN

Adult *Catocyclotis adelina.*

Catocyclotis adelina pinned. Dorsal view.

Napaea umbra umbra (Riodinidae: Riodininae)
Quilted Metalmark

0–1300 m

FWL: 18-20 mm. **Description:** DFW is grayish-brown with irregular curved dark brown lines that form bubbles in the basal half, a longitudinal line in medial area, and a white spot in sub-apical area. DHW is same as DFW but has two longitudinal lines, one post-medial, the other sub-marginal. **Similar species:** *Periplacis laobotas* has more than one white spot on DFW, the spots are post-medial instead of sub-apical, and the brown basal lines do not form bubbles. **Habitat:** Generally seen flying along trails and forest edges of secondary and primary forest; sometimes seen in open areas feeding from small white inflorescences. Localities include Alajuela, Garabito, Sarapiquí, Puerto Jiménez, and Santa Cruz. **Distribution:** Mexico to Costa Rica. **Seasonality:** Present all year. **Natural history:** This is a fast-flying butterfly, which makes it difficult to identify in flight, but it can be observed when it perches on underside of leaves with the wings wide open. Females fly lower than males, at 1-4 m, while males perch at 5 m, or higher in light gaps, but always on underside of leaves. The species is relatively common but seldom collected. **Early stages:** Eggs are white. Larva is densely covered in long, soft white setae, which are arranged in four tufts on each body segment; body is greenish-white with black spiracles and a faint brown longitudinal dorsal line. It rests on underside of host plant's leaves. Pupa is white on dorsal side of abdomen, which is flattened but thickens until it generates a prominent crest on segment A_2; from there, the crest decreases and is brown; lateral area of segments A_6–A_8 are also brown; wing pads, dorsal side of thorax, and head are white. **Host plants:** *Bromelia pinguin*, *Ananas comosus*, *Guzmania* (Bromeliaceae).

Larva of *Napaea umbra umbra*.

Pupa of *Napaea umbra umbra*.

Adult *Napaea umbra umbra*.

Napaea umbra umbra pinned. Dorsal view.

Calephelis schausi (Riodinidae: Riodininae)
Schaus's Calephelis

0–3400 m

FWL: 10–12 mm. **Description:** DFW is somewhat variable, from blackish to reddish-brown with a faint, broken darker band running longitudinally from costal to anal margin; silver lines parallel to the darker band, one (broken) in post-medial area, and another (not broken) in sub-marginal area; and distal area reddish in sub-marginal and marginal area, with a row of black dots between the two silver lines. DHW has same pattern as DFW. VWs are reddish-brown with discontinuous small black longitudinal lines in basal and medial areas, followed by a broken silver line and a continuous sub-marginal silver line. **Similar species:** Very similar to *Calephelis browni* and *C. sodalis* but difficult to distinguish from all other Costa Rican *Calephelis* species. The first feature to check is FWL; few species are larger. Second, the FW is not falcate (sickle-shaped) in *C. schausi*. Third, the ventral coloration is reddish-brown in *C. schausi*.

Larva of *Calephelis schausi*.

Pupa of *Calephelis schausi*.

Calephelis schausi pinned. Ventral view.

Adult *Calephelis schausi*.

Once these characteristics are matched, compare dots and line patterns to confirm. **Habitat:** Open areas such as roadsides, gardens, croplands, and cattle ranches, mainly on the Pacific slope, although sometimes present on the Caribbean side. It is common in localities such as Santa Rosa, Palo Verde, Carara, Manuel Antonio, Los Quetzales, and Corcovado National Parks. **Distribution:** Honduras to Panama. **Seasonality:** Present all year but much more common during rainy season. **Natural history:** These active small butterflies fly fast and erratically. Both sexes are active from mid-morning to late afternoon when the sun shines. They visit many flowers of the Asteraceae family, with a special preference for species with small white inflorescences such as *Bidens pilosa* and *Melanthera nivea*. They also visit other flowers, including *Alternanthera pubiflora* (Amaranthaceae). During courtship, both sexes fly extremely fast in a spiral until the female crashes to the ground; the male rapidly engages, and then both fly together to the underside of a nearby leaf. Very little is known about this genus in the tropics, although it is common in many habitats and very diverse in distribution. **Early stages:** Eggs are white, sausage-shaped, with reticulate surface, and are laid singly in joints of petioles or stalks of host plants. Young caterpillars are light green with few setae and are first skeletonizers on undersides of leaves. Mature larva, which also rests on leaf underside, is heavily covered in long, soft white setae, arising from small warts in four tufts per segment; there are two faint longitudinal bands of blackish dots on dorsolateral surface. Pupa is oval, whitish, with black dots on dorsum of each abdominal segment; one larger spot each in segments T_3 and A_1; dorsolateral black dots in segments T_1, T_2, and A_2; and small dorsal lines on head and T_1. Pupa is covered in a light layer of the old caterpillar setae and is attached to substrate by a silk girdle across segment A_1. **Host plant:** *Cosmos sulphureus* (Asteraceae).

Brush-footed Butterflies (Nymphalidae)

Nymphalidae is one of the most interesting and diverse butterfly families. Its members inhabit any place that has vegetation at least part of the year, from tundra to tropics—where the highest diversity is found. There are some very small species (*Microtia elva*) and some very large ones, such as *Morpho polyphemus catalina*, the largest butterfly in Costa Rica. Nymphalidae contains around 6000 species worldwide (Ackery et al. 1999, Peña & Espeland 2015), of which at least 460 inhabit Costa Rica. This large family presents diverse behavior and biology. Some species are nectar feeding (*Heliconius*), others feed from rotting fruits, feces, or animal carcasses (*Prepona* and *Hypanartia*). Some genera are powerful canopy fliers (*Historis*), while others live in the understory (*Ithomia*), moving slowly through the shadows. The longest insect migration of the world is performed by the nymphalid butterfly *Danaus plexippus*, the Monarch.

Nymphalid larvae feed on host plants from dozens of families including, among others, Lauraceae, Euphorbiaceae, Urticaceae, Passifloraceae, Heliconiaceae, Fabaceae, Poaceae, and even spore-dispersed plants such as Selaginellaceae. Eggs may be spherical, semispherical, with or without ridges, barrel-shaped, ovoid, or keeled. The larvae can be hairless or fuzzy, bearing rigid spines, soft spines (setae), or fleshy projections—including scoli, small fleshy protuberances with a group of rigid setae on top—and in some the spines can sting. Eversible glands on the ventral side of the prothorax, which function as ant or parasitoid deterrents, are present in some genera, such as *Eryphanis*, *Caligo*, and *Morpho*. Larvae of some species have two-forked tails, and others bear many horns on the head capsule. Pupae are always attached to the substrate by the cremaster, usually hanging, but in some genera, such as *Hamadryas* and *Catonephele*, pupae are horizontal or even pointing up. Adult nymphalids show an amazing variety of wing shapes, from squared (*Siproeta*) or triangular (*Morpho* and *Danaus*) to long and narrow (*Heliconius* and *Mechanitis*) to tailed (*Marpesia* and *Consul*).

Nymphalidae contains twelve subfamilies (M. Zhang et al. 2008), of which ten are present in Costa Rica: Apaturinae, Libytheinae, Danainae, Charaxinae, Satyrinae, Nymphalinae, Biblidinae, Limenitidinae, Cyrestinae, and Heliconiinae. These subfamilies share some common characteristics. All have modified forelegs, which are smaller than those of other butterfly families and have olfactory functions. In female nymphalids, these forelegs bear small spines enabling them to break the cuticle of the leaf, releasing the plant's chemical compounds, which indicate to the butterfly whether or not it is the right host plant on which to lay its eggs. The family's common name, brush-footed butterflies, refers to the forelegs' brushlike modification. Another important characteristic of nymphalid butterflies is the presence of ventral longitudinal keels on their antennae (Murillo-Hiller 2008).

Danaus plexippus plexippus (Nymphalidae: Danainae)
Monarch

0–2500 m

FWL: 40–48 mm. **Description:** DWs and VWs are orange with black veins; marginal areas are black with white spots. *Female:* Black veins are thicker, resulting in less orange area. *Male:* Black veins are thinner; orange is brighter; and DHW has patches of scent scales (androconia) on vein Cu$_2$. **Similar species:** *Danaus gilippus thersippus* does not have black veins on DWs and has a medial line of three or four white spots on DFW. *D. eresimus montezuma* has white sub-apical dots on DFW of the same size as the sub-marginal ones, while *D. plexippus plexippus* has much larger sub-apical spots than sub-marginal ones and often also has pink or orange blotches in sub-apical area. **Habitat:** Open areas, gardens, fields, and crops. This butterfly prefers low secondary vegetation, as is found in abandoned cattle ranches, and also rocky peaks and savannas. It is a common inhabitant of backyards in suburban areas of the Central Valley. **Distribution:** North and Central America to northern South America, Cocos Island, Philippines, Australia, western Europe. **Seasonality:** Present all year. In Central Valley it seems to be more abundant during dry season. **Natural history:** The Monarch is one of the most-studied butterflies in the world, owing to the magnificent migration it performs every year in North America; in consequence, it has become an icon for insect conservation. In Costa Rica, its encounters with people are common, as adult and caterpillar. Its host plants are planted as ornamentals and also grow widely because of wind-blown seed dispersal. Adults feed from flowers such as *Lantana camara, Stachytarpheta, Ageratum conyzoides, Tithonia, Asclepias curassavica*, and many others. This species is often cited as a model of aposematic coloration, its bright colors serving as a warning signal of unpalatability. However, preliminary experiments (DeVries 1987) have shown that individuals from early wet season were eaten more by magpie-jays (*Calocitta*) than the ones from later the same year. More studies should be conducted to determine whether it was a

Larva of *Danaus plexippus plexippus.*

Pupa of *Danaus plexippus plexippus.*

Adult *Danaus plexippus plexippus.*

Danaus plexippus plexippus pinned. Dorsal view.

matter of very hungry birds or differences in the concentration of toxins (cardiac glycosides) in host plants related to water quantity. Larvae are parasitized by flies (*Lespesia archippivora*, *Hyphantrophaga*, *Archytas*, Tachinidae). Females of *Archytas* deposit their eggs on leaves of the Monarch's host plants; the eggs are then swallowed by the young butterfly larva and develop inside it; the flies usually abandon the Monarch in its pupal stage. Caterpillars are also attacked by wasps (*Brachimeria*, Chalcididae), which oviposit directly on the mature larva's body or on the fresh pupa. The Green Wall Lizard (*Sceloporus malachiticus*) preys on adults (pers. obs.). **Early stages:** Eggs are creamy-white, short, bottle-like, and laid singly on underside of mature leaves or in new shoots of host plants. Larvae are cannibalistic: in the first instar, a larger one will try to eat the smaller ones, so usually there are very few larvae per plant; but when there are few host plants, many larvae will coexist together. Larvae are transversely striped in black, white, and yellow, though some can be largely black, lacking white bands. Additionally, each larva has four prominent fleshy warts, two anterior on segment T_2, and two posterior on A_8; they serve as sense organs and also as decoys, since predators will confuse head with tail. Pupa is jade green, rounded, with a gold and black half ring on dorsal side of segment A_3, eight gold spots dispersed throughout head and thorax area, and two more on wing pads. **Host plants:** *Asclepias curassavica*, *A. oenotheroides*, *Matelea*, *Gomphocarpus physocarpus*, *Gonolobus edulis* (Apocynaceae).

Tithorea tarricina pinthias (Nymphalidae: Danainae)
Cream-spotted Tigerwing

0–1500 m

FWL: 38–45 mm. **Description:** DFW is black with eleven or twelve small yellow spots in medial and sub-marginal areas. DHW is orange with marginal area black. Both DWs have very small white spots in margins. VFW is same as DFW but with an extra row of white sub-marginal spots. VHW is same as DHW but also has two or three larger light orange spots in sub-apical area. **Similar species:** Both sexes are distinguished from *Eresia ithomioides eutropia* (Nymphalinae), *Heliconius hecale* (Heliconiinae), and *Callithomia hezia* (Danainae) by the number, size, and location of the yellow spots on DFW. Male *T. tarricina pinthias* has androconial hairs on costal margin of DHW, lacking in males of other butterfly groups, including other nymphalids. **Habitat:** Both Pacific and Caribbean slopes, on forest trails, roadsides along small dirt roads through forest, and riversides where abundant vegetation grows. Found in well-preserved habitats such as Cahuita, Braulio Carrillo, Arenal Volcano, and Rincón de la Vieja National Parks, and, on the Pacific side, in Carara, La Cangreja, and Corcovado National Parks. Generally absent in Guanacaste lowlands but sporadically found in very wet years during rainy season. **Distribution:** Honduras to Panama. **Seasonality:** Present all year. **Natural history:** This species inhabits preserved forest and prefers small sunny places inside the forest or along rivers. It flies 1–4 m above the ground.

Larva of *Tithorea tarricina pinthias*.

Pupa of *Tithorea tarricina pinthias*.

Tithorea tarricina pinthias pinned.
Dorsal view.

Adult *Tithorea tarricina pinthias*.

Males set up territories and stay there for several days; they perch on leaves that are hit by sunrays and chase other butterflies that pass by. Females are more associated with darker areas in the forest interior where they search for ovipositing sites. Adults feed from flowers such as *Psychotria*, *Eupatorium*, *Epidendrum paniculatum*, *Stachytarpheta*, *Lantana*, *Hamelia patens*, and *Ageratum conyzoides*. Pupae are attacked by parasitoid wasps (Chalcididae). **Early stages:** Eggs are small, creamy-white, rounded and bottle-like, and are laid singly on undersides of leaves. Larvae are similar during all stages. Fifth instar is bright-colored and transversely striped, anterior and posterior parts black with yellow, middle part black with white; two long fleshy warts arise from segment T_2 and probably serve as sense organs; head capsule is smooth and striped black and yellow. Pupa is bright, shiny gold-silver, and rounded, with concave wing pads, six rows of black spots along abdomen, and some black lines along wing pads, thorax, and head; bears two small anterior peaks in area of eyes. **Host plants:** *Prestonia portobellensis*, *P. longifolia* (Apocynaceae).

Mechanitis polymnia isthmia (Nymphalidae: Danainae)
Polymnia Tigerwing

0–1600 m

FWL: 35–39 mm. **Description:** A variable species with long wings. DFW is black with yellow streaks and bands in medial and post-medial areas, a round orange spot on tornus, and sometimes orange in basal area and anal border. DHW is orange with a transverse black band, variable in size or sometimes almost absent; and a black margin. **Similar species:** Orange spot on DFW tornus, which is always isolated and somewhat round, distinguishes *Mechanitis p. isthmia* from *M. menapis*, *M. lysimnia*, and other Ithomiini species. **Habitat:** Common on almost all forest trails. In Central Valley, abundant along rivers and in nature reserves in the surrounding mountains. In rural areas can be observed feeding from flowers in parks and gardens. **Distribution:** Costa Rica and Panama. **Seasonality:** Present all year but much more abundant during rainy season. **Natural history:** This is one of the commonest members of Danainae in Costa Rica. It is present in all forested habitats at suitable elevations. Like many other members of its subfamily, it prefers to inhabit the dark understory of the forest, but it is not unusual to find individuals looking for flowers in open areas. They can be seen early in the morning (around 6:00 a.m.) feeding on the nectar of *Ageratum conyzoides* in open fields. In the morning, males establish territories in the forest, and females oviposit. Males will also look for bird droppings or dead insects to feed on. Larvae are attacked by a parasitoid fly (*Houghia aurifera*, Tachinidae). **Early stages:** Eggs are white, bottle-like, and laid in clusters of about thirty on underside or upper side of leaves. Caterpillars rest and eat together, congregating on underside of the leaf they are eating. Larva is bluish-white with a light yellow dorsal line, white sides with fleshy prolongations from segments A_1–A_8, and shorter, forward-projecting fleshy prolongations from T_1. Pupae are formed together, generally on undersides of host plant's leaves. Pupa

Fifth instar larvae of *Mechanitis polymnia isthmia.*

Mechanitis polymnia isthmia orange form pinned. Dorsal view.

Pupae of *Mechanitis polymnia isthmia.*

Adult *Mechanitis polymnia isthmia* (black band form) feeding from *Ageratum* sp. CC

is longitudinally elongate but rounded, and most of its body is very bright, chrome-like silver, with three red longitudinal lines across wing pads and T$_1$, black spiracles, a concave red keel on thorax, two small spikes on head, and black cremaster. **Host plants:** *Solanum aturense, S. hayesii, S. hazenii, S. hispidum, S. jamaicense, S. lanceifolium, S. rudepannum, S. rugosum, S. schlechtendalianum, S. torvum* (Solanaceae).

Greta morgane oto (Nymphalidae: Danainae)
Darkened Rusty Clearwing

400–1700 m

FWL: 28–30 mm. Description: Wings are mostly transparent with brown-orange borders and brown veins. FW has a thick black bar at distal end of discal cell, followed by a white band and finally a thick reddish-brown apex. FW anal border and HW costal border are brown. **Similar species:** The clearwing butterflies, including *Greta morgane oto*, *G. andromica lyra*, *Hypoleria lavinia cassotis*, and *Oleria vicina*, are generally hard to identify, often requiring a specimen in hand. Venation arrangement helps distinguish them to genus level and color pattern to species. In the case of *Greta*, compare HW venation with that of other species shown in Fig. 26. *G. m. oto* is then distinguishable by well-differentiated dark brown triangle at distal margin of vein Cu$_1$ together with thick black FW apex. **Habitat:** This is one of the commonest clearwings in Costa Rica, found in any forested and moderately well-preserved habitat, such as the mountains surrounding the Central Valley. Found along forest trails and rivers in San Rafael de Heredia, Monteverde, Cerro de la Muerte, Tapantí, Guayabo National Monument, San Ramón in Alajuela, Las Cruces Biological Station, Rincón de la Vieja, and many other places. **Distribution:** Mexico to Panama. **Seasonality:** Present all year in higher montane habitats and the Caribbean slope but absent from the Pacific dry forest during dry season. **Natural history:** This species presents some very interesting behaviors. Males look for flowers, such as *Ageratum conyzoides*, *Heliotropium*, and others, to drink their nectar and sequester from it a poisonous compound (pyrrolizidine alkaloids) that the male transfers in small sacs to the female during copulation. In order to expose themselves

Fifth instar larvae of *Greta morgane oto*.

Pupae of *Greta morgane oto*.

Greta morgane oto
pinned. Dorsal view.

Greta

Hypoleria

Oleria

Fig. 26. Hind-wing venation of *Greta morgane oto* and two other genera.

Greta morgane oto feeding from *Tournefortia hirsutissima.*

to females, males establish leks in places with the right conditions, where they congregate to release pheromones and display their wings. Females find the many males together and select one. This species is intolerant of low humidity; apparently it makes altitudinal migrations on the Pacific slope to escape the dry season, which could allow individuals to live up to 5 months (G. Styles, pers. comm.); however, no records have been collected for more than 3 months under controlled conditions in the LAREBUB. **Early stages:** Eggs are small, whitish, rounded, and are laid on underside of host plant's leaves. Young larva is light green; it rests on underside of leaves, always gregariously. Mature larva is bluish-white, with a smooth surface and two dorsolateral rows of black and yellow spots; head capsule is white with two prominent black spots on top, and stemmata are also surrounded by black. Pupa is rounded dorsally, especially on wing pads, which are very convex. It is light green, crossed by thin chrome-gold lines, with disperse black spots (variable in number and size) on abdomen, wing pads, and head area; thorax has a black-tipped ventral spike, pointing upward, and a black-tipped spike arises from the area of each eye. **Host plants:** *Cestrum glanduliferum, C. microcalyx, C. tomentosum, Lycianthes, Solanum hayesii* (Solanaceae).

Adelpha fessonia fessonia (Nymphalidae: Limenitidinae)
Band-celled Sister

0–1000 m

FWL: 25–32 mm. **Description:** A brown butterfly with a broad white band longitudinally crossing both FW and HW through the middle; this band extends from anal section of HW tornus to costal margin of FW, just outside of distal part of discal cell. DFW has large orange sub-apical spot; DHW has smaller orange spot in tornus. VWs have mosaic of longitudinal light brown, dark brown, white, orange, and gray lines, variable in thickness. **Similar species:** *Adelpha f. fessonia* is distinguishable from *Doxocopa druryi acca* female (p. 148), *D. pavon* female, and other *Adelpha* species by the white band, which ends in the DFW right at the sub-costal vein and not before it. **Habitat:** This species likes sunny places associated with forest patches in deciduous habitats. Not commonly seen in urban areas, but in rural areas can be seen along forest trails in light gaps, on coffee and timber plantations, along rivers, and on forest edges. Locations include Liberia, Nicoya, Santa Cruz, Abangares, Ciudad Colón, and, more rarely, in the rainy southern Pacific. **Distribution:** Mexico to Panama. **Seasonality:** Present all year but most common during rainy season. **Natural history:** Generally, this species likes to fly fairly close to the ground, although males can be seen setting up and patrolling territories 3–4 m above the ground in open areas and second-growth habitats, and feeding from flowers of *Cordia*, *Croton*, and arborescent Asteraceae such as *Tithonia*. Females stay closer to the ground and look for young host plants on which to oviposit. Occasionally, both sexes feed from rotting fruits such as figs on the forest floor. Caterpillars are attacked by a parasitoid fly (*Zizyphomyia arguta*, Tachinidae). **Early stages:** Eggs are small, whitish, rounded, very ornamented, and are laid on underside of leaves. After eating its eggshell, the young larva eats the leaf, leaving only the central vein, and builds up the central vein by mixing excrement with silk; it rests on the central vein. Mature larvae, always gregarious, rest on top of leaves with the anterior half of the body erected, somewhat resembling a piece of moss. Mature larva is green alternating with very light grayish-blue and brown, with many large fleshy projections with thick, sharp spines; head capsule is also very spiny. Pupa looks like a dried leaf. It has a flat projection from each eye area at a 90° angle with the body; from dorsal side of abdomen a large projection arises and folds over, touching another projection and forming a loop with a hole, which disrupts the pupal form, adding to its camouflage. **Host plants:** *Randia aculeata*, *R. armata*, *R. echinocarpa*, *R. monantha* (Rubiaceae).

Fifth instar larva of *Adelpha fessonia fessonia*.

Pupa of *Adelpha fessonia fessonia*.

Adult *Adelpha fessonia fessonia*.

Adelpha cocala lorzae (Nymphalidae: Limenitidinae)
Lorza's Sister

0–800 m

FWL: 25-30 mm. **Description:** A small to medium-size butterfly with brown background color to DWs. DFW is crossed in middle by a longitudinal orange band. DHW is crossed in middle by a longitudinal white band, appearing as a continuation of DFW's orange band. VFW has a variegated pattern of light orange, white, and brown longitudinal bands; VHW is similar but with conspicuous white longitudinal band. **Similar species:** *Adelpha* species are often difficult to identify—even more so if the specimen is damaged or in flight. In comparison with *A. heraclea heraclea* and *A. boeotia oberthurii*, *A. cocala lorzae* must first be distinguished by the longer orange segment of the DFW orange band at veins M_2-M_3; once that is found, there must be little to no notch in the inner margin of the orange bar at M_3-Cu_1. **Habitat:** Common in secondary and mature forests such as those in Cahuita, Tortuguero, Carara, and Corcovado National Parks. Always associated with rainforests and generally seen in forest edges, light gaps, or along trails. **Distribution:** Mexico to Venezuela and Ecuador. **Seasonality:** Abundant during rainy season and early part of dry season. Rare and in reproductive diapause during dry season. **Natural history:** Male perches 2–3 m high in light gaps and will stay on a certain perching spot for more than a day, spending the hottest hours of the day chasing other butterflies, no matter the species, out of its territory. Female flies in late morning or midday, wandering around the forest searching for host plants, inspecting them, and laying single eggs on plants 20 cm to 2 m high. Both sexes feed at rotting fruits and dung. **Early stages:** Eggs are gray-green, round, and generally laid on dorsal tip of leaf. Female lands on any part of the plant and then walks on it in reverse, touching the plant with the abdomen tip until it finally finds a leaf tip. First to third instars rest on a frass chain, a small veinlike structure that the caterpillar builds using its own frass (excrement) stuck together with silk, generally located at tip of host plant leaves. Fourth and fifth instars rest on top of leaf, close to the petiole, on central vein of leaf they are eating. Mature larva resembles a piece of moss, its body covered with many fleshy projections bearing spiny setae on

Fifth instar larva of *Adelpha cocala lorzae*.

Pupa of *Adelpha cocala lorzae*.

Adult *Adelpha cocala lorzae*.

Adelpha cocala lorzae pinned. Ventral view.

segments T_2, T_3, A_2, A_7, and A_8, in pale yellow and tones of brown and green; head capsule has a double pale yellow spiny crown. Pupa is somewhat polychromatic, ranging from dark to light brown to green; head bears flat flanges projected to the sides; dorsally it has two abdominal keels, one projected toward the posterior part and the other to the anterior, which combine with a thoracic keel to create the effect of a hole in a piece of dry leaf. **Host plants:** *Calycophyllum candidissimum, Chimarrhis parviflora, Pentagonia donnell-smithii, Chomelia, Genipa, Uncaria, Psychotria* (Rubiaceae).

Euptoieta hegesia meridiania (Nymphalidae: Heliconiinae)
Mexican Fritillary

0–1300 m

FWL: 30–35 mm. **Description:** DWs are orange, both with a row of black spots in sub-marginal area; DFW has two black rings on discal cell and some black lines in sub-medial and medial area. VWs have a complex variegated pattern of brown and black lines on an orange and gray background. **Similar species:** *Euptoieta poasina* has a white sub-marginal line on VHW, absent in *E. hegesia meridiania*. **Habitat:** Open areas such as pastures, cattle ranches, and semi-urban areas countrywide. **Distribution:** Mexico to Argentina. **Seasonality:** Present all year. **Natural history:** This common species likes to fly close to the ground in open, sunny areas. It flies slowly but when alarmed can be very fast. Both sexes are active during the hottest hours of the day, looking for flowers such as *Lantana, Duranta erecta, Turnera*, and various Asteraceae genera including *Ageratum* and *Tithonia*. In Mexico and USA, this species has been reported to migrate, but this has not been observed in Costa Rica. **Early stages:** Eggs are green and laid singly on underside of small leaves. Mature larvae is dark brown with a longitudinal line of white spots on dorsum, then a lateral reddish band, followed by another line of white spots; each segment has six prominent spiny scoli of similar size, except for T_1, on which the two dorsal ones are much longer and

Fifth instar larva of *Euptoieta hegesia meridiania*.

Pupa of *Euptoieta hegesia meridiania*.

Adult *Euptoieta hegesia meridiania*.

round-tipped; head capsule is smooth, black and red, and unlike most related species, lacks horns. Pupa is dark brown with silver patterns, and dorsal rings of brown, silver, and red spines on all abdomen and thorax segments. **Host plants:** *Passiflora biflora, P. foetida, Turnera ulmifolia* (Passifloraceae).

Actinote anteas anteas (Nymphalidae: Heliconiinae)
Anteas Actinote

0–1400 m

FWL: 30–35 mm. **Description:** Wings are black with a large orange patch on DHW and a yellow medial band and other smaller patches on DFW. VWs are similar. **Similar species:** Yellow medial band on DFW, which is wider, brighter yellow, and has an irregular distal border, distinguishes *Actinote a. anteas* from *A. melampeplos* and all other Costa Rican *Actinote* species. **Habitat:** This is a species of open areas associated with secondary or early successional habitats. It is found on riversides, in pastures, and in croplands in the Central Valley and at mid-elevations of the central mountain ranges on both Pacific and Caribbean slopes. Sporadically it occurs in parks, gardens, and urban areas. **Distribution:** Mexico to Panama. **Seasonality:** Present all year but more common during dry season. **Natural history:** Usually glides above vegetation looking for flowers such as *Lantana, Emilia fosbergii, Asclepias, Tithonia, Eupatorium,* and *Senecio.* Its slow and gliding flight is probably explained by the species' capacity to sequester secondary compounds from its larval host plants, making it unpalatable to predators; also, when caught, adults release an orange fluid from prothoracic glands that probably has a bad taste. This species has been introduced in South Africa, Indonesia, and Sumatra as a bio-control agent to address overgrowth of the invasive weed *Chromolaena odorata,* but it has not produced successful results. **Early stages:** Eggs are yellow, laid in large groups on underside of host plant's leaves. Larvae eat together, and when disturbed they fall from the plant and hang by a silk string. Early instars are

Fifth instar larvae of *Actinote anteas anteas.*

Pupa of *Actinote anteas anteas.*

Adult *Actinote anteas anteas.*

Actinote anteas anteas
pinned. Dorsal view.

whitish. Fifth-instar larva is dark brown; each body segment bears four long, black scoli, which emerge from a white ring; head capsule is smooth and shiny black. Pupa is white with five pairs of black spines on dorsal side of abdomen, black lines on wing pads, and small black markings all over body. **Host plants:** *Mikania micrantha*, *Chromolaena odorata* (Asteraceae).

Dione juno huascuma (Nymphalidae: Heliconiinae)
Juno Longwing

0–1400 m

FWL: 30–40 mm. **Description:** An orange butterfly with distinctive long wings. DFW has two black bars: one inside discal cell and another on its distal end, the latter one continuing as a thin black line to distal margin of wing. DHW has thick black margin. VFW is orange at base, with black bands in medial area and silver spots in apical area. VHW is dark orange and brown with many silver spots. **Similar species:** *Eueides aliphera gracilis* and *Dryas iulia moderata* (facing page) lack any silver spots on VWs. *Dione moneta poeyii* has two orange tones (one basal, the other distal) in FWs and HWs, while *D. j. huascuma* has a single orange tone. **Habitat:** This is a species of open areas and secondary forest. It has become more abundant in urban ecosystems, and is commonly seen in Central Valley gardens, backyards, and parks, since *Passiflora miniata*, one of its host plants, was introduced from Ecuador and Peru as an ornamental. Also occurs in rural areas with cultivation of passionfruit (*Passiflora edulis*), of which it is an important pest. **Distribution:** Mexico to Panama. **Seasonality:** Present all year but more common during rainy season. **Natural history:** Likes to glide along forest edges or from flower to flower in semi-open and early successional habitats to feed from species such as *Lantana*, *Stachytarpheta*, *Ageratum*, *Comaclinium montana*, and *Zinnia*. This species has become a pest of gardens and crops, where its voracious larvae often reduce their host plants to stalks. Rufous-collared Sparrow (*Zonotrichia capensis*, Emberizidae) is reported to be an avid predator of the larvae (E. Salas, pers. comm.). Larvae are parasitized by flies (*Patelloa*, Tachinidae). **Early stages:** Eggs are bright yellow and laid in groups of up

Fifth instar larvae of *Dione juno huascuma*.

Pupa of *Dione juno huascuma*. CRU

Dione juno huascuma pinned. Dorsal view.

Dione juno huascuma pinned. Ventral view.

Adult *Dione juno huascuma.*

to 100. All larval instars are highly gregarious, resting, moving, and feeding in groups. Young larvae are reddish with shiny black heads. Mature larva is black with multiple small orange dots on dorsum and four to six black spines on each segment; head capsule is shiny black with a conical projection on each side. Pupa is black with dark reddish-brown markings, convex wing pads, and a keel on dorsal side of thorax. **Host plants:** *Passiflora alata, P. biflora, P. edulis, P. pedata, P. pittieri, P. platyloba, P. vitifolia* (Passifloraceae), *Erblichia odorata* (Turneraceae).

Dryas iulia moderata (Nymphalidae: Heliconiinae)
Julia

0–1800 m

FWL: 40–45 mm. **Description:** Wings are long and narrow. DWs are mostly orange; DFW has variable amount of black, sometimes with black margins and a transverse black line from costal vein to distal margin in sub-apical area. VWs are variegated orange and whitish-orange, with thin black distal margins. **Similar species:** *Eueides aliphera gracilis* is very similar but much smaller and has thin black lines on veins in distal area of DHW. *Dione juno huascuma* (opposite page) is differentiated by silver spots on VHW. In flight, *Marpesia petreus petreus* (p. 160) is very similar; once it lands, check to see if it has tails: *M. petreus petreus* does, *D. iulia moderata* does not. **Habitat:** Common and persistent in most habitats, from open areas and crops to primary forest. Usually found in open areas with early successional vegetation. **Distribution:** Mexico to Panama. **Seasonality:** Present all year. **Natural history:** Seen in gardens and fields when the sun shines, visiting flowers such as *Lantana, Cosmos, Zinnia, Stachytarpheta,* and *Jatropha.* If there are not many flowers, it passes overhead, flying fast at high altitude, which makes it difficult to observe. Males can be seen puddling on riverbanks or even drinking from eye and nose fluids of reptiles along rivers. Larvae are attacked by parasitoid flies (*Myothyriopsis,* Tachinidae). **Early stages:** Eggs are bright yellow and corncoblike. Mature larva is light brown with creamy patches and black transverse lines, the amount of color in each area very variable; every segment has four to six long, black branching spines; head capsule is red with black and yellowish lines and two well-developed, upward-raised horns. When larva is resting, the scoli of segments T_1 to A_1 aggregate together, pointing forward. In contrast to most other Heliconiinae species, the larva of *D. iulia moderata* builds itself a refuge by cutting part of a host plant leaf, resulting in an area of dry leaf where the larva can hide and seek safety from predators such as ants. This behavior has been noticed also in *Dryadula* and *Philaethria*, closely related genera. Pupa resembles a dried leaf; it is elongate, with wing pads more prominent than abdomen; variegated white, brown, and gray; and has chrome-silver dorsal spots from T_3 to A_2 and very small horns on head area.

Host plants: *Passiflora adenopoda, P. auriculata, P. bicornis, P. biflora, P. capsularis, P. coriacea, P. costaricensis, P. foetida, P. pittieri, P. platyloba, P. suberosa, P. talamancesis, P. vitifolia* (Passifloraceae).

Fifth instar larva of *Dryas iulia moderata*.

Pupa of *Dryas iulia moderata*.

Adult *Dryas iulia moderata*.

Dryas iulia moderata pinned. Dorsal view.

Dryadula phaetusa (Nymphalidae: Heliconiinae)
Banded Longwing

0–1500 m

FWL: 37–43 mm. **Description:** DWs are black with orange transverse bands; DHW has a sub-marginal row of orange spots. VWs are also black but with yellow-orange bands and a yellow row of sub-marginal spots on VHW. **Similar species:** None; *D. phaetusa* is distinguished by black-and-orange-banded pattern on dorsal wings in combination with its size. **Habitat:** Semi-open and open areas such as parks and early successional habitats, as well as secondary forests and open fields close to forest edges. In the Central Valley it can be seen along rivers associated with preserved forest patches such as Santa Ana, Ciudad Colón, and Palmares. In Guanacaste it may be found in forested patches along coasts and rivers. In rural areas, it may visit flowers in backyards and croplands. **Distribution:** Mexico to Brazil. **Seasonality:** More common during rainy season, but a few individuals can be found in humid spots persisting during dry season. **Natural history:** This butterfly wanders along rivers searching for host plants and flowers in semi-open areas. Group roosting behavior under leaves has been documented. Adults feed mainly from nectar of flowers such as *Lantana*, *Asclepias*, and various Asteraceae species; sometimes they can be seen feeding from bird droppings. **Early stages:** Eggs are yellow, corncoblike, and laid

Fifth instar larva of *Dryadula phaetusa*.

Pupa of *Dryadula phaetusa*.

Adult female *Dryadula phaetusa*.

singly on dried leaves or tendrils. Young larvae build shelters by cutting up pieces of leaves and then letting them dry. Mature larva is dull purple with long orange scoli on all body segments; head capsule is bright orange (DeVries 1987 incorrectly states it is dark purple) with two black raised horns. Pupa is whitish with gold dorsal marks on segments T_3, A_1, and A_2, and a dark long projection on head at area of each eye; it lacks spines. **Host plants:** *Passiflora biflora, P. costaricensis, P. talamancensis,* and possibly other *Passiflora* species (Passifloraceae).

Eueides isabella eva (Nymphalidae: Heliconiinae)
Isabella's Longwing

0–1500 m

FWL: 37–43 mm. **Description:** This butterfly typically has a "tiger-mimic" pattern: a black background with transverse orange lines combined with yellow spots and bands, covering all wing surfaces, both dorsal and ventral. A melanistic form has an almost completely black DFW with only a few small yellow spots, and borders of black bands on DHW are irregular and have some contact areas among them, characters the common form never shows. **Similar species:** The tiger-mimic pattern is shared by many butterfly species, constituting a mimicry ring, and it can be very difficult to distinguish the various species. In general terms, *Eueides isabella eva* is smaller than *Heliconius ismenius clarescens*, same size as *Mechanitis* species (e.g., p. 130), and larger than *Eresia ithomioides alsina* and *E. eunice mechanitis*. On *E. isabella eva*, a yellow teardrop-shaped spot is located in DFW medial area. **Habitat:** This species is common in many habitats,

Fifth instar larva of *Eueides isabella eva.*

Pupa of *Eueides isabella eva.*

Adult *Eueides isabella eva* orange form.

Adult *Eueides isabella eva* black form.

from croplands to primary forest to gardens, backyards, and parks with abundant flowers, as well as second-growth habitats and rivers and creeks, where it glides along edges of vegetation. Common in Central Valley and towns in rural areas. **Distribution:** Mexico to Panama. **Seasonality:** Present all year but more common during rainy season on the Caribbean slope; absent or rare during dry season in Guanacaste lowlands. **Natural history:** This species flies when the sun shines, visiting flowers such as *Lantana*, *Stachytarpheta*, and *Pentas*. As with many members of its subfamily, it sequesters toxic secondary compounds from its host plants, becoming unpalatable to predators. Many other unpalatable and palatable species mimic its pattern, making it almost impossible to identify them, unless they are perched or still. This species' flight is usually slow, but it can fly very fast and high if alarmed. The melanistic form is more common in some populations than in others, although never abundant. **Early stages:** Eggs are yellow, corncoblike, and laid singly on underside of leaves. Larva rests alone on leaf underside along central vein. Young larva has a longitudinal creamy-yellow lateral band. Mature larva retains that band, but dorsal area is variable in color and pattern, from variegated gray and brown to red, and the last two segments sometimes are bright red; all body segments have four to six long, branched spines (scoli), which can vary from black to white; head capsule varies in coloration from shiny black to black with gray lines, and bears two long, variably colored, upward-projected scoli. Pupa is gray with pink bands; all abdominal and thoracic segments have thick projections, most short but the ones on A_3 and A_4 very long; two long projections extend from head. This striking pupal form may resemble a dead insect attacked by fungus of the genus *Cordyceps*. **Host plants:** *Passiflora serratifolia*, *P. adenopoda*, *P. ambigua*, *P. biflora*, *P. capsularis*, *P. pedata*, *P. platyloba* (Passifloraceae).

Heliconius charithonia vazquezae (Nymphalidae: Heliconiinae)
Zebra Longwing

0–1350 m

FWL: 40–48 mm. **Description:** A black butterfly with three parallel yellow bands on FW, and one band and a line of dots, both yellow, on HW. **Similar species:** *Heliconius hewitsoni*, *H. pachinus*, and *H. sara theudela* do not have three yellow bands on DFW. **Habitat:** Open areas and secondary forest. Likes to fly where bushes provide security; can be seen on flowers along croplands or cattle ranches or in forest edges. **Distribution:** USA to Panama. **Seasonality:** Present all year but most common during rainy season. **Natural history:** *H. charithonia vazquezae* is common in most disturbed habitats, especially bushy patches.

Fifth instar larva of *Heliconius charithonia vazquezae*.

Pupa of *Heliconius charithonia vazquezae*.

Adult *Heliconius charithonia vazquezae*.

Adults feed from a variety of flowers such as *Lantana, Stachytarpheta, Hamelia,* and *Tithonia*. Males actively search for host plants, seeking pupae of their own species, and then stay close to the found pupae. A few days before they emerge, the female pupae release a pheromone, prompting one to four males to land on it and await the emergence of the female. The male in the best position will mate with the female within the first seconds and in doing so transfer a spermatophore, a nutrient-rich packet important for egg production and the vitality of the female. This species also displays roosting behavior, with many adults resting together in the same place for many nights in a row. The reason for this behavior is not definitively known, but one possibility is that a butterfly sleeping in the forest can easily be startled and take flight, a huge risk because it is blind in the dark. Any individual encountering a group resting together might intuit that the roosting place is safe; and night after night, an increasing number of individuals may thus gather. In nature, the sex ratio in the roost is about 2:1, male to female. In the case of this species, which practices pupal mating, this could be because males obtain information from other males. Once a male knows where female pupae are, the new males may wait until morning to follow him and perhaps obtain better chances of finding female pupae and increasing their probability of reproducing. Larvae of this species are attacked by parasitoid flies (*Myothyriopsis*, Tachinidae). **Early stages:** Eggs are yellow, corncoblike, and laid singly or in small groups in new shoots of host plant. Mature larva is white with disperse black dots all over the body, and long, branched black spines arising from some of these dots; head capsule is white with black patches in front and around stemmata, and a long black scolus on each side. Resembling a dry leaf, pupa is a mosaic of brown, gray, and black (some pupae are darker than others) with silver-chrome dorsal patches on segments T_1, A_1, A_2, and A_3; it is long, with convex wing pads, flat dorsal projections that end with black spines on abdominal segments, and a keel-like dorsal projection on each side of the abdomen with a black spine at the end on A_4, A_5, and A_6; head has a long, corrugated, downward-pointing leaflike projection on each eye. **Host plants:** *Passiflora adenopoda, P. apetala, P. bicornis, P. biflora, P. lobata, P. menispermifolia* (Passifloraceae).

Heliconius erato demophoon (Nymphalidae: Heliconiinae)
Heliconius erato petiverana
Heliconius erato hydara
Crimson-patched Longwing

0–1600 m
demophoon

0–1600 m
petiverana

0–1600 m
hydara

FWL: 30–37 mm. **Description:** A black butterfly with a red transverse band from costal to distal border on FWs, and a yellow transverse band from costal to anal border on HWs. *H. erato demophoon* has thicker yellow band on DHW than *H. e. petiverana*; *H. e. hydara* does not have the yellow band. **Similar species:** *Heliconius melpomene rosina* is very similar, but yellow band on its VHW does not touch costal margin, as it does in *H. erato*. **Habitat:** Secondary forest and old-growth forest, along rivers and forest edges; generally less common in primary forest. May be seen visiting flowers in gardens or abandoned cattle ranches. It is common in Central Valley river systems and surrounding rural areas and coffee plantations. **Distribution:** *H. e. demophoon*, Mexico to Panama (Pacific side); *H. e. petiverana*, Mexico to Panama (Caribbean side); *H. e. hydara*, Costa Rica (close to Panama border) to Brazil. **Seasonality:** Present all year but more common during rainy season on Pacific slope. **Natural history:** This is the commonest Costa Rican *Heliconius* species. It prefers secondary and old-growth forest edges over primary forest. Both sexes feed from flowers such as *Lantana*, *Stachytarpheta*, *Hamelia*, and many others, but they always search for *Psiguria*, *Gurania*, *Anguria*, *Palicourea*, and *Cephaelis* plants to obtain the amino acid proline, which is important for increasing their life span (to up to 6 months), flying power, egg production (females), and spermatophore production (males). This species is also known to practice roosting and pupal mating (as described in *H. charithonia vazquezae*, p. 143), usually roosting 1 m above the ground, lower than other species. It tends to stay close to its roosting site and favorite flower source, so its home range is often no more than 1 km². **Early stages:** Eggs are yellow, corncoblike, and laid singly in new shoots and young leaves of host plant. Mature larva is whitish with a dark yellow longitudinal band on each side, black dots over entire body (largest and closest to each other on the lateral band), and long, black scoli on each segment; head capsule is white with black stemmata, two triangular black patches in front, and two long black scoli. Pupa is dark brown and gray with silver-chrome dorsal patches on segments T_1, A_1, A_2, and A_3, nine pairs of short dorsal spines, and a keel-like dorsal projection with a black spine at the end on each side of abdomen on A_4 and A_6; head has a long, corrugated downward-pointing leaflike projection on each eye. **Host plants:** *Passiflora auriculata*, *P. bicornis*, *P. biflora*, *P. caerulea*, *P. coriacea*, *P. costaricensis*, *P. pittieri*, *P. talamancensis*, *P. vitifolia* (Passifloraceae).

Fifth instar larva of *Heliconius erato demophoon*.

Pupa of *Heliconius erato demophoon*.

Heliconius erato demophoon pinned.
Dorsal view.

Adult *Heliconius erato demophoon* roosting.

Heliconius erato petiverana pinned.
Dorsal view.

Adult *Heliconius erato petiverana*.

Heliconius erato hydara pinned.
Dorsal view.

Adult *Heliconius erato hydara*.

Heliconius cydno galanthus (Nymphalidae: Heliconiinae)
Heliconius cydno chioneus
Cydno Longwing

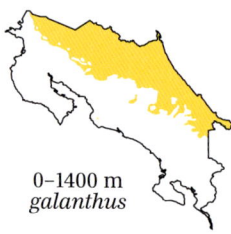

0–1400 m
galanthus

FWL: 38–44 mm. **Description:** Wings black with variable white markings. VHW has red transverse lines, variable in size. *H. c. galanthus*: DHW is entirely black except for few small white dots on apex. *H. c. chioneus*: DHW has large, conspicuous white marginal band. **Similar species:** *Heliconius sapho leuce* has white area on veins M_2–M_3 from discal cell to distal margin of FWs. **Habitat:** Secondary and primary rainforest on Caribbean slope, along trails, light gaps, riversides, forest edges, and roads. Common visitor of flowers in farms and gardens with nearby forest patches. Locations include Cahuita, Braulio Carrillo, Tortuguero, and Arenal Volcano National Parks. *H. c. chioneus* is found on Caribbean slope only from Manzanillo southward, while *H. c. galanthus* occurs there and also north of that point. **Distribution:**

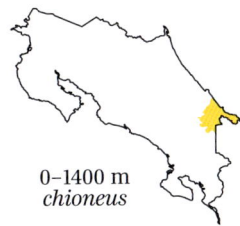

0–1400 m
chioneus

H. c. galanthus, Mexico to Costa Rica; *H. c. chioneus*, Costa Rica (close to Panama border) to Colombia. **Seasonality:** Present all year. **Natural history:** This conspicuous butterfly is common throughout Caribbean slope. It likes to fly from forest understory to sub-canopy and is very dominant when visiting flowers of *Psiguria* and *Gurania*, expelling any other butterfly trying to get nectar from them. Like most *Heliconius* butterflies, it obtains toxic compounds from its host plants, which are stored in its body, both as larva and adult, making it distasteful to predators. This has led to very complex rings of Müllerian mimicry among different species, making them very similar and difficult to distinguish under field conditions. This species naturally hybridizes with *Heliconius melpomene rosina* in certain Caribbean localities and with *H. pachinus* on passes through mountains between Caribbean and Pacific slopes. **Early stages:** Eggs are yellow, corncoblike, and laid singly in new shoots of host plants. Mature larva is white with many black dots dispersed throughout body and four to six long black scoli on each segment; head capsule is light orange with black stemmata and two black scoli. Pupa is brown with chrome-silver dorsal spots on segments T_3, A_1, and A_2, very concave wing pads, long black dorsal spines rising from T_2 and A_3–A_7, a conspicuous row of short black spines from dorsal head area to ventral wing-pad borders, and two short, down-pointing protuberances from head. **Host plants:** *Passiflora alata*, *P. ambigua*, *P. apetala*, *P. auriculata*, *P. biflora*, *P. caerulea*, *P. capsularis*, *P. coriacea*, *P. costaricensis*, *P. menispermifolia*, *P. oerstedii*, *P. pittieri*, *P. quadrangularis*, *P. vitifolia* (Passifloraceae).

Fifth instar larva of *Heliconius cydno galanthus*.

Pupa of *Heliconius cydno galanthus*.

Adult *Heliconius cydno galanthus*.

Adult *Heliconius cydno chioneus*.

Heliconius cydno galanthus
pinned. Dorsal view.

Heliconius cydno chioneus
pinned. Dorsal view.

Doxocopa druryi acca (Nymphalidae: Apaturinae)
Silver Emperor

0–800 m

FWL: 35–40 mm. Description: Both sexes are brown with a longitudinal crossband on both FW and HW; VWs are shiny or glossy silver; HWs have small spike-like tails. *Female:* Crossband is all white and extends from near costal margin of FW to anal margin of HW; DFW has large orange circular patch in sub-apical area. *Male:* Crossband on DW is mostly orange, except for posterior end, which is white, and orange part of band fuses with orange sub-apical circular patch; iridescent purple occurs on DHW and anal margin of DFW. **Similar species:** Many *Adelpha* species (e.g., pp. 133-136) are similar, but none are shiny ventrally. Female *Doxocopa pavon theodora* is similar but much smaller. Male *D. linda plesaurina* has no iridescent purple on DHW, and female has half-white, half-orange DFW crossband. **Habitat:** Secondary and primary forest on both slopes but more common in Pacific deciduous habitats. Rare individuals are seen along rivers or forest edges in Guanacaste and Puntarenas, perching 2–4 m high. Occasionally, both sexes can be seen puddling on sand of riverbanks or dirt roads. In some years, this species becomes very abundant, and both sexes visit flowers. **Distribution:** USA to Panama. **Seasonality:** Present all year on both slopes; on Pacific slope more common during rainy season. **Natural history:** This fast-flying species is generally very wary and flies away when approached. It is not very common but can be seen regularly in some preserved localities. Both sexes visit sap, dung, rotting fruits, and flowers of *Croton draco, Lasianthaea fruticosa, Ageratum,* and *Cordia.* It is not known why this species is so similar to *Adelpha.* Pinheiro et al. (2016) contend that *Adelpha* and *D. d. acca* are part of a mimetic ring of palatable butterflies with great facility for escaping from danger; predators (such as birds) learn that the cost is too high relative to benefit to try to catch these butterflies and therefore ignore them. Larvae are attacked by a parasitoid wasp (*Microcharops tibialis,* Ichneumonidae). **Early stages:** Eggs are light green, rounded, with faint longitudinal ribs, and are

Fifth instar larva of *Doxocopa druryi acca.*

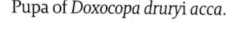

Pupa of *Doxocopa druryi acca.*

Adult *Doxocopa druryi acca.*

Doxocopa druryi acca pinned. Dorsal view.

laid on the margin and underside of leaves. Mature larva is green with short pubescence, many very small white warts dispersed all over body, and a fused, two-forked tail; segments A_2, A_4, and A_7 have circular, gray to yellow dorsolateral patches (in fifth instar, patch on A_7 almost disappears), which can be accompanied by a neighbor patch (variable in brightness, but in younger instars brighter and yellowish); head capsule has bluish-gray and white band in front, short green lateral spikes, and two long, upraised, green and black horns with some lateral spikes and ending in a black two-forked tip. Pupa is triangular and pale green with diagonal darker markings, large dorsal protuberances on A_3, A_5, A_6, and A_7, and a two-forked head. **Host plant:** *Celtis iguanaea* (Cannabaceae).

Doxocopa laurentia cherubina (Nymphalidae: Apaturinae)
Turquoise Emperor

400–1600 m

FWL: 35–40 mm. **Description:** Both sexes have glossy light brown VWs with black dots and disperse small lines; and small spike-like tails on HW. *Female:* DWs are brown with a longitudinal band in medial area of both wings; band on DFW is orange, turning to white as it approaches anal border and variable in thickness, sometimes diminishing to spots; band on DHW varies from white to blue or both together; VHW has longitudinal dark brown line from costal to anal margin. *Male:* DWs have broad blue band from discal cell of DFW to anal margin of DHW.

Similar species: Male *Doxocopa excelsa* has almost all-blue patch in DFW discal cell; male *D. l. cherubina* has very reduced blue patch, plus has two or three white dots on sub-apical area of DFW. Female *D. excelsa* has one blue-centered eyespot on veins M_3–Cu_2 on VFW. **Habitat:** Well-preserved mid-elevation forest patches on both slopes; always associated with rainforest habitats. **Distribution:** Mexico to Colombia. **Seasonality:** Present during rainy season. **Natural history:** This glorious species is present only in undisturbed habitat, and adults can be found only during rainy season. On hot days, males may be seen puddling on dirt roads or riverbanks. Both sexes can also be seen perching 5–6 m high in vegetation of forest edges or riversides. Adults feed from dung, rotting fruits, and flowers of trees such as *Croton draco*. Male is very territorial; it perches on high exposed leaves with closed wings and energetically chases away any other butterfly that passes by. Female stays in forest canopy and rarely flies down. This species seems to represent a case of escaping mimicry (see pages 25 and 292) of palatable butterflies, since it strongly resembles species of *Prepona*, *Archaeoprepona*, and *Adelpha* (females). **Early stages:** Eggs are light green, round, with faint longitudinal ribs, and are laid on underside of leaves, generally toward side margin of leaf. Mature larva is green with short pubescence, disperse small white warts, a pair of small circular light yellow and bluish-gray dorsolateral patches on segments A_2, A_4, and A_7, and a forked tail; head capsule is white with a longitudinal bluish band in front, black stemmata and mouthparts, and two long, upward projections that end in a black two-pronged fork. Pupa is green and triangular, with a large point on dorsal side of segment A_3, sharp edges, and two short horns on head. It is formed on underside of leaf, with ventral area very close to the leaf, giving it the appearance of a leaf protuberance. **Host plant:** *Celtis iguanaea* (Cannabaceae).

Fifth instar larva of *Doxocopa laurentia cherubina*.

Pupa of *Doxocopa laurentia cherubina*.

Adult *Doxocopa laurentia cherubina*.

Adult *Doxocopa laurentia cherubina*.

Doxocopa laurentia cherubina pinned. Dorsal view.

Dynamine dyonis (Nymphalidae: Biblidinae)
Blue-eyed Sailor

400–1300 m

FWL: 18–23 mm. **Description:** *Female:* DWs are dark brown with a wide white band from vein Cu_1 on DFW to anal margin of DHW, a smaller white patch from DFW costal margin to M_3, a smaller white spot on sub-marginal area (Cu_1–Cu_2) of DFW, and other, smaller disperse white markings; basal costal margin of DFW is light brown. *Male:* DWs are metallic greenish-yellow with brown marginal borders and a brown square from M_2 to M_3 on distal margin of DFW; basal costal margin of DFW is yellow. Both sexes: VWs have pattern of wavy white and brown bands and patches, some metallic blue lines, and two prominent eye-spots with blue scales in the center on medial area of VHW. **Similar species:** *Dynamine artemisia glauce* and many other Costa Rican *Dynamine* species are similar, but male *D. dyonis* can be distinguished by yellowish coloration and by arrangement of brown DW markings, especially brown square on DFW at M_2–M_3, which only touches the distal brown margin. *D. artemisia glauce* has very reduced

VHW eyespots, which separates it from female *D. dyonis*. **Habitat:** Open areas and secondary forest edges along Pacific slope. Both sexes are usually seen puddling on wet mud during hot mornings; females tend to stay close to host plants in early successional habitats such as abandoned fields and crops. **Distribution:** USA to Panama. **Seasonality:** Present all year but more common during dry season. **Natural history:** This is a very localized species, highly dependent on presence of host plant patches. Both sexes are powerful fliers and perch on emergent vegetation 2–5 m high. They fly only during the hottest hours of the day. Adults have been reported to feed from flowers of various species of Asteraceae and *Lantana* as well as from rotting fruits. **Early stages:** Eggs are green, with prominent longitudinal ribs, and are laid singly on underside of leaves (not in floral structures as with many other *Dynamine* species). Mature larva is green with faint longitudinal white lines and bands; body is covered in profuse green scoli, each branched with around ten spines tipped with rounded vesicles. Fifth-instar head capsule is green and smooth (no projections at all) with black stemmata. Pupa is green with small white streaks on abdominal segments, brown markings on wing pads and thorax, a pronounced brown dorsal protuberance on segment A_2, and a smaller one on T_2; segments A_3–A_6 are brown with white dorsal blotches; head has two short spikes. **Host plant:** *Tragia volubilis* (Euphorbiaceae).

Fifth instar larva of *Dynamine dyonis*.

Pupa of *Dynamine dyonis*.

Adult male *Dynamine dyonis*.

Dynamine dyonis pinned.
Dorsal view.

Dynamine dyonis pinned.
Dorsal view.

Catonephele numilia esite (Nymphalidae: Biblidinae)
Blue-frosted Banner

0–1500 m

FWL: 38–42 mm. **Description:** *Female:* DWs are black; DFW has a diagonal light yellow band from costal border to distal margin but not touching the edge, and some small yellow spots and a red spot in sub-apical area; DHW is all black with two sub-marginal rows of yellow spots almost forming a line. VWs are a mosaic of gray, black, and brown wavy patches, with same yellow band on VFW as on DFW. *Male:* DWs are velvet-black with two large, conspicuous orange patches on DFW and one on DHW, which also has blue apical and tornus margins. VWs are grayish-brown and black, with a light yellow basal patch on VFW. **Similar species:** No other butterfly in Costa Rica has six circular orange patches on the wings. **Habitat:** This is a forest species, common in secondary and well-preserved forest habitats and always associated with rainy places. It may be seen along forest edges, rivers, open trails, and roads through forests and abandoned plantations, in Sarapiquí, Guápiles, Ciudad Colón, Jacó, Dominical, and other rainy lowland localities. **Distribution:** Mexico to Ecuador and Venezuela. **Seasonality:** Present all year but more common during dry season. **Natural history:** Males are fast and powerful fliers that perch in clearings, 2–20 m high. They protect territories by chasing away other butterflies and returning to the same area (not necessarily the same leaf). Females spend more time in forest interior and semi-open habitats looking for host plants and laying eggs on leaves 1–4 m high. Both sexes feed from rotting fruits and tree sap, and when resting, they perch on tree trunks and leaves with wings closed and head pointing downward. **Early stages:** Eggs are greenish-white, barrel-like, with prominent longitudinal ribs, and are laid singly on underside of mature leaves. After hatching, young larva goes to end of a leaf vein and eats the leaf all around it, then starts extending the vein with small frass balls, attaching them at the end with silk. First- and second-instar larvae rest concealed at the end of this frass chain. Third and fourth instars rest on top of leaves with body in an S position. Third-instar larva is green with a

Fifth instar larva of *Catonephele numilia esite*.

Pupa of *Catonephele numilia esite*.

Adult *Catonephele numilia esite*.

Catonephele numilia esite pinned.
Dorsal view.

black rectangular dorsal patch on T_2–A_2 and another on A_7–A_{10}; it is covered in long, compound scoli. Mature larva is all green with longer scoli; head capsule is green except for lateral margins, which are black. All instars except the first have compound scoli on head, which are short in second instar and very long in subsequent ones. Pupa is different tones of green with a beanlike yellowish patch, which may resemble an eye, on each side of body at dorsal border of wing pads. Generally formed on top of leaf in an erect posture, the pupa is more or less similar to a reptile head, which could be intimidating to small predators. **Host plants:** *Alchornea costaricensis, A. latifolia* (Euphorbiaceae).

Biblis aganisa (Nymphalidae: Biblidinae)
Red Rim

0–1100 m

FWL: 30–35 mm. **Description:** Wings are black. DHW has a sub-marginal row of large red square patches; DFW has a sub-marginal reddish shadow. VWs are similar, but patches on VHW are smaller and pink instead of red. **Similar species:** In female *Lyropteryx lyra cleadas* (Riodinidae), red reaches margin of DHW and also extends along DFW to costal margin. **Habitat:** Disturbed and open habitats such as cattle ranches, abandoned crops, and farms on both slopes; more common on Pacific side. May be seen flying close to the ground along bushes and sides of roads and trails. Locations include Ciudad Colón, La Garita of Alajuela, Dominical, and other places but always in very local populations associated with presence of host plant. **Distribution:** USA to Panama. **Seasonality:** Present during rainy season, but an occasional old individual can be seen wandering during dry season. **Natural history:** This beautiful butterfly has a slow, fluttering flight. Its bright colors and behavior suggest that it tastes bad to predators; and its host plants tend to produce toxic secondary compounds, which may be sequestered by its caterpillars. Furthermore, its similarity to female *L. l. cleadas* also suggests unpalatability, since when sexual dimorphism is present, the female tends to mimic unpalatable species. In the LAREBUB adults live up to 70 days. Caterpillars are attacked by parasitoid wasps (*Diolcogaster*, Braconidae) and flies (*Pseudosturmia*, Tachinidae). **Early stages:** Eggs are light yellow, round, and laid singly on underside of mature leaves. Females can lay around 500 eggs (T. Fox, pers. comm.). Young caterpillars are black with dark yellow on dorsum, and rest on top of leaves. Mature larva is very cryptic, with brown, green, and whitish pattern of lines and bands; all segments bear four pairs of light brown, long, compound scoli; head capsule is creamy-colored with many brown dots on sides and black in front, many short spines, and two long, brown zigzag scoli ending in a spiny rosette. Mature larvae are very difficult to find because they usually do not rest on the host plant but move to adjacent vegetation or structures to hide. Pupa resembles a small dry leaf, colored different tones of green and brown, with dorsal protuberances on segments A_2 and T_2 (a feature shared by *Dynamine dyonis*; p. 150) and two small spines on head; pupa is broader when seen in dorsal view, with lateral projections on abdomen and thorax. **Host plants:** *Tragia volubilis, Gitara nicaraguensis* (Euphorbiaceae), *Gouania lupuloides* (Rhamnaceae; W. Haber, pers. obs.).

Fifth instar larva of *Biblis aganisa*.

Pupa of *Biblis aganisa*.

Adult *Biblis aganisa.*

Myscelia cyaniris cyaniris (Nymphalidae: Biblidinae)
Whitened Bluewing

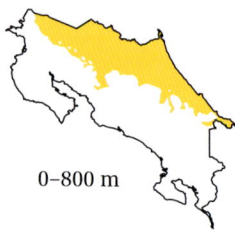

0–800 m

FWL: 29–33 mm. **Description:** Male DWs are very iridescent blue; female lacks iridescence. Both sexes have a beaklike shape to apical area of FW, and black DWs with bluish transverse bands on both DFW and DHW. DFW has a white band on discal cell, and white blotches in medial and sub-apical area. VWs are grayish with white bands, and VFW has white blotches. Female has more black and disperse dark areas on VHW than males. **Similar species:** Male is the only Costa Rican *Myscelia* species with bright blue iridescence. Female has only a long white band in discal cell of DFW, which distinguishes it from *M. ethusa* (two blue lines) and *M. leucocyana* (one blue line and two white dots). **Habitat:** Primary and well-preserved forest on Caribbean slope. Rare individuals may be seen flying 2–8 m high, perching in light gaps and along riversides, dirt roads, and forest edges. Localities include Braulio Carrillo, Cahuita, Arenal Volcano, Tortuguero, and Barbilla National Parks, and Hitoy-Cerere Biological Reserve. **Distribution:** Mexico to Panama. **Seasonality:** Present all year. **Natural history:** This butterfly is not common in the wild because it needs well-preserved habitats with local presence of its host plants; however, it has been bred extensively in butterfly houses and is often seen in live-butterfly exhibitions. Male perches on tree trunks with wings closed and head pointing downward. It does not present marked territorial behavior, but it is keen to chase females of its own species. When they find each other, the female will fly slowly while beating her wings very fast, and the male does the same, just a few centimeters above her, in a courtship that can last around five minutes, after which they land on a leaf to mate or to repeat the process. Female perches on high branches of trees neighboring the host plant and lays many eggs on single plant. Both sexes visit rotting fruits, and males puddle during hot days. **Early stages:** Eggs are white, with longitudinal ribs, and are laid singly on underside of mature and young leaves. First-instar larva is pale green and rests on a frass chain in central vein of leaf. Fifth-instar larva is green with many disperse white dots and yellowish dorsolateral markings, and its body is covered in long, green and black compound scoli; head capsule is mostly black with white front, long green spines at sides, and two long black scoli with many spines along shaft and at tip. Fifth-instar larvae rest in middle of upper side of leaf in a humped position, so that parts of body are not in contact with the surface. Pupa

Fifth instar larva of *Myscelia cyaniris cyaniris*.

Pupa of *Myscelia cyaniris cyaniris*.

Adult *Myscelia cyaniris cyaniris*.

is formed on upper side of leaf in an erect position; it is green with brown at sides of wing pads and head, a reddish dorsolateral patch on A_1–A_2, a prominent dorsal projection on T_2, and two short, conical spines on head. **Host plants:** *Adelia triloba*, *Dalechampia triphylla* (Euphorbiaceae).

Hamadryas amphinome mexicana (Nymphalidae: Biblidinae)
Red Cracker

0–1500 m

FWL: 35–40 mm. **Description:** DWs have a complex pattern of bluish-gray zigzag lines and dots on a bluish-black background; DFW has a transverse medial white band composed of seven closely spaced blotches. VFW is black with same characteristic white band as on DFW, and an orange basal area; VHW has large orange patch extending from basal area to sub-marginal area close to anal margin and to medial area at costal margin; both VWs have marginal white dots. **Similar species:** *Hamadryas arinome arienis* does not have orange patch on VHW. **Habitat:** Ground level in second-growth habitats; also in canopy

Fifth instar larva of *Hamadryas amphinome mexicana*.   Pupa of *Hamadryas amphinome mexicana*.

Adult *Hamadryas amphinome mexicana*.

Hamadryas amphinome mexicana pinned.
Ventral view.

and sub-canopy in mature forest. It is more easily seen on the Pacific slope, in rural areas near coffee plantations, abandoned fields, and riparian forest patches around Liberia, Abangares, Tárcoles, Dominical, and Osa Peninsula. On the Caribbean slope it is more difficult to see since it tends to perch higher in trees; it occurs around cattle ranches and in forest areas of Guápiles, Sarapiquí, and San Carlos. **Distribution:** Mexico to Colombia, Cuba. **Seasonality:** Present all year on both slopes, but on the Pacific slope much more common during rainy season. **Natural history:** As with many (but not all) species of its genus, *H. a. mexicana* has the ability to produce sounds for intra- and interspecific communication that are audible to humans; males make a cracking sound to deter other males from a territory or to court a female. Sound is produced by the slight bending of the sub-costal vein of the FW when the wing is beating, which produces a vibration that is then amplified in a spiral organ inside the same vein with a vibratory membrane (Vogel's organ), which releases the sound to the environment. Many *Hamadryas* species are palatable to predators, but this one is thought to be unpalatable; in experiments, Jacamars (*Galbula ruficauda*) have been shown to decline to eat live individuals of *H. a. mexicana*. **Early stages:** Eggs are white, round, with longitudinal ribs, and are laid on underside of leaves in pendant chains of up to fifteen eggs. Young larvae feed in groups; mature larvae are also gregarious. Young larvae are dark and covered in long compound scoli; head capsule is red with long black horns. Mature larva is black with red lateral patches on and around prolegs on segments A_2-A_5, yellow dorsal lines that transcribe the letters *O, T, O* across each of the same segments, and compound scoli that are black except in the same segments, where the spines are orange; head capsule is similar to that of younger instars but larger. Pupa is elongate, resembling a curled-up leaf, variable in coloration, from mostly brown to yellowish on wing pads (but always with a brown mark in the middle), with red longitudinal lines; head bears two long, twisted leaflike projections. **Host plants:** *Dalechampia scandens, D. websteri* (Euphorbiaceae).

Epiphile adrasta adrasta (Nymphalidae: Biblidinae)
Common Banner

200–1700 m

FWL: 26–30 mm. **Description:** *Female:* DHW is orange with black apex. DFW is orange in basal area; the rest is black with a yellow band from costal margin to tornus, and a white spot in sub-apical area. *Male:* DWs are velvet-black with two wide longitudinal orange bands surrounded by bluish-purple iridescence; DFW apical area has orange beaklike margin. Both sexes: VHW is brown with blurred eyespots in medial area and a conspicuous white triangular patch on costal margin; VFW has a yellow basal area, a transverse band in medial area, and two or three small sub-apical eyespots. **Similar species:** Although many butterflies are black-and-orange-banded, this species' beautiful simplicity, size, and wing shape make it hard to mistake. It is distinguishable by the medial orange area from male *Temenis pulchra*, in which the area is more red and does not touch the wing margin. **Habitat:** On both slopes, this species flies in light gaps and forest edges in secondary and primary forest areas in montane places such as Sabanilla in Montes de Oca, San Rafael de Heredia; and Escazú. It is sometimes seen in riparian forest along rivers in Central Valley, and also occurs in Monteverde, Orosí Valley, San Vito in Coto Brus, and rarely in Guanacaste mountain range. **Distribution:** USA to Panama. **Seasonality:** Present all year but more common during rainy season. **Natural history:** Males are powerful fliers that glide swiftly from one perch to another at sub-canopy level in direct sunshine in early mornings; they do not stay for more than a day or two in the same light gap (pers. obs.). Females fly low, at ground level, in mid-morning, searching for host plants at forest edges or trails. Adults feed from dung, decaying matter, sap, and rotting fruits at sub-canopy and ground level. **Early stages:** Eggs are white, barrel-like, with longitudinal ribs, and are laid singly on underside of leaves. Young larvae rest on frass chains; mature larvae rest in middle of upper side of leaves. Mature larva is green with many thin, yellow longitudinal lines and gray lateral patches on abdominal segments, and small, short orange and black dorsolateral scoli from T_2 to A_{10}, plus a dorsal, slightly larger scolus on A_9; head capsule is green with spines, black stemmata,

Fifth instar larva of *Epiphile adrasta adrasta.*

Pupa of *Epiphile adrasta adrasta.*

Adult male *Epiphile adrasta adrasta.*

Epiphile adrasta adrasta pinned. Dorsal view.

and two long, dark green, backward-curved horns. Pupa is erect, generally formed on top of host plant leaves; it is mostly whitish-green, though darker green dorsally from T_1 to A_2, and has a dorsal keel on T_2–T_3 and bright silver patches on lateral area of wing pads at T_3 level; head is two-forked, silver on dorsal side and green on ventral. **Host plants:** *Paullinia costaricensis, Cardiospermum grandiflorum, Serjania atrolineata* (Sapindaceae).

Diaethria astala astala (Nymphalidae: Biblidinae)
Astala Eighty-eight

700–1600 m

FWL: 23–25 mm. **Description:** *Female:* DWs are dark brown; DFW has a metallic-green band from sub-costal vein almost to tornus margin, and a small white band in sub-apical area. *Male:* DWs are black with iridescent dark blue; DFW has a greenish-silver band from discal cell to tornus that does not touch wing margin, and a white spot in sub-apical area. Both sexes: VFW is red in basal area, black in medial area, and has two white bands in sub-apical area. VHW is white with four concentric black lines, the ones in the inner part with black spots in the center, forming a variable *88* or *89* figure; one of the sub-marginal black lines has a thin red parallel line, and costal area has red markings. **Similar species:** *Diaethria anna anna* has continuous red costal margin on VHW, while in *D. astala astala* the red is discontinuous. *D. clymena marchalli* has much more red in VFW than *D. astala astala*, and *D. gabaza eupepla* has much less. **Habitat:** Found in primary and disturbed habitats such as coffee plantations, riversides, and abandoned fields, perching from 2 m high to canopy, on both Pacific and Caribbean slopes. It is common in Central Valley in suburban areas near forested patches; males can be seen puddling on sidewalks or curbside gutters. **Distribution:** Mexico to Colombia. **Seasonality:** Present all year but more abundant during rainy season. **Natural history:** This splendid butterfly is a common inhabitant of Central Valley suburban areas. Both sexes prefer to fly during hottest hours of morning and midday. Males perch in light gaps or forest edges along rivers and puddle on wet soil with closed wings; they will fly away very fast if alarmed but return close by in a few minutes. Females lay eggs on host plants 1–2 m high, infrequently as high as 16 m, from 10:00 a.m. to midday. Both sexes feed on rotting fruits, sap, and dung. The bright coloration of this butterfly is poorly understood; it might be unpalatable to predators, or this might be a case of good advertising, with its bright coloration training predators to remember not to pursue it because it is so fast and hard to catch, as proposed by Pinheiro et al. (2016). **Early stages:** Eggs are light green, barrel-shaped, with conspicuous longitudinal ribs, and are laid singly on underside of leaves. Young larvae are light green and rest on frass chains; mature larvae rest in middle of upper side of leaves. Mature larva is green with many disperse white and yellow warts; from A_2 to A_{10} it has dorsolateral orange scoli, each with one short black spine, except for the one on A_{10}, which has two; head capsule is green with two very long spiny horns. At rest, these horns point forward, facing the leaf; if disturbed, the larva uses the horns to attack. Pupa is erect, formed on upper side of leaves (usually on host plant); it is green with disperse whitish dots, a small, white-tipped dorsal keel on T_2, and a small whitish line on dorsolateral part of wings pads; head has white patches on dorsal area and is slightly two-forked. **Host plants:** *Allophylus racemosus*, *Cardiospermum*, *Paullinia*, *Serjania* (Sapindaceae).

Fifth instar larva of *Diaethria astala astala*.

Pupa of *Diaethria astala astala*.

Adult *Diaethria astala astala*.

Diaethria astala astala pinned. Dorsal view.

Diaethria astala astala pinned. Dorsal view.

Diaethria astala astala pinned. Ventral view.

Marpesia petreus petreus (Nymphalidae: Cyrestinae)
Ruddy Daggerwing

0–1600 m

FWL: 37–42 mm. **Description:** FW has a prominent beaklike apex; HW has a long brown tail (vein M$_3$), a smaller tail (vein Cu$_2$), and a notched apex. DWs are orange with parallel longitudinal black stripes; DHW has a brown distal margin. VW pattern is a brownish-gray mosaic of parallel dark lines and black dots. **Similar species:** *Marpesia berania fruhstorferi* lacks beaklike apex on FW and notch on apex of HW. *Dryas iulia moderata* (p. 139) lacks tails on HW. **Habitat:** More common on Pacific slope than Caribbean. Can be found in secondary and disturbed habitats, flying swiftly in open areas such as gardens, cattle ranches, and croplands, or visiting tree flowers around swimming pools, where it sometimes puddles, as it does on beaches, dirt roads, and riverbanks. Localities include Puntarenas, Limón, and Guanacaste coastlines; Central Valley parks and riparian forests. **Distribution:** USA to Bolivia. **Seasonality:** Present all year. **Natural history:** This species greatly resembles *Dryas iulia moderata* when flying since the black stripes and tails are not very visible when it is moving. Both species feed from very toxic host plants, but it is uncertain whether they are palatable to predators or if their similarity is a case of Müllerian, Batesian, or escaping mimicry. Based on larval and adult coloration and ecology, *M. petreus petreus* is likely unpalatable. Adults feed on flowers of several Asteraceae plants; trees and shrubs such as *Inga, Cordia, Croton draco, Lantana,* and *Hamelia patens*; and vines such

Fifth instar larva of *Marpesia petreus petreus.*

Pupa of *Marpesia petreus petreus.*

Adult *Marpesia petreus petreus.*

as *Mikania* and *Tournefortia hirsutissima*. **Early stages:** Eggs are cream-colored, barrel-shaped, with reduced but apparent longitudinal keels, and are laid on underside of leaves. Mature larva rests on upper side of leaf on central vein. It has four long, erect black dorsal spines on segments A_2, A_4, A_6, and A_8; dorsal segments from T_1 to A_1 are orange with four black dots; segments A_3, A_5, A_7, and A_8 have black diagonal dorsolateral bands surrounded by cream color; lateral area is half red and half whitish; a conspicuous black dot marks the side of each proleg. Head is bright orange with black stemmata and two long black horns. Pupa is pale yellowish-green with dorsolateral row of black dots along all abdominal and thoracic segments; a row of black dots on dorsum, some with small spikes, and those on segments A_2 and A_7 with long black projections, the one on A_2 two-forked; long, thin, black lateral projections on wing-pad base continued from black line on wing pad; and two black, curved projections arising dorsally from head. **Host plants:** *Ficus crocata*, *F. cahuitensis*, *F. cotinifolia*, *F. obtusifolia*, *F. crassinervia*, *F. tinctoria*, *F. pertusa*, *F. costaricana* (Moraceae), *Anacardium* (Anacardiaceae).

Smyrna blomfildia datis (Nymphalidae: Nymphalinae)
Blomfild's Beauty

0–1800 m

FWL: 32–40 mm. Description: *Female:* DFWs are brown in basal half and black in distal half, with three sub-apical white spots, and a diagonal white band separating the brown and black areas. DHW is mostly brown except for black apical area. *Male:* Similar to female, but DFW has bright orange where the female is brown and lacks diagonal white band. Both sexes: HW has a small spike-like tail (vein Cu_2). VFW has a brown basal area, yellow medial area, then a black area with the same three white sub-apical spots as in DFW, plus some other white lines; VHW has a complex design of white concentric and parallel lines on a brown background, and a row of four eyespots in sub-marginal area. **Similar species:** *Pycina zamba zelys* has four or five white spots on DFW, plus an extra one in apical area, instead of just three as in *S. blomfildia datis*. **Habitat:** In some years, this species is extremely common in the northern Pacific from coastline to mountains and is seen in gardens and cattle ranches that are close or connected to any kind of natural vegetation area, from an abandoned field to a primary forest. It is easy to see along rivers or in backyards, attracted to fruits or perching on tree trunks. It is rarer on the Caribbean side. In Central Valley it is common where there are some forest patches, such as Santa Ana, La Carpintera, Alajuela, and Escazú. **Distribution:** USA to Panama. **Seasonality:** Present all year but much more common during rainy season. **Natural history:** Both sexes fly rapidly from ground level to the canopy. Males perch on tree trunks, so generally are in semi-shaded places; they rest with wings closed and the head pointing downward. During dry season in the dry forest, adults undergo diapause, sheltering from sunlight between tree buttresses, 1–5 m above the ground, at sides of dry rivers or creeks. They spend up to 5 months there and will fly only if rotting fruit, plant sap, or fresh dung appears nearby. **Early stages:** Eggs are green, spherical, with longitudinal ribs, and are laid singly on underside of host plant's leaves. Young

Fifth instar larva of *Smyrna blomfildia datis*.

Pupa of *Smyrna blomfildia datis*.

Smyrna blomfildia datis
pinned. Dorsal view.

Smyrna blomfildia datis pinned.
Ventral view.

Adult *Smyrna blomfildia datis.*

larva builds a frass chain to rest on; mature larva cuts base of main leaf veins in order to build a tentlike shelter, then rests singly on leaf underside on central vein. Mature larva is black (amount of black variable) with a cream-colored lateral band; two rows of white spots form a double line along dorsum, and disperse spots appear on dorsolateral areas; most segments have seven long, compound cream-colored scoli, the dorsal ones with black-tipped spines, the lateral ones entirely cream-colored; head capsule is covered in short spines and is dark orange with black stemmata, black and white mouthparts, and two long, thick, spiny orange horns. Pupa is bulblike in shape and shows different tones of dull brown with five longitudinal rows of cream-colored dorsal spots, a large black patch at base of each wing pad, and a small dorsolateral one between segments T_3 and A_1, the pair of which may resemble eyes, probably to deter predators. **Host plants:** *Myriocarpa longipes, M. bifurca, Urera caracasana* (Urticaceae).

Colobura dirce dirce (Nymphalidae: Nymphalinae)
Dirce Beauty

0–1600 m

FWL: 33–38 mm. **Description:** HW has a short square projection on veins Cu_2–A_2. DFW is brown with a diagonal cream-colored band from costal margin to tornus. DHW is all brown except for white costal margin. VWs have black-and-white-striped pattern of parallel and concentric lines. **Similar species:** *Baeotus beotus* has three thin tail projections on HW instead of one square projection. *Tigridia acesta* has three sub-apical white spots on DFW, and DHW is orange instead of brown. *Colobura annulata* is extremely similar, distinguished by details of the wing pattern (Fig. 27). **Seasonality:** Present all year but more abundant during rainy season. **Habitat:** Secondary- and primary-forest habitats, along trails, riversides, forest edges, and light gaps. Sometimes visits gardens surrounded by forest or close to

Fifth instar larva of *Colobura dirce dirce*.

Pupa of *Colobura dirce dirce*.

Fig. 27. (A) *Colobura annulata*. (B) *C. dirce dirce*.

Adult *Colobura dirce dirce*.

Colobura dirce dirce pinned. Dorsal view.

riparian vegetation. Localities include Guanacaste, along the coast; Caribbean slope, along rivers and on cattle ranches; Osa Peninsula; and Central Valley, in Ciudad Colón and near rivers of downtown San José. **Natural history:** This species is common in many habitats, flying during most hours of the day, 1–10 m high. Both sexes perch on tree trunks with the wings closed and the head pointing downward. The HW projections have a kind of eyespot, which might lure predators to the end of the body pointing up, rather than the butterfly's head. Both sexes feed from tree sap, rotting fruits, and dung. **Early stages:** Eggs are green, spherical, with longitudinal ribs, and are laid singly or in small groups of up to fifteen on underside of leaves. Younger instars build frass chains alongside their siblings, separated by 1 cm from each other. Fourth and fifth instars rest gregariously on underside of leaves near where the petiole attaches to the leaf; there the larvae chew the leaf veins, making a tentlike structure to rest inside. Mature larva is black with lateral yellow patches along body; each segment is armed with a ring of compound scoli, which are variable on thorax from white to yellow color and elsewhere all yellow; head capsule is black with lateral black and yellow spines and compound yellow horns. Pupa resembles a stick of wood; it is grayish-brown and long, thin, and cylindrical, with three pairs of short, upward-pointing dorsal spines on abdomen, a short thorax protuberance, and a two-forked head. **Host plants:** *Cecropia obtusifolia, C. peltata* (Urticaceae).

Anartia jatrophae luteipicta (Nymphalidae: Nymphalinae)
White Peacock

0–1300 m

FWL: 27–30 mm. **Description:** A whitish-gray butterfly with some longitudinal orange-brown wavy parallel lines, bolder in sub-marginal area and DFW discal cell, lighter on rest of DWs. DFW has a conspicuous black medial spot between veins Cu$_1$ and Cu$_2$. DHW is similar, with extreme variation in extent of orange-brown markings, and black spots at M$_1$-M$_2$ and Cu$_1$-Cu$_2$. HW has a small tail (vein M$_3$). VWs are similar to DWs but lighter. **Similar species:** None in Costa Rica. **Habitat:** Common in most disturbed habitats such as cattle ranches, croplands, suburban areas, and gardens, generally seen at ground level, flying among grasses and flowers. Occurs on all coasts and in grass fields and other open areas and is a common inhabitant of early successional habitats. **Distribution:** USA to Panama. **Seasonality:** Present all year but more abundant during rainy season. **Natural history:** This species is easily seen and identified. When alarmed, it can fly very fast and may be confused with *Protographium epidaus epidaus* (Papilionidae; p. 68) because of the colors and velocity, but it is easily recognized when stationary. The high variability in adult coloration is thought to be caused by environmental conditions during development. It has the widest distribution of its genus and also has the most reported host plants. Adults feed from flowers such as *Ageratum, Lantana, Bidens*, and many others. **Early stages:** Eggs are light green, spherical, with longitudinal ribs, and are laid singly in new shoots and underside of young

Fifth instar larva of *Anartia jatrophae luteipicta.*

Pupa of *Anartia jatrophae luteipicta.*

Adult *Anartia jatrophae luteipicta*.

leaves. Larvae rest on underside of leaves. Mature larva is mostly black with many disperse small white spots; each segment has a ring of long, spiny compound scoli, which are all black in the four anterior segments but have an orange base on the remaining segments; head capsule is shiny black with two long, hairy, forward-pointing horns with globular tip. Pupa has a rounded-cylindrical shape and is light green with many small disperse black dots. **Host plants:** *Blechum pyramidatum*, *Ruellia simplex* (Acanthaceae), *Bacopa* (Plantaginaceae), *Lindernia* (Linderniaceae), *Lippia bracteosa* (Verbenaceae).

Junonia genoveva (Nymphalidae: Nymphalinae)
Tropical Buckeye

0–1500 m

FWL: 25–31 mm. Description: DWs are mostly brown with prominent post-medial eyespots on DFW at M_1–M_2 and Cu_1–Cu_2 and on DHW at R_5–M_3 and Cu_1–Cu_2. DFW has orange-red bands on discal cell and sub-marginal areas, and a pale orange diagonal band running from post-medial sub-costal margin to distal sub-marginal area. VWs are brown with faint darker lines; basal area of VFW is orange, with eyespots very reduced or absent except for the one at Cu_1–Cu_2, which is always present. **Similar species:** *Junonia evarete* is identical, but ventral part of male's antennal club is pale, while on *J. genoveva* it is dark. **Habitat:** Open areas such as fields, cattle ranches, and gardens. It occurs on both slopes but is more common in areas of the north Pacific such as the Guanacaste coastline. Found in disturbed habitats of Puntarenas, Cañas, Parrita, Osa Peninsula, and Limón. **Distribution:** Mexico to Panama. **Seasonality:** Present all year but more abundant during rainy season. **Natural history:** This species is one of the most resilient to prolonged dry periods and is among the few butterflies that can be seen in the hot dry season in dry forest. It flies low, at ground level, resting with open wings on vegetation or puddling. Both sexes feed from flowers such as *Ageratum*, *Lantana*, *Comaclinium montana*, and *Asclepias curassavica*. It is the only butterfly that can live in mangrove habitats in Costa Rica; other species seen there are generally only passing by.

Fifth instar larva of *Junonia genoveva*.

Pupa of *Junonia genoveva*.

Adult *Junonia genoveva*.

Early stages: Mature larva is variably dark, often black, with a ring of long scoli on each body segment; each scolus has a small iridescent blue ring at the base, and the lateral ones also have orange at base; head capsule is black, with orange in front area (in some individuals back of head is orange), many small black spines and long black setae, and two short horns covered with setae. Pupa is cylindrical, with short dorsal spikes on abdomen, and black with white stains on dorsal side of segments T_2, A_1, A_2, A_4, A_7, and A_8, and white and gray marking on wing pads and head area. **Host plants:** *Blechum pyramidatum*, *Hygrophila costata*, *Avicennia germinans*, *Dyschoriste quadrangularis*, *Ruellia simplex* (Acanthaceae), *Stachytarpheta jamaicensis*, *S. frantzii* (Verbenaceae).

Siproeta stelenes biplagiata (Nymphalidae: Nymphalinae)
Malachite

0–1500 m

FWL: 44–49 mm. **Description:** DWs are dark greenish-brown with large light yellow-green band in medial area from anal margin of DHW to anal margin of DFW, which then splits into separate blotches, one inside discal cell and others in sub-apical area; there is a row of large spots of same color in sub-marginal area of DHW, then two spots on DFW sub-margin, and remaining spots reduced to small white stains. Both wings have slightly wavy borders, and HW has a small tail (vein M_3). Same color pattern appears on VWs but in shades of light brown and pale green, plus a few more disperse white markings. **Similar species:** *Philaethria diatonica* (Heliconiinae) has FW longer than HW, while in *S. stelenes biplagiata* the wings are about the same length; also, *P. diatonica* lacks tail from M_3 vein on HW. **Habitat:** Present on both slopes but more common on the Pacific coast, in primary, secondary, and riparian forest as well as gardens and semi-open areas such as croplands and cattle ranches. In Guanacaste, commonly seen flying low in the understory along forest trails and roadsides. In Central Valley can be seen around Alajuela, La Garita, and La Guácima. **Distribution:** USA to Amazon basin. **Seasonality:** Present all year but more common during rainy season. **Natural history:** With its bright greenish-yellow color and great

Fifth instar larva of *Siproeta stelenes biplagiata*.

Pupa of *Siproeta stelenes biplagiata*.

Adult *Siproeta stelenes biplagiata*.

abundance in many places, this species one of the more conspicuous and well-known butterflies in Costa Rica. It has been proposed that this species is a mimic of *P. diatonica*, but this is improbable since the two species do not share the same habitat most of the time. Also, the palatable *S. stelenes* is much more common than the unpalatable *P. diatonica*, and in the mimicry model, the distasteful species should be the more common. **Early stages:** Eggs are green, barrel-shaped, with prominent longitudinal keels, and are laid in tips of new shoots or base of young petioles. Young larva is all black and spiny. Mature larva is black, covered in long, compound, reddish scoli, those in two dorsolateral rows with dark orange base; head capsule is black with many setae and two long, curved horns covered in setae. Pupa is green with a row of short, yellow dorsal spines at segments $A_3–A_4$; rest of abdomen covered in small black dots; a small black dorsal spike on T_2; and head two-forked, with each fork black dorsally. **Host plants:** *Blechum pyramidatum, Justicia, Ruellia inundata, R. terminalis* (Acanthaceae).

Chlosyne janais janais (Nymphalidae: Nymphalinae)
Crimson-patch Checkerspot

0–1400 m

FWL: 23–29 mm. Description: Both DWs are black with a row of white spots on distal margins; DFW has many disperse small white spots throughout, and DHW has a large red patch, variable in size, from basal to medial area. VFW is similar to DFW. VHW is black with a large yellow basal patch filled with small black markings; a row of large red rectangular patches at distal border of yellow patch; a row of white spots in black medial area; and a sub-marginal row of yellow blotches that are variable in size and number. Both VWs have a row of white spots on distal margins. **Similar species:** *Chlosyne janais janais* has four or more white dots on basal area of DFW inside and beneath discal cell, while *C. lacinia lacinia* has none or sometimes one, and *C. lacinia saundersi* is all black in the same area or has orange patches but never white dots. *C. lacinia* is an extremely variable species, and some individuals very closely resemble *C. janais janais*. **Habitat:** Open areas and gardens associated with urban and agricultural habitats including cattle ranches and rural roads, beaches, and rivers. Seen in habitats with abundant shrubs and herbaceous plants in Central Valley, Limón, Siquirres, San Carlos, Puntarenas, Osa Peninsula, and Guanacaste. **Distribution:** Mexico to Colombia. **Seasonality:** Present all year but more common during dry season, especially on the Caribbean slope. **Natural history:** This common species inhabits human-disturbed habitats since its many larval host plants are often used as ornamentals in gardens and landscaping. It flies only when the sun shines, with a slow, heavy, gliding flight. Both sexes feed from flowers such as *Ageratum, Bidens, Alternanthera pubiflora, Comaclinium montana, Cosmos, Lantana, Coffea,* and many others. **Early stages:** Eggs are yellow, barrel-shaped, with longitudinal ribs, and are laid in groups of up to thirty on underside of host plant's leaves. Young larvae feed synchronously on underside of leaves; at first they are all dark and spiny but change to reddish-brown with black spines. Mature larva is greenish-brown, segmented with black rings bearing long, setae-covered scoli; head capsule is orange-red with many black setae and black stemmata. When alarmed, larvae drop off the host plant to escape danger. Pupa is grayish-green, covered in small black spots and lines on wing pads, thorax, and head; it is cylindrical, with a 90° angle from cremaster to body, which allows it to pupate on vertical surfaces and still hang down. **Host plants:** *Graptophyllum pictum, Odontonema tubiforme, Pachistachys lutea, Pseuderanthemum cuspidatum, Ruellia simplex, Megaskepasma erythrochlamys* (Acanthaceae).

Fifth instar larva of *Chlosyne janais janais.*

Pupa of *Chlosyne janais janais.*

Adult *Chlosyne janais janais.*

Chlosyne janais janais pinned. Ventral view.

Anthanassa atronia (Nymphalidae: Nymphalinae)
Brown Crescent

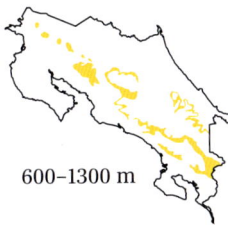

600–1300 m

FWL: 16–23 mm. **Description:** *Female:* DFW is brown with a diagonal row of three large white dots in medial area, followed by four or five white dots; basal area is light brown; DHW has large white medial band from veins R_5 to 2A, followed by a short, thin grayish sub-marginal line and another grayish line parallel to it from anal to costal margin; VFW is mostly brown with grayish apical area and a few small white sub-marginal spots; VHW is grayish-brown with sub-marginal row of faint eyespots, framed distally and basally by thin parallel brown lines.
Male: Much smaller than female; DFW has a diagonal row of three light brown dots in medial area, followed by four or five whitish dots, and profuse light brown scales in basal area; DHW is brown in costal half and in anal half is covered basally in light brown scales and has three thin parallel yellow lines toward distal margin; VWs are similar to female's but with many fewer white markings. **Similar species:** Male *Anthanassa otanes otanes* lacks the three thin yellow lines *A. atronia* male has on DHW. Female *A. atronia* has a single white spot on DFW from Cu_2 to 2A, distinguishing it from female *A. o. otanes* and both sexes of *A. ardys*, which have two spots. Also, VWs of *A. o. otanes* have yellowish shades, while VWs of *A. atronia* lack yellow but have an irregular white medial band. **Habitat:** Secondary habitats and open areas, including riversides and coffee plantations, on both slopes but more common on the Pacific side. Found in mid-elevations such as Central Valley, Alajuela, Atenas, Tilarán, and San Vito in Coto Brus. **Distribution:** Mexico to Costa Rica. **Seasonality:** Present all year but more common during dry season. **Natural history:** This species is a common inhabitant of open areas, where both sexes fly actively, visiting flowers of a large variety of Asteraceae species such as *Emilia fosbergii*, *Bidens pilosa*, *Melanthera nivea*, and *Alternanthera pubiflora*. Males visit mud on hot days, and females oviposit on sunny mornings (9:00–10:00 a.m.) inside the forest interior on host plant saplings. Males are very territorial and will use the same perch for several days in a row, chasing away any other *Anthanassa* butterfly that flies near it. When protecting the territory, the male's wings remain open. Perches are generally exposed leaves in direct sunshine, about 1 m high. **Early stages:** Eggs are yellow, spherical, and laid in large clusters of about forty on underside of leaves. Larvae feed gregariously until fifth instar, when they separate. Mature larva is dark brown with many disperse beige spots, a beige dorsolateral line on abdominal segments, and a lateral band along entire body; each segment has a ring of short, setae-covered scoli, which are dark brown except on segment T_1, where they are brownish-orange; head capsule is shiny black with brown patches. Pupa is light brown with two dorsal rows of small warts on A_4-A_7 and two larger spikes on T_2. **Host plant:** *Megaskepasma erythrochlamys* (Acanthaceae).

Fifth instar larva of *Anthanassa atronia*.

Pupa of *Anthanassa atronia*.

Adult male *Anthanassa atronia*.

♂

Anthanassa atronia pinned. Dorsal view.

Anthanassa atronia pinned.
Dorsal view.

♀

Anthanassa atronia pinned.
Ventral view.

♀

Consul fabius cecrops (Nymphalidae: Charaxinae)
Tiger-striped Leafwing

0–1200 m

Elevation: 0–1200 m. FWL: 35–40 mm. **Description:** FW has a very acute apex and a larger beaklike projection in distal margin; VW has a long, spatulate tail (vein M_3) and a smaller tail (vein 2A). DWs have a general tiger-striped pattern of yellow, orange, and black. DFW has a black background with a medial diagonal yellow band, which does not touch distal margin, followed by a large, transverse orange band from basal area almost to distal margin, barely touching it. DHW has orange basal, medial, and costal areas, and remaining area black with a sub-marginal row of four or five dorsally compressed yellow spots. VWs have a complex, cryptic pattern of different tones of brown, with gray and orange patches and wavy lines. **Similar species:** Immediately distinguished from *Heliconius ismenius clarescens* (Heliconiinae), *Melinaea lilis scylax* (Danainae), and many other tiger-stripe-patterned species by the spiky FW projections and tailed HW plus the cryptic ventral pattern. **Habitat:** Forest trails, understory, and riparian forest in secondary and primary habitats with a pristine understory. It is common on both slopes, in localities such as Cahuita, Braulio Carrillo, Arenal Volcano, La Cangreja, Corcovado, and Santa Rosa National Parks. In Central Valley it may be seen in Ciudad Colón and nearby areas such as Orotina and Acosta. **Distribution:** Mexico to Bolivia. **Seasonality:** Present during rainy season. **Natural history:** This species' behavior and appearance are unusual among members of Charaxinae. It has a slow and floppy flight and inhabits forest understory, precisely where *H. i. clarescens*, *M. l. scylax*, and other similar butterflies fly. These species represent a great example of color, shape, and behavioral convergence driven by Batesian mimicry. But when *C. f. cecrops* is resting, its cryptic underside, providing it infallible protection, is typical of Charaxinae. Adults feed on rotting fruits and tree sap from ground level up to 8 m and rarely higher. Females oviposit during late morning to midday on plants in light gaps or semi-open understory. Larvae are attacked by a parasitoid wasp (*Bracon alejandromasisi*, Braconidae) and a fly (*Winthemia*, Tachinidae). **Early stages:** Eggs are pale green, spherical, with smooth surface, and

Fifth instar larva of *Consul fabius cecrops*.

Pupa of *Consul fabius cecrops*.

Adult *Consul fabius cecrops*.

Consul fabius cecrops pinned. Dorsal view.

are laid singly on underside of leaves. From first to third instar, larva rests on a frass chain at tip of host plant and resembles a piece of dry leaf, with black, dark brown, and gray colors distributed in patches, and a hump from segment T_2 to A_3. Mature larva builds a live leaf roll to rest inside, usually on leaf tip but sometimes on a side of the leaf. Mature larva is green, a little darker in segments T_2, A_9, and A_{10}, with two black dorsolateral dots on A_2 and A_5, and disperse light yellow dorsolateral markings on T_1–T_3, and is covered in very short whitish setae; head capsule is mostly black with many yellow lines in front, many lateral yellow spikes, two short, wartlike projections, and two short, black spiny horns. The armed head presumably serves as a shield at the entrance of the leaf roll, blocking access to the vulnerable body to ants, parasitoids, or predators such as vespid wasps. Pupa is light green, conical, with a slightly two-forked head area, a dorsal shelflike shape on segment A_3, light brown lines on dorsal side of thorax and anterior of abdomen, and a light brown spot on each wing pad; segments A_8–A_{10} are shiny black. **Host plants:** *Piper aduncum, P. linearifolium, P. amalago, P. arboreum, P. colonense, P. lanceifolium, P. multiplinervium, P. auritum, P. jacquemontianum, P. peltatum, P. phytolaccifolium, P. reticulatum, P. sancti-felicis, P. trigonum, P. tuberculatum, P. umbricola, P. marginatum* (Piperaceae).

Archaeoprepona amphimachus amphiktion (Nymphalidae: Charaxinae)
White-spotted Prepona

0–1900 m

FWL: 50–57 mm. **Description:** DWs are velvet-black with a wide longitudinal greenish-blue band in medial area from vein M_3 on FW to 2A on HW, and two small greenish-blue patches (often fused) on costal margin of DFW at medial level. **Similar species:** Dorsally, it is almost impossible to distinguish among Costa Rican *Archaeoprepona* species. Ventrally, this species is distinguished by a straight line that runs longitudinally from VFW costal border at distal part of discal cell to vein Cu_2 and continues on VHW from costal border in medial area to tornus margin. *A. meander megabates* has two different brown tones on VWs, a pale brown from distal margin to post-medial area, a very dark brown in medial area, and pale brown again in basal area (on both VFW and VHW), and the division between the marginal light brown and the medial dark brown area is straight (making a curve only as it reaches the costal margin). VWs of *A. amphimachus amphiktion* sometimes, but not always, have two tones, darker distally and paler basally; the division of these areas in *A. amphimachus amphiktion* is wavy and diffuse. All *Prepona* species in Costa Rica have large eyespots on VHW at R_5–M_1 and Cu_1–Cu_2, lacking in *Archaeoprepona* species. **Habitat:** Well-preserved rainforest habitats on both slopes; rarer in dry forest. It is a canopy species but may be seen on rotting fruits in light gaps. Sometimes female individuals fly lower along trails or riparian forest, feeding from tree sap or decomposing matter or looking for host plants. Along dirt roads at lower elevations, males sometimes perch on exposed leaves on roadside vegetation as low as 3–4 m. **Distribution:** Mexico to Colombia. **Seasonality:** Present all

Fifth instar larva of *Archaeoprepona amphimachus amphiktion*.

Pupa of *Archaeoprepona amphimachus amphiktion*.

Adult *Archaeoprepona amphimachus amphiktion.*

Archaeoprepona amphimachus amphiktion pinned. Ventral view.

year but more common during rainy season. **Natural history:** These powerful fliers generally spend most of their time in the canopy, where males set up territories and perch on leaves exposed to direct sunshine and vigorously chase away any other butterfly that passes by. Females fly through forest understory in late morning to midday searching for host plants. They prefer to lay eggs on young trees 1–4 m tall and generally lay more than one egg on each plant. Mature larvae are attacked by parasitoid flies (*Winthemia*, Tachinidae), which lay their eggs directly on the larval body, generally on sides of thorax. Janzen and Hallwachs (2019) recorded a nearly 8-month period for one individual *A. a. amphiktion* from egg to pupal formation, at 1150 m elevation. **Early stages:** Eggs are pale green, spherical, with smooth surface, and are laid on underside of host plant's leaves. First- and second-instar larvae rest singly on a frass chain at tip of leaves; third instar rests at leaf tip, but not on a frass chain, and is greatly camouflaged by its dried-leaf appearance, even depositing pieces of real dried leaves to enhance crypsis; mature larva rests on top of host plant leaves or on thin branches. Mature larva has a highly disruptive shape: T_1 and T_2 are very small, segments T_3, A_1, and A_2 have a pronounced hump, and body tapers from segment A_6 to posterior tip, ending in a long, thin, two-forked tail. It is dark green, gray, and brown, with a diagonal lateral dark line from tip of hump on segment T_3 to tip of A_{10}; dorsolateral areas of A_2 have large eyespots, and hump of T_3 ends in a triangular tip, creating the appearance of a large head; head capsule is dark brown in front, pale green on sides, and has two long, thick, curved horns. Pupa is conical with a two-forked head, very large, and green with grayish transverse bands. **Host plants:** *Cinnamomum brenesii, C. triplinerve, C. hammelianum, Nectandra salicifolia, N. smithii, N. umbrosa, Ocotea atirrensis, O. insularis, Persea povedae, P. veraguasensis, Mespilodaphne macrophylla, M. veraguensis* (Lauraceae).

Memphis proserpina elara (Nymphalidae: Charaxinae)
Proserpina Leafwing

400–1800 m

FWL: 40–45 mm. **Description:** FW anal margin is indented, giving tornus a distinctive shape. *Female:* HW has small spatulate tail (vein M_3). DFW is black with metallic blue in basal area and in irregular sub-apical band that touches costal margin. DHW has same metallic blue on most of wing except darker sub-marginal area; distal margins are also blue. VWs have an intricate pattern of mottled reddish-brown with grayish areas. *Male:* DFW is similar to female's but with inner blue area smaller and darker; sub-apical blue band is present and fades into a marginal iridescent bluish wake. DHW has smaller and darker blue patch than female, but more area of the wing shows bluish iridescence; distal margin is also blue. VWs have same coloration as female's but with smaller grayish areas. **Similar species:** Male and female *Memphis ambrosia ambrosia* have a sub-apical band on DFW that is twice as wide, larger, and closer to apex

Fifth instar larva of *Memphis proserpina elara*.

Pupa of *Memphis proserpina elara*.

Adult *Memphis proserpina elara*.

Memphis proserpina elara pinned.
Dorsal view.

Memphis proserpina elara
pinned. Ventral view.

Memphis proserpina elara pinned.
Dorsal view.

Memphis proserpina elara pinned.
Ventral view.

than that of *M. p. elara*. Male and female *M. beatrix* have DFW sub-apical band composed of two blue spots arranged together and a third one closer to costal margin, separated and a little more basal than the other two. Male *M. laura laura* can be distinguished from male *M. p. elara* by absence of DFW sub-apical blue band; only the marginal–sub-marginal blue wake is present. **Habitat:** Premontane forest habitats on both Pacific and Caribbean slopes of central mountain ranges. Found in well-preserved forest patches in localities such as Bosque de la Hoja in San Isidro de Heredia; Braulio Carrillo, Rincón de la Vieja, and Turrialba Volcano National Parks. Also occurs in Las Cruces Biological Station, Monteverde, and Cerro de la Muerte. **Distribution:** Mexico to Panama. **Seasonality:** Present all year but more common during rainy season. **Natural history:** This conspicuous species is a fast and powerful flier. Males are active only during bright sunshine and can be seen with binoculars in the forest canopy, where they perch on treetops (including treetops as low as 3–4 m, as in some montane habitats) or at lower altitudes if there are light gaps. Females are more commonly seen flying from 10:00 a.m. to midday in forest understory searching for host plants on which to oviposit. They lay their eggs on small plants in forest interior during rainy season (T. Fox, pers. comm.). Both sexes feed from rotting fruits. Larvae are attacked by a parasitoid fly (*Houghia longipilosa*, Tachinidae). **Early stages:** Eggs are pale green, spherical, with smooth surface, and are laid singly on underside of leaves. Young larvae are green with dark brown patches; they build a frass chain and rest there through third instar; fourth and fifth instars build a leaf-roll shelter in leaf tip. Mature larva is cylindrical, black, with many disperse white scoli bearing short setae, a dorsolateral row of white spots, and a lateral row of larger spots; head capsule is black with white warts bearing white setae, and two white lines from front area to antennal base. Pupa is globular, pale green with reddish-brown markings on wing pads, dorsal and lateral sides of abdomen, dorsal side of thorax, legs, and head; a reddish-brown line crosses dorsally from one spiracle to the other on segments A_3 and A_4; and segments A_9 and A_{10} are black. **Host plants:** *Mollinedia costaricensis*, *M. viridiflora* (Monimiaceae).

Memphis oenomais (Nymphalidae: Charaxinae)
Edge Leafwing

0–1100 m

FWL: 27–32 mm. **Description:** FW has acute apex (much more acute in male), and anal border of tornus is excavated. HW has spatulate short tail (vein M_3). *Female:* DFW is light brown with basal half metallic blue and an irregular blue sub-marginal band from costal border almost to M_3. DHW is almost all metallic blue except for brown distal sub-marginal and marginal areas. VWs are grayish, mottled with brown and reddish shades; very cryptic. *Male:* DWs similar to female's but with blue area darker and less extensive. On DFW, sub-apical blue band similar to that of female or absent; some individuals have a small, additional sub-marginal area of blue from veins Cu_1 to 2A. DHW can have a variable amount of blue on distal border. VWs are reddish-brown mottled with gray and white; also very cryptic. **Similar species:**

Fifth instar larva of *Memphis oenomais*.

Pupa of *Memphis oenomais*.

Female *Memphis oenomais*.

Memphis oenomais pinned.
Dorsal view.

Memphis oenomais pinned.
Ventral view.

Memphis oenomais pinned.
Ventral view.

HW tail distinguishes male *Memphis oenomais* from males of *M. mora orthesia* and other similar *Memphis* species, except for *M. moruus boisduvali*, which is also tailed and has identical coloration on DWs. However, *M. oenomais* has a straight whitish line from apex to Cu_1 on VFW, which *M. moruus boisduvali* lacks or has in a very weak form. Females of *Memphis* species cannot be distinguished by color pattern. **Habitat:** Second-growth habitats, disturbed areas such as riverside vegetation on cattle ranches, light gaps in forested areas, and along trails and rivers on both slopes. Common in central and southern Pacific slope but rare in Guanacaste dry forest. Localities include San Carlos, Guápiles, Ciudad Colón, and San Isidro del General. **Distribution:** Mexico to Argentina. **Seasonality:** Present all year but more common during rainy season. **Natural history:** This is a common species in most places where at least small patches of natural vegetation remain or are growing back; it can inhabit living fences and semi-urban localities. Males stay mostly in the forest canopy and seldom descend. Females are easier to see since they fly lower (1–4 m above the ground) searching for host plants on which to oviposit. Both sexes feed on rotting fruits and tree sap. In the LAREBUB, individuals have lived for 2 months. **Early stages:** Eggs are pale green, spherical, with smooth surface, and are laid singly on underside of leaves. Young larva builds a frass chain at tip of host plant leaves and rests there until fourth instar, then builds a conical leaf-roll shelter in the same place, where it remains until pupation. Mature larva has a thick, longitudinal orange-brown dorsal band from middle of segment T_1 to A_7; remainder of abdomen and lateral areas are black; body is covered in many disperse wartlike white scoli, many with a short white seta; head capsule is black with many white scoli with setae, and four small, short protuberances. Pupa is bulb-shaped and white with reddish-brown lines on dorsal side of abdomen and thorax, head, legs, and wing pads; last abdominal segments and cremaster are black. **Host plants:** *Croton niveus, C. schiedeanus* (Euphorbiaceae).

Fountainea eurypyle confusa (Nymphalidae: Charaxinae)
Pointed Leafwing

0–900 m

FWL: 27–30 mm. **Description:** HW has tail (vein M₃). *Female:* DFW is dark brown with orange in basal area and a thick, irregular sub-apical orange band from costal margin to M_3; DHW is orange; VWs are brownish-orange splashed with whitish dots and patches. *Male:* Same color pattern as female but with blue iridescence. **Similar species:** *Fountainea eurypyle confusa* male is twice the size of *F. halice chrysophana* and can be distinguished from *F. ryphea ryphea* by latter's lack of HW tail. Female is distinguished from other Costa Rican *Fountainea* species by presence of a fairly straight, solid (never broken) dark line in medial area of VHW from vein R_5 to 2A. **Habitat:** Secondary forest and old abandoned fields, along living fences and forest edges, in areas such as Santa Ana, Ciudad Colón, Puntarenas, Osa Peninsula, Guápiles, and Guanacaste. **Distribution:** Mexico to Panama. **Seasonality:** Present all year but more common during rainy season. **Natural history:** Although not rare, this butterfly is difficult to see because of its cryptic underside, its preference for forest canopy, and its fast flight. Adults do not fly much, generally sitting quietly with wings folded and becoming very cryptic. Males perch in the canopy on tree leaves with head pointing down, chasing other butterflies that pass by. Females are easier to see in late morning and midday, flying swiftly along forest edges and light gaps looking for host plant saplings. Both sexes feed from tree sap and rotting fruits. Caterpillars on host plants are easier to find than adults. Larvae are attacked by parasitoid flies (*Eujuriniodes assimilis*, *Winthemia*, Tachinidae). **Early stages:** Eggs are pale green, with smooth surface, and are laid singly on underside of host plant's leaves. Young larvae are green and build a frass chain at tip or side of host plant leaves; in first instar, head capsule is smooth; in second instar, small protuberances and yellow lines appear on head.

Fifth instar larva of *Fountainea eurypyle confusa.*

Pupa of *Fountainea eurypyle confusa.*

Adult *Fountainea eurypyle confusa.*

Fountainea eurypyle confusa pinned. Dorsal view.

Mature larva builds a conical leaf-roll shelter on tip or side of leaf. When attacked by ants or parasitoids, the caterpillar retracts its head, sealing the shelter's entrance and leaving exposed only its well-armed head capsule. Mature larva is pale green with darker green longitudinal lines; a dorsolateral row of black lines; black spiracles; dark dorsal saddles on segments A_2, A_5, sometimes A_6, and again on A_7; and scattered whitish-yellow spots and small dense hairs all over body. Head capsule is variably pale to dark green with greenish-yellow parallel lines and black stemmata, and armed with around sixteen large yellow spikes and two short black horns. Pupa is compact and pale green with a yellowish dorsal shelf on abdomen, yellowish anal margin of wing pads, and a slight keel continuing around head from one wing pad to the other. **Host plants:** *Croton schiedeanus, C. billbergianus, C. draco, C. verapazensis, C. niveus* (Euphorbiaceae).

Morpho helenor narcissus (Nymphalidae: Satyrinae)
Common Morpho
Morpho helenor marinita
Marinita Morpho

0–1800 m
narcissus

0–1800 m
marinita

FWL: 62–79 mm. **Description:** *M. h. narcissus*: Large blue butterfly with DW margins black on male and dark brown on female, both sexes with variable amount of sub-marginal white spots. *M. h. marinita*: Large brown butterfly with variable amount of blue. In Costa Rican central Pacific, it has only a patch of blue on medial area of DFW; southward it has more blue on DHW medial area, and on Osa Peninsula it has a complete blue medial band on both DFW and DHW. VWs of both subspecies are identical: dark brown with many cream lines and bands, with three or four medial eyespots on VFW and four or five medial eyespots on VHW. **Similar species:** *M. helenor narcissus* is extremely similar to *M. deidamia polybaptus*, but the latter has a notable greenish iridescence to the blue coloration and also has bright, conspicuous greenish lines on VWs, while *M. h. narcissus* has paler, cream-colored, more diffuse lines on VWs and has a small black mark on costal border of DFW from distal end of discal cell, which *M. d. polybaptus* lacks. *M. h. marinita* has no similar species in Costa Rica. **Habitat:** Common on both Pacific and Caribbean slopes in many habitats, always associated with forest patches, not necessarily very well-preserved ones. It flies along roads and rivers lined with forest or in gardens close to forest patches. **Distribution:** *M. h. narcissus*, Nicaragua to Panama; *M. h. marinita*, Costa Rica to Panama. **Seasonality:** Present all year but more abundant during rainy season. **Natural history:** This morpho butterfly is nearly always seen flying, making no stops, giving an observer just a few seconds to appreciate it. When perched it is seldom seen, thanks to its great camouflage. Only in early morning or late afternoon is it possible to

Fifth instar larva of *Morpho helenor narcissus*.

Pupa of *Morpho helenor narcissus*.

Adult male *Morpho helenor narcissus*.

Adult *Morpho helenor marinita*.

Adult *Morpho helenor marinita*.

Morpho helenor narcissus pinned. Dorsal view.

Morpho helenor marinita pinned. Dorsal view.

Morpho helenor marinita pinned. Dorsal view.

see it perching with open wings in direct sunshine. Males patrol territories that are exposed to direct sunlight during the morning, such as riversides, roads, gardens, or light gaps. It is not clear whether they fly in circles or just fly to a certain point and then come back, or if it is the same male that passes by an area now and again for several days. Females wander along forest edges and light gaps from mid-morning to midday, searching for trees from 2 m to 6–7 m tall on which to oviposit. Both sexes feed from rotting fruits in the understory; they stay still when eating and are almost invisible. The great adaptability of this species has allowed it to live in many different habitats from Santa Rosa dry forest to Monteverde pre-montane forest to Tortuguero swamp forest. In consequence, some populations are more seasonal than others. For instance, mature larvae of the dry forest population decrease their metabolism during dry season and eat only very small pieces of leaves each day. In this way they use less water from already dehydrated leaves, save the host plant's few remaining leaves, and delay becoming butterflies, thus avoiding adverse weather. Murillo-Hiller and Canet (2018) reported that the fifth-instar larvae lasted 76 days during dry season, compared with 26 days during rainy season (Paniagua & Murillo-Hiller 2015). Larvae are attacked by parasitoid flies (*Hyphantrophaga morphophaga*, *Winthemia*, Tachinidae). **Early stages:** Eggs are semispherical, pale green with a crown of black dots appearing at 3–4 days, and are laid singly on upper side of leaves. Young larvae rest on underside of host plant leaves; they are reddish-brown with whitish dorsal patches and are covered in black setae. Mature larvae usually rest on tree trunk at base of branches or at base of tree; if larval density is large, they will rest together, side by side, concealing themselves with bark. Mature larva is very hairy and brown, with two dark yellow dorsal patches on segments A_2–A_3 and A_5–A_7, a lateral veined pattern of dark yellow areas with brown and red lines, a pair of small dorsolateral red and white tufts of hairs on A_4, a smaller pair on A_5, and a larger pair on A_7, and a short two-forked tail; head capsule is reddish-brown, covered in many red setae, with front of head varying from brown to white, and small horns on each side of top of head. Pupa is rhomboid, green with white spiracles, and has two small, downward-pointing spikes on head. **Host plants:** *Mucuna monticola*, *Dalbergia retusa*, *Machaerium kegelii*, *M. biovulatum*, *M. salvadorense*, *M. seemannii*, *M. floribundum*, *Lonchocarpus oliganthus*, *L. macrophyllus*, *Andira inermis*, *Pterocarpus rohrii*, *P. officinalis*, *Platymiscium parviflorum*, *Inga punctata* (Fabaceae), *Heteropterys laurifolia* (Malpighiaceae), *Guarea glabra tuerckheimii* (Meliaceae), *Geonoma ferruginea* (Arecaceae), *Dichapetalum morenoi*, *D. grayumii* (Dichapetalaceae).

Morpho menelaus amathonte (Nymphalidae: Satyrinae)
Menelaus Morpho

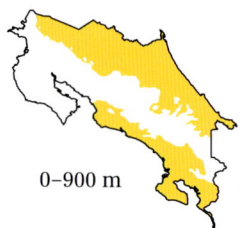

0–900 m

FWL: 73–79 mm. **Description:** A huge iridescent blue butterfly. *Female:* DWs are blue in basal area, with wide brown margins containing two rows of white spots. *Male:* DWs are all blue except for black apex on DFW. Both sexes: VWs are brown with pale white lines and seven or eight eyespots on medial area. **Similar species:** None in Costa Rica. **Habitat:** Common in well-preserved rainforest habitats of both slopes but absent from Pacific side of Guanacaste. Localities include Cahuita, Tortuguero, Carara, Corcovado, and Rincón de la Vieja National Parks; and areas around San Carlos, Upala, Uvita, and Sarapiquí, among others. **Distribution:** Nicaragua to Panama. **Seasonality:** Present all year on both slopes, but on Pacific side more common in rainy season. **Natural history:** This spectacular butterfly has a slow, floppy flight, which together with its enormous size makes it impossible to ignore. Males fly in understory, 1–3 m high, along rivers, roads, or forest edges; they invariably fly in early morning (6:30–9:00 a.m.), establishing territories in semi-open and open areas, and then stay hidden or feed. They may fly later to search for decomposing fruit at ground level. Females are more difficult to see because they fly inside the forest and for less time per day; they may be seen in light gaps or along trails in forest interior, looking for host plants from mid-morning to midday. It is unclear whether this butterfly is represented in Costa Rica by one or two species or, if it is one species, whether it is *M. amathonte centralis* or

Fifth instar larva of *Morpho menelaus amathonte*.

Pupa of *Morpho menelaus amathonte*.

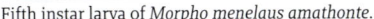

Adult male *Morpho menelaus amathonte*.

Morpho menelaus amathonte pinned. Dorsal view.

Morpho menelaus amathonte pinned. Dorsal view.

M. menelaus amathonte. There are important differences in the pupa between Caribbean and Pacific populations and, according to one study (Nield 2008), also in adults, suggesting two species may occur in the country. Main predators are jacamars (Galbulidae) and other birds; larvae are attacked by a parasitoid fly (*Hyphantrophaga anacordobae*, Tachinidae). **Early stages:** Eggs are semispherical, with smooth surface, and pale green, with a black ring appearing on each egg after 3 days; they are laid in small groups of three to ten on upper side of leaves. Young larvae rest on leaf undersides; they are covered in reddish setae, with two yellow dorsal patches. Mature larvae rest in small groups and build shelters of two or three leaves held together with silk. Mature larva has large, bright yellow dorsal patches at segments A_2–A_4 and A_5–A_7; sides are veined with yellow, red, black, and brown stripes; and they have two-forked tails; head capsule is orange and covered in long setae. Pupa is rhomboid, with white spiracles and two-forked head; in Pacific population, remainder of pupa is green; in Caribbean, pupa has prominent white band from side to side on dorsal side of abdomen at A_4. **Host plants:** *Pterocarpus officinalis, P. rohrii, P. hayesii, Lonchocarpus macrophyllus, L. oliganthus, Dioclea malacocarpa, Machaerium seemannii* (Fabaceae), *Dichapetalum grayumii* (Dichapetalaceae), *Prestoea decurrens* (Arecaceae).

Morpho polyphemus catalina (Nymphalidae: Satyrinae)
White Morpho

700–1300 m

FWL: 75–82 mm. **Description:** An enormous white butterfly. *Female:* DWs are mostly white; DFW has brown apex and brown distal end of discal cell; DHW has a sub-marginal row of brown spots in distal area followed by small brown marginal spots. VFW is similar to DFW but has an extra brown band inside discal cell and small, yellow-margined black eyespots in post-medial area on veins M_1–M_2, M_3–Cu_1, and Cu_1–Cu_2. VHW has brownish, longitudinal zigzag lines in medial area, followed by four yellow-margined black eyespots on R_5–M_1, M_1–M_2, M_3–Cu_1, and Cu_1–Cu_2, and has faint sub-marginal brown lines on distal margin. *Male:* Wings are like female's but the dark markings on DWs are smaller and darker, and VWs lack brownish zigzag medial lines. **Similar species:** The similar *Morpho theseus heraldica* has a conspicuously serrate to short-tailed HW distal margin and lives only in the southern extreme of the country, while *M. polyphemus* lives only in the north. **Habitat:** Secondary and primary forest at mid-elevations where evergreen forest begins on the Pacific slope; flies along rivers and roads as well as in forest light gaps. Localities include Tenorio Volcano, Miravalles Volcano, Rincón de la Vieja, and Orosí Volcano in Guanacaste National Park; **Distribution:** Costa Rica (endemic). **Seasonality:** Present only during August and September. **Natural history:** Males can be seen in sunshine from 9:00 a.m. to early afternoon, gliding along 4–20 m above the ground in open areas of forest. They set up territories, which they constantly patrol; when one passes by a certain point, in about 15–20 minutes it will pass again. Females start flying in late morning to early afternoon, with a heavy, floppy

Fifth instar larva of *Morpho polyphemus catalina*. RC

Pupa of *Morpho polyphemus catalina*. RC

Adult *Morpho polyphemus catalina*. RC

Morpho polyphemus catalina pinned. Dorsal view.

flight, along the forest interior from 2 m to 4–5 m above the ground, searching for small host plant trees on which to oviposit. Both sexes feed on a large variety of rotting fruits but are often seen on decomposing parts of *Monstera* fruits that are still attached to plants high on tree trunks. This is the only *Morpho* species in Costa Rica with an annual life cycle (univoltine). Larvae are attacked by parasitoid flies (*Hyphantrophaga*, Tachinidae). **Early stages:** Eggs are pale green, semispherical, and laid singly on upper side of host plant's leaves. Larvae last around 10 months, feeding from the same plant. Mature larva is grayish-pink with two large yellow dorsal patches; its body is covered in short, dense white setae but has six dorsolateral tufts of black setae; head capsule is gray and densely covered in long white, black, and red setae. The mature larva builds a shelter with old (but still alive) dry leaves and lives in it alone. Pupa is formed around July; it is green, globular, and has three large, dorsolateral white spots on segments A_1, A_2, and A_3; head bears two small spikes. **Host plants:** *Inga punctata*, *Zygia palmana* (Fabaceae). Larvae can also feed on *Z. longifolia* (R. Cubero, pers. comm.) and *Forchhammeria trifoliata* (Capparaceae)

Opsiphanes fabricii fabricii (Nymphalidae: Satyrinae)
Split-banded Owl-Butterfly

0–1500 m

FWL: 35–41 mm. **Description:** A robust butterfly with mostly brown DWs. DFW has an orange medial band from tornus to costal margin that splits when it reaches the discal cell, with one bar going inside the discal cell and the other outside; two small white spots are in sub-apical area. DHW has a sub-marginal orange band from apex to vein M_3 in female and extending in males, sometimes faintly, to 2A. **Similar species:** In male *Opsiphanes jacobsorum*, the DFW orange bar inside the discal cell never touches the medial band; in female *O. jacobsorum*, the single band is white instead of orange and does not split or enter discal the cell. In female *O. periphetes*, the orange band in DHW is divided into spots, lacking a clear straight edge as in *O. fabricii fabricii*; in male *O. periphetes*, the DHW sub-marginal orange band is shorter, thinner, with irregular margins, and does not reach vein 2A (Piovesan et al. 2022). Around Central Valley, *O. fabricii fabricii* can be mistaken for *O. cassiae*, but the latter is larger and much less common. **Habitat:** Common all over the country and in all habitats. Flies at dusk in gardens, parks, and streets of the Central Valley, even in cities. In rural areas it flies along forest edges, in fields, or even in the forest interior, where it is rarer since it prefers disturbed habitats; it is common in Guanacaste, Puntarenas, Osa Peninsula, and throughout the Caribbean slope. **Distribution:** Nicaragua to Panama. **Seasonality:** Present all year but more common during rainy season. **Natural history:** Both sexes fly at dusk, 2–3 m high. Because this butterfly takes wing near dark and is a powerful and fast flier, it is generally hard to see its characteristics unless it is perched. But because it is very common, in certain places such

Fifth instar larva of *Opsiphanes fabricii fabricii*.

Pupa of *Opsiphanes fabricii fabricii*.

Adult *Opsiphanes fabricii fabricii*.

Opsiphanes fabricii fabricii pinned.
Dorsal view.

Opsiphanes fabricii fabricii pinned.
Ventral view.

as urban areas, identification can be accurately made even when it is flying. Both sexes of *O. fabricii* feed on rotting fruits and sometimes carrion. Males are very territorial and perch on exposed leaves and chase other males out of their space. Females fly swiftly along streets or other open areas looking for host plants, laying several eggs on the same plant and flying quickly from one branch to another. Larvae are attacked by parasitoid flies (*Anoxynops auratus*, *Winthemia*, Tachinidae). **Early stages:** Eggs are whitish, spherical, with faint longitudinal ribs, and are laid singly or in small groups of two or three eggs on underside of leaves or in stalks of host plant. Young larvae rest parallel with veins on underside of leaves. Mature larvae build a shelter by folding the leaflet to the underside and fastening it with abundant silk, making a tunnel. Mature larva is bright green with longitudinal lines of different tones, and has long two-forked tail; head capsule has alternating longitudinal bands of two green tones and is armed with six backward-pointing horns, each ringed first in yellow, then pale green, ending in a black tip, all variable in width. Pupa resembles a young leaf; it is green, curved ventrally, with thin, longitudinal darker green lines, slightly keeled wing-pad margins, a faint dorsal keel running longitudinally from abdomen to thorax, and two short spikes on head. It has a small, eye-like, silver lateral patch, which together with the pupal shape makes it resemble a reptile head when seen from certain angles. **Host plants:** *Chamaedorea costaricana*, *Dypsis lutescens*, *Cocos nucifera*, *Acrocomia aculeata*, *Bactris guineensis*, *B. mayor* (Arecaceae).

Eryphanis lycomedon (Nymphalidae: Satyrinae)
Split-spotted Owl-Butterfly

0–1200 m

FWL: 55–62 mm. **Description:** *Female:* DFW is dark brown with a dull blue basal area followed by a brownish-orange medial patch with a little iridescent blue area close to anal margin; post-medial area is dark brown with longitudinal brownish-orange band. Basal area of DHW is dull brown followed by an iridescent-blue medial band and a dark brown margin. *Male:* DFW is dark brown with a large iridescent dark purple patch from basal to sub-marginal and sub-apical areas; a brownish-orange sub-marginal band (variable in brightness) from sub-apical area to tornus; and a brownish-orange distal margin. DHW is mostly dark brown with a continuation of iridescent purple on medial area from sub-apical and sub-marginal distal border to around vein Cu_1. Both sexes: VWs have a complex pattern of different brown tones veined with thin white and black lines; medial area of VHW has a small eyespot on M_3–Cu_1 and a larger one on Cu_1–Cu_2. **Similar species:** Both sexes can be distinguished from *Eryphanis bubocula* by eyespots on VHW; in *E. bubocula* they are larger, and their external black ring extends more than 1 mm into Cu_2–2A area, which never occurs in *E. lycomedon*. **Habitat:** The easiest places to find this difficult-to-see species are thickets of *Chusquea* or other bamboos, sitting quietly, 1–2 m

Fifth instar larva of *Eryphanis lycomedon*.

Pupa of *Eryphanis lycomedon*.

Adult male *Eryphanis lycomedon*.

Eryphanis lycomedon pinned.
Dorsal view.

Eryphanis lycomedon pinned.
Dorsal view.

Eryphanis lycomedon pinned.
Ventral view.

above the ground. Localities include Sarapiquí, Guápiles, Limón, San Carlos, Upala, Parrita, Ciudad Colón, and Osa Peninsula. **Distribution:** Guatemala to Brazil. **Seasonality:** Present all year. **Natural history:** A fast and erratic flier with very cryptic colors, this species is hard to see in nature, not least because it spends most of its time perched, flying less than a half hour each day. Its activity periods in the LAREBUB are roughly 6:00–6:15 a.m. and 5:30–5:45 p.m. Females fly at dawn to lay eggs, and males are more active in evening, when they set up territories in dark places close to host plant patches. Originally, this species probably used only *Chusquea* as a host plant, but with the expansion and use of different species and genera of bamboos as ornamental plants, the species has become more common in disturbed and human-made landscapes. **Early stages:** Eggs are white, spherical, with longitudinal ribs, and are laid singly on underside of leaves. Young larvae rest on upper side of leaves along strips of dead leaf tissue where their length and brownish colors make them very hard to see. Mature larva is elongate and patterned with a longitudinal dark green band on dorsum and light green sides, generating a disruptive shape. It is covered in a dense layer of short setae and has many small granulations on dorsolateral and lateral areas but not on dorsum, where instead it has five long, thin soft spine-like structures on segments A_2–A_6 (which earlier instars do not have); segment A_{10} is forked into two long tails that, when alarmed, the larva opens and folds dorsally to the front, in an intimidating posture. Head capsule is light green, slightly striped longitudinally, and has six long, straight, backward-pointing horns close to thorax. Before pupation, the larva becomes entirely yellow. Pupa is elongate and grayish-green with faint darker longitudinal lines; head has two very long, fused, downward-pointing projections. **Host plants:** *Guadua angustifolia*, *Chusquea* (Poaceae).

Caligo telamonius memnon (Nymphalidae: Satyrinae)
Caligo telamonius menus
Pale Owl-Butterfly

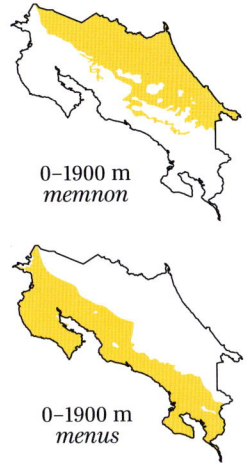

0–1900 m
memnon

0–1900 m
menus

FWL: 65–78 mm. Description: Sexes are very similar, differing in blue iridescence of DHW. *C. t. memnon:* DFW has black distal margins, gray basal area, and cream-white medial area from costal margin of discal cell to vein 2A; DHW is gray in basal area, and the rest is black, which in male has blue iridescence. VFW is veined with black and gray fine lines and is cream-colored with brown markings in medial area; VHW has large owl-like eyespot on veins M_3–2A. *C. t. menus:* Very similar; DFW medial area is more yellow than cream-white and does not touch medial costal area (as in *C. t. memnon*), which has a gray margin; DHW of male *C. t. menus* has much more blue iridescence than *C. t. memnon*. VWs are very similar to those of *C. t. memnon* but with many more brown markings inside cream medial area of VFW. **Similar species:** In *Caligo oedipus oedipus* and *C. oileus scamander*, "shadow" surrounding eyespot on VHW runs from anal margin to costal margin, even wrapping around smaller eyespot on $Sc+R_1$–R_5; in *C. telamonius*, the shadow diminishes as it approaches the sub-costal eyespot. In *C. oedipus fruhstorferi* and *C. brasiliensis sulanus*, DFW is dark gray in medial area between Cu_2 and 2A, while in *C. telamonius* it is yellow or cream-white. **Habitat:** Both subspecies are common in disturbed habitats, secondary forest, croplands, and rural areas. They can be seen flying at dusk and dawn along rivers and forest edges; in coffee plantations in Central Valley; and along coastlines and beaches. *C. t. menus* is common in tourist areas of Guanacaste; *C. t. memnon* is common in swamp forests such as in Tortuguero and Cahuita National Parks, where generally it is seen perched on a tree trunk or flying fast through forest understory. In rural areas, it often enters houses or perches on decks and gazebos. **Distribution:** *C. t. memnon*, Mexico to Panama; *C. t. menus*, Costa Rica to Colombia. **Seasonality:** Present all year but more common during rainy season. **Natural history:** This iconic species is the commonest and most widespread owl-butterfly in Costa Rica. It flies for short periods in very early morning, when it usually eats, and then again late at dusk, when males fly fast and erratically, chasing each other and establishing territories. The male searches for an exposed leaf, lands, and waits for another male to strike; if a female passes, he will chase her vigorously for several meters. Generally, males remain in the area the next day but do not necessarily establish their territory in the same spot. Females lay eggs from 3:00 to 6:00 p.m., depending on weather; on dark or rainy days they will fly more often. Both sexes feed from rotting fruits at ground level. This species has lived as an adult in the LAREBUB for as long as 110 days. Great-tailed Grackle (*Quiscalus mexicanus*) has been observed preying on adults at dusk

Fifth instar larva of *Caligo telamonius menus*.

Pupa of *Caligo telamonius menus*.

(pers. obs.), and the Giant Toad (*Rhinella marina*) is an important predator of mature larvae in northwestern Costa Rica (S. Hiller, pers. comm.). Eggs are parasitized by wasps (*Ooencyrtus*, Encyrtidae); larvae are attacked by parasitoid wasps (*Cotesia*, Braconidae; *Hyposoter*, Ichneumonidae) and a fly (*Winthemia pinguis*, Tachinidae), while vespid wasps (*Mischocyttarus*, Vespidae) attack fifth-instar larvae to obtain food for their own larvae. **Early stages:** Eggs are white, spherical, with longitudinal ribs, and laid singly or in small groups of up to ten. Females deposit them on any part of the plant, even on hanging dry leaves. Larvae are very cryptic and rest at day and eat at night. Young caterpillars are light green and rest in groups (sometimes formed from different broods) on undersides of leaves, lined up along central vein. Mature larvae resemble bark and rest together side by side at base of host plant stalk. Fifth-instar larva is light brown with darker lines and has a longitudinal dark brown band on segments A_1–A_7, from which soft black spines arise; an elliptic patch on A_4; and a long two-forked tail on A_{10}. Head capsule is striped and bears six curved cream-colored horns covered in long setae. Larvae have an eversible ventral gland on segment T_1 that releases a spray of acid to repel ants and parasitoids. Pupa is dark to light brown, veined with many thin darker lines; it is wider at the sides at wing-pad level and has a prominent keel on dorsal thorax, black setae patches on dorsal side of abdomen and in head area, and two small lateral silver spots at margin of wing pads. **Host plants:** *Heliconia latispatha*, *Heliconia imbricata* (Heliconiaceae), *Calathea lutea*, *C. macrosepala*, *Thalia geniculata*, *Goeppertia villosa* (Marantaceae), *Canna indica* (Cannaceae), *Musa acuminata*, *M. sapientum* (Musaceae).

Caligo telamonius memnon pinned. Dorsal view.

Adult female *Caligo telamonius menus*.

Caligo telamonius memnon pinned. Ventral view.

Caligo telamonius menus pinned. Ventral view.

Caligo atreus dionysos (Nymphalidae: Satyrinae)
Banded Owl-Butterfly

0–1700 m

FWL: 70–86 mm. Description: DWs are mostly black; DFW has a longitudinal blue medial band with a large iridescent area and two small white sub-apical spots on costal margin; DHW has a wide yellow longitudinal sub-marginal band that sometimes has blue iridescence. VWs have a complex veined pattern of brick-brown and black lines; VFW has a sub-apical eyespot; VHW has one eyespot on medial costal margin and a larger one in medial area from M_3 to a little beyond Cu_2. **Similar species:** *Caligo atreus dionysos* is twice the size of female *Catoblepia orgetorix championi* and can also be distinguished by yellow band on DHW, which in *C. orgetorix championi* reaches the wing margin and in *C. atreus dionysos* never does. **Habitat:** Rainforest, on both slopes. It is absent from driest northern Pacific areas of Costa Rica, having its boundary in Orotina, but is easy to see throughout Caribbean and Osa Peninsula, along forest trails, light gaps, riversides, and forest edges. It is usually seen perched 1–4 m high on tree trunks, rocks, or vines. Sometimes enters houses or gazebos and is common in gardens or croplands if they are close to well-preserved forest, which is necessary for supporting a population of *C. atreus dionysos*. **Distribution:** Nicaragua to Panama. **Seasonality:** Present all year. **Natural history:** This huge crepuscular species is a powerful flier. Specimens have lived up to 2.5 months in captivity in the LAREBUB. Males are very territorial; they like to perch on exposed leaves, at around 3 m high to have a good view, from which they will chase any other *Caligo* that flies by, expel it, and come back in few seconds; they do not use the same perch day after day, since in the early morning they look for rotting fruits to eat and generally have to fly away from their previous perch. Females feed from rotting fruits

Fifth instar larva of *Caligo atreus dionysos*.

Pupa of *Caligo atreus dionysos*.

Adult *Caligo atreus dionysos*.

Caligo atreus dionysos pinned. Ventral view.

from 6:00 to 8:00 a.m. and then rest until 5:00 or 6:00 p.m., when they spend about a half hour laying eggs. Mature caterpillars are attacked by parasitoid flies (*Winthemia*, Tachinidae). **Early stages:** Eggs are white, spherical, with longitudinal ribs, and are laid on any part of host plant such as stalks or dead or live leaves. Young caterpillars up until fourth instar are green and rest on upper side of leaves on the central vein where the petiole meets the leaf. Fifth-instar larva is dark brown with black zigzag stripes on each segment, a longitudinal whitish lateral band, four or five soft (false) spines on dorsum of abdomen, and a two-forked tail, and the body is covered in dense short hairs; head capsule is cream-colored with few longitudinal lines, covered in dense creamy-white hairs, and has six horns, two side pairs and a central pair with a globular tip. It rests either in the same place as previous instars or at base of host plant's stalk with head pointing up. Pupa is yellowish-brown with many darker markings, a dark brown dorsal line running longitudinally from cremaster to head, and two large chrome-gold spots on medial edge of wing pad; it is wider at segment A_3 on wing pads, which are slightly keeled, and has a prominent keel on dorsal side of T_2. Unlike pupae of other *Caligo* species, it is hairless. **Host plants:** *Costus scaber* (Costaceae), *Hedychium coronarium* (Zingiberaceae), *Heliconia latispatha*, *H. tortuosa* (Heliconiaceae), *Hylaeanthe hoffmannii*, *Calathea crotalifera* (Marantaceae), *Musa acuminata*, *Musa sapientum* (Musaceae), *Asterogyne martiana*, *Chamaedorea tepejilote* (Arecaceae).

Cithaerias pireta pireta (Nymphalidae: Satyrinae)
Rusted Clearwing-Satyr

0–1800 m

FWL: 30–33 mm. **Description:** A clear-winged butterfly. DFW has two thin, pink longitudinal medial lines and pink margins. DHW has pink margins, wider pink sub-marginal bands on apex and tornus, and a black sub-marginal eyespot between veins M_1 and M_2. **Similar species:** *Haetera macleannania* is larger and has two eyespots on HW instead of one. *Dulcedo polita* lacks pink on HW tornus and has a conspicuous blue iridescence overall. **Habitat:** Well-preserved forest; common in humid and swampy forest habitats, and intolerant of dryer Guanacaste habitats, finding its northern Pacific limit around Carara National Park. Although it can be found at pre-montane elevations, it is much more common at lower elevations. It prefers dark areas where disperse sunrays hit forest floor. Localities include many national parks including Braulio Carrillo, Cahuita, Tortuguero, Barbilla, Corcovado, and Arenal Volcano. **Distribution:** Guatemala to Colombia. **Seasonality:** Present all year. **Natural history:** This species may be seen along trails traveling close to the ground in a fluttery flight. Both sexes fly just 20–30 cm above the ground and perch on top of leaves with closed wings. Males set up territories about 10 m long from which they chase away intruders, and unless expelled by another male will stay there for many weeks. Both sexes feed from rotting fruits and other decomposing organic matter at ground level. It is widely accepted that members of Satyrinae are palatable to predators and therefore bear cryptic or mimic

Fifth instar larva of *Cithaerias pireta pireta*.

Empty pupa of *Cithaerias pireta pireta*.

Adult *Cithaerias pireta pireta*.

coloration. However, in this case, no Batesian mimic models are known, and it has been proposed that the pink coloration signals that *C. pireta* is actually unpalatable, obtaining secondary compounds from its host plants. **Early stages:** Eggs are pale green, spherical, with smooth surface, and are laid singly on underside of host plant's leaves. Young larva is light green with black head capsule. Mature larva is dark brown with two dorsolateral bands of triangular light brownish-orange spots, highly accentuated in segments A_3–A_6, and a short two-forked tail; head capsule is cream-colored with brown patches and two long, whitish, dorsolateral horns. It rests on underside of leaf at side of central vein. Pupa is dark brown with disperse grayish markings, two short dorsolateral protuberances on A_3, and a dorsal keel as well as short lateral projections on T_2; head is slightly two-forked. **Host plant:** *Philodendron herbaceum* (Araceae).

Manataria maculata maculata (Nymphalidae: Satyrinae)
White-spotted Satyr

0–2500 m

FWL: 40–45 mm. Description: DFW is brown in basal area and black in distal area with a medial diagonal row of white spots and an extra white spot in sub-apical area. DHW is all brown with a thin, darker sub-marginal line. VFW is brown in basal half with same row of medial white spots as in DFW; sub-apical area is grayish with an eyespot. VHW is grayish with a post-medial row of eyespots and elsewhere filled with concentric, thin black and brown curved lines. **Similar species:** None; distinguished by medial diagonal row of white spots on DFW together with post-medial row of eyespots on VHW. **Habitat:** Pacific slope, in many types of habitats from dry tropical forest and disturbed areas to montane oak forest. Frequently attracted to lights in Central Valley, it can be found resting on roofs and walls. **Distribution:** Mexico to Colombia. **Seasonality:** Present all year but actively flies during rainy season. **Natural history:** *M. maculata maculata* flies at dusk and dawn, so generally is seen as a dark, medium-size butterfly quickly zigzagging along open trails and roads. However, if hungry, it will fly to find rotting fruits or

Fifth instar larva of *Manataria maculata maculata*.

Pupa of *Manataria maculata maculata*.

Manataria maculata maculata pinned. Ventral view.

Adult *Manataria maculata maculata*.

other decaying matter at any time of day. It perches on tree buttresses in shaded places with wings folded and head downward. This species reproduces during the early rainy season at low elevations from sea level to around 1000 m in the central and north Pacific; the offspring migrate northward to higher elevations and go into reproductive diapause. A few to several dozen individuals will establish daytime roosts in dark places such as hollow trees and shaded cliffs; during this time they sporadically feed. At dusk, they fly to the forest canopy, where they spend the night perched on leaves and branches; the shift in roosting location may help them avoid predators. This behavior continues until the following year, when at the end of dry season they fly back southward, reproduce, and die. For most of this annual (univoltine) cycle, they are adults; therefore they live around 10 months. This species can perceive ultrasound signals of bats and use them to avoid attacks at night. Larvae are attacked by a parasitoid fly (*Patelloa xanthura*, Tachinidae). **Early stages:** Eggs are creamy-white, spherical, with smooth surface, and are laid in clusters of about thirty on underside of leaves. Young larvae are pale green and feed gregariously. They rest together on underside of leaves until third instar, when they start to disperse; larvae then rest alone or in pairs on central vein of leaves on either upper side or underside. Mature larva is longitudinally striped: dorsal area is dark green with two longitudinal rows of creamy-green dots; lateral area has three longitudinal creamy-green bands alternating with two dark green bands. Its body is covered in a dense layer of short whitish setae, and segment A_{10} is forked with two long tails; head capsule varies from light green to dark brown and has small yellow spikes on sides, reddish stemmata, and two short, spiny yellow horns bearing long, black spatulate setae. Pupa is globular, pale green with no markings of any kind, covered in a fine waxy layer, and angled at 110° from cremaster to rest of body. **Host plants:** *Bambusa vulgaris, Rhipidocladum pittieri, Lasiacis, Guadua angustifolia, Chusquea* (Poaceae).

Oxeoschistus tauropolis mitsuko (Nymphalidae: Satyrinae)
Yellow-patched Satyr

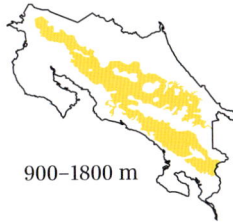

900–1800 m

FWL: 35–41 mm. **Description:** DFW is brown with a creamy-white patch from vein Cu_1 to 2A, and a row of post-medial cream spots. DHW has a large post-medial creamy-white patch from costal margin to Cu_2, a faint eyespot on Cu_1–Cu_2 and another on Cu_2–2A, and reddish tornus margin. VWs are reddish-brown with cream lines. VFW has same pattern as DFW. VHW has a post-medial longitudinal row of faint eyespots from costal margin to tornus between two whitish lines. HW margin is serrate. **Similar species:** Distinguished from *Oxeoschistus cothon* by sub-marginal longitudinal white band on VHW, which is very sharply toothed in *O. tauropolis mitsuko* and less so in *O. cothon*; and medial white band running from costal margin to discal cell on VHW, which is at least twice as thick in *O. cothon* as in *O. tauropolis mitsuko*. **Habitat:** Forests in all mountains of Costa Rica. Common in Central Valley in riparian forest where bamboo thickets are present and in forest patches in mountains surrounding the valley. **Distribution:** Costa Rica (endemic). **Seasonality:** Present and common all year. **Natural history:** This common species flies 1–4 m high with a fast and bouncy flight in forest interior or along trails. It likes to perch where sunrays hit low vegetation. Males are territorial and stay in the same territory for several days, always very close to *Chusquea* or other bamboo thickets. Both sexes are active all day and feed from rotting

Fifth and fourth instar larvae of *Oxeoschistus tauropolis mitsuko*.   Pupa of *Oxeoschistus tauropolis mitsuko*.

Adult *Oxeoschistus tauropolis mitsuko*.

Oxeoschistus tauropolis mitsuko pinned.
Dorsal view.

Oxeoschistus tauropolis mitsuko pinned.
Ventral view.

fruits and any other kind of decaying matter. **Early stages:** Eggs are small, whitish, spherical, with smooth surface, and are laid on underside of host plant's leaves. Mature larvae rest on upper side of the leaves and stalks; they are very cryptic, resembling dead plant tissue. Mature larva is light brown with darker lines, with posterior half of dorsum slightly darker than anterior half, and has a long, conical, fused two-forked tail; head capsule has two long, forward-pointing horns with same streaked color pattern as body. Pupa is elongate, greenish-gray with many faint dark streaks, and has two small dorsolateral protuberances on segment A_3 and a dorsal one on T_2; head has two very long, downward-pointing beaklike projections that are fused together. **Host plant:** *Guadua angustifolia* (Poaceae).

Megeuptychia antonoe (Nymphalidae: Satyrinae)
Cramer's Satyr

0–1500 m

FWL: 30–35 mm. **Description:** DWs are all brown, DFW with two faint, darker sub-marginal lines. VFW is also brown, with two reddish longitudinal medial lines, one or two small eyespots in sub-apical area, and three parallel sub-marginal lines on distal border. VHW is same color as VFW and also crossed by two reddish longitudinal medial lines, and has a row of five post-medial eyespots; the eyespots on veins M_1–M_2 and Cu_1–Cu_2 are much larger than the others, and the latter one has a double white pupil. There are two reddish parallel sub-marginal lines and another one on distal margin. **Similar species:** Many satyrs are hard to distinguish at species level. With this species, consider a combination of characters: first the size, then the lack of serrate HW margins, and finally the double white pupil on Cu_1–Cu_2 eyespot of VHW. **Habitat:** Rainforest in localities such as Carara, Corcovado, Barbilla, and Arenal Volcano National Parks; Hitoy-Cerere Biological Reserve; and Guápiles and Sarapiquí forested areas. **Distribution:** Mexico to Brazil. **Seasonality:** Present all year on the Caribbean slope and only during rainy season on the Pacific slope. **Natural history:** Adults are seldom seen since they spend most of their time in the forest canopy. Females are more often seen flying in forest understory along trails or in light gaps. This butterfly is poorly known, probably because of its canopy habitat and flying behavior; it is not generally seen flying in mornings and perhaps flies at dusk and dawn as do many other satyrs. Adults are reported to feed on fruits in the canopy. The species' host plants, however, are in forest understory, and larvae are always seen no more than 2 m high, so females must fly low at times. **Early stages:** Eggs are pale green, spherical, with smooth surface, and are laid in groups of up to thirty-five on underside of leaves. Young larvae rest on underside of leaves, aggregated in rounded masses. They are bluish-green, smooth, with a short, yellow two-forked tail; head capsule is same color as body and bears two short, thick horns. Mature larva, around 4 days before pupating, becomes pale yellow, brighter in anterior quarter and tail, with dark longitudinal dorsal shading that suggests the digestive tract can be seen through the skin; head

Fifth instar larvae of *Megeuptychia antonoe*.

Pupae of *Megeuptychia antonoe*.

Adult *Megeuptychia antonoe*.

Megeuptychia antonoe pinned.
Ventral view.

capsule is pale pink with a granulated texture. Larvae eat gregariously at any time of day or night; when they move, they walk one behind the other, as if in a procession. To pupate they leave the plant together in a line and then disperse in smaller groups of around four individuals. Pupa is round and either pink or pale yellow. **Host plant:** *Calathea lutea* (Maranthaceae).

Taygetis thamyra (Nymphalidae: Satyrinae)
Thamyra Satyr

0–1500 m

FWL: 32–37 mm. **Description:** DWs are all brown with a slightly darker sub-marginal brown line on both wings; HW has a strongly serrate distal margin. VWs are brown, with a longitudinal shift in medial area to grayish-brown, then a post-medial longitudinal row of eyespots, followed by a sub-marginal and marginal dark brown area. Eyespots on both wings are large, and most are yellowish-brown inside, but the one on Cu_1–Cu_2 is generally black inside with a very thin yellow border; in some cases, other eyespots can be very dark too. **Similar species:** Distinguished from *T. uzza* by the medial line that divides the proximal brown area from the post-medial clearer area, which in *T. uzza* is followed by a noticeable whitish line with very clearly defined edges (edges are diffuse in *T. thamyra*); in the VHW that line is strait while in *T. thamyra* it is curved, following the shape of the wing. There is intense debate about the validity of *T. thamyra* in Costa Rica; the species *T. inconspicua* is very similar, and in the future *T. thamyra* may be changed to *T. inconspicua*. **Habitat:** Common in well-preserved forested habitats as well as secondary and disturbed habitats on both slopes, although not very common in open areas. In Guanacaste, it is seen along forested riversides and in forest patches surrounding mangroves. Other localities include Guápiles, Osa Peninsula, Ciudad Colón, and La Fortuna. **Distribution:** Mexico to Brazil. **Seasonality:**

Fifth instar larva of *Taygetis thamyra*.

Pupa of *Taygetis thamyra*.

Adult *Taygetis thamyra*.

Taygetis thamyra pinned.
Dorsal view.

Taygetis thamyra pinned.
Ventral view.

Present all year. **Natural history:** This species is generally seen along forest trails, no more than 1 m high. It is often found in the morning, perched on top of leaves close to the ground, but is skittish and hard to get close to. It does not fly much, traveling a short distance of about 2–3 m; then it lands and walks, pausing and partially opening its wings. Both sexes feed from rotting fruits and fungi at ground level. *T. inconspicua* is most active at dusk and is perhaps also nocturnal. DeVries (1987) states that Guanacaste populations may enter reproductive diapause during dry season. **Early stages:** Eggs are pale green, spherical, with smooth surface, and are laid singly on underside of host plant's leaves. Young larvae are light green and rest on underside of leaves, while mature larvae rest on stems. Mature larva is green with thin, lighter green, parallel longitudinal lines and a two-forked tail; head capsule has two short, spiny horns, which like the capsule itself are light green dorsally and dark green ventrally, as well as two very small spikes on sides of head. Pupa is green when formed, becoming darker with age, with small, disperse whitish and brown spots; it is ventrally curved and has a dorsal hump on dorsal thorax, a ridge on sides of wing pads that continues to sides of head, and a slightly two-forked head. **Host plant:** *Acroceras zizanioides* (Poaceae).

Cyllopsis hedemanni vetones (Nymphalidae, Satyrinae)
Stub-tailed Gemmed-Satyr

800–2100 m

FWL: 20–25 mm. **Description:** DFW is all brown with a very faint, longitudinal, post-medial orange band. DHW is also brown with a very faint longitudinal post-medial orange band and a black sub-marginal eyespot from vein M_3 to Cu_1. VFW is variegated brown. VHW has the same black eyespot as DHW and a lighter longitudinal post-medial line from costal to anal margin. Cu_1 vein of HW is slightly prolonged, forming a smooth, small tail. **Similar species:** Distinguished from all other small brown Costa Rican Satyrinae butterflies, such as *Hermeupthychia* and other *Cyllopsis*, by HW tail, which gives it a distinctive wing shape.

Habitat: Pre-montane and montane forest habitats as well as secondary forest or disturbed places where host plants are present. Occurs on both Pacific and Caribbean slopes but associated with mountain ranges. Localities include San Isidro and San Rafael de Heredia, foothills of Irazú Volcano, and Central Valley, around riverside vegetation. **Distribution:** Costa Rica and Panama. **Seasonality:** Present all year but more common during dry season. **Natural history:** This localized species can be seen flying swiftly around thickets of host plants and in light gaps from early morning to early afternoon, 1–3 m above the ground. It patrols its own territory for several days, landing on top of leaves and chasing other small satyr butterflies that approach it. Both sexes feed from rotting fruits and flowers such as *Mikania micrantha* and probably many others as well. Females fly along the forest understory searching for places to lay eggs from 10:00 a.m. to midday. **Early stages:** Eggs are whitish, rounded, with smooth surface, and are laid singly on underside of leaves. Larva is long and thin with a two-forked tail; it is light green with a light blue longitudinal dorsal band, a light yellowish lateral band that is broken into segments from A_2 to A_4, and a longitudinal yellowish ventrolateral line; body and head are covered in a faint layer of short setae and have a granulated texture; head capsule is green with four whitish lines, black stemmata, and two long, straight horns. Pupa is green with a streaked pattern; it is elongate and slightly curved ventrally with a faint dorsal hump on thorax and two long, acute horns, stuck together, on head. **Host plants:** *Lasiacis* (Poaceae).

Fifth instar larva of *Cyllopsis hedemanni vetones.*

Pupa of *Cyllopsis hedemanni vetones.*

Adult *Cyllopsis hedemanni vetones.*

Cyllopsis hedemanni vetones pinned.
Ventral view.

Higher Ditrysia Moths

This group is represented by around 20 families, a smaller number than lower Ditrysia (pp. 33-43), which comprises around 110 families. (The numbers of families are imprecise because of unresolved taxonomic issues.)

This chapter begins with the small to medium-size moths of the families Pyralidae and Crambidae, which are among the most diverse families, with 5900 and 9600 species, respectively, distributed worldwide. In Costa Rica, around 650 species of Crambidae and close to 200 of Pyralidae have been reported. Both families are recognized by the presence of a pair of tympanic organs in the ventral area of the first abdominal segment, a characteristic unique to them among Lepidoptera. Remarkably, some pyralids have aquatic larvae that live in rivers. One species of Tortricidae is treated next; this family comprises 10,300 species worldwide, close to 400 of which occur in Costa Rica. Tortricid larvae are commonly known as "leaf-rollers" because they roll up live leaves of their host plant to create a shelter. Lasiocampidae, another family present on all continents, contains around 1500 described species, of which about 130 occur in Costa Rica. They are small to medium-size, robust moths with a very short or absent proboscis. The larvae build shelters with leaves and abundant silk, and can be solitary or gregarious. Eupterotidae is represented worldwide by around 300 species, of which only 3 inhabit Costa Rica. Their adult proboscis and other mouthparts are reduced or absent. Saturniidae is a family of medium-size to huge moths; there are 1500 species in the world and 118 in Costa Rica. The proboscis is absent or very reduced, although adults still drink water. Saturniid larvae can be very large and may be gregarious (*Hylesia continua*) or solitary; some have conspicuous spikes (Ceratocampinae) while others are smooth-skinned (*Titaea tamerlan*). Species of the family Sphingidae are medium-size to very large. They have a very well-developed proboscis, which in some cases can reach 25 cm long, adapted to reach the nectar in long, tubular flowers. Their body is robust, and their abundant musculature allows them to beat their wings many times per second and buzz like hummingbirds. About five species in this family are diurnal and are often confused with hummingbirds. Sphingids are solitary as larvae, which are recognized by a dorsal horn on segment A8. There are 1200 species worldwide, and 145 occur in Costa Rica.

Included in this book is the only Costa Rican diurnal species of the family Uraniidae, which contains 41 species in the country and 700 worldwide. The non-diurnal species of the family are medium-size to small, very cryptic, and often have folds and notches in their wings. In the same superfamily (Geometroidea) is Geometridae; with 23,000 species (1500 in Costa Rica), it is the second-most diverse family in the order Lepidoptera. Its larvae are popularly known as inchworms for their manner of movement. Most species are very cryptic in adult as well as caterpillar stage, but some are conspicuous, flashing aposematic coloration, and even day-flying. Another huge group of moths, the Noctuoidea, is recognized by the position and structure of a metathoracic tympanal organ. It is composed of six families, of which four are included herein: Notodontidae has 3800 species worldwide, of which about 500 live in Costa Rica. Its larvae are smooth-skinned (lacking setae), often have humps or notches, and are colorful. Noctuidae has about 12,000 species worldwide, and larvae are generally smooth, though some species have disperse setae. Nolidae has 1740 species, of which about 90 live in Costa Rica; their larvae are lightly covered in setae.

Erebidae is the most diverse family of the Lepidoptera, with almost 25,000 species world-wide; caterpillars may be either smooth-skinned or very fuzzy.

These families can be distinguished from one another by venation characters called trifine and quadrifine conditions (see figs. 18b and 18c in Introduction, p. 21). Erebidae presents the quadrifine condition in its hind wing, which differentiates it from Noctuidae, with trifine hind-wing venation. Notodontidae is distinguishable by its trifine forewing venation from Noctuidae, Erebidae, and Nolidae, which have quadrifine forewing venation.

Nolidae is distinguished by the bar-shaped retinaculum (a tissue cavity at the base of the costal area of the VFW, into which the spine-like frenulum is inserted, anchoring the HW with the FW to make the wings move synchronously; see Fig. 15, p. 20) from all other groups with the exception of the subfamily Arctiinae (Erebidae). Arctiinae is distinguished by the position of the post-spiracular counter-tympanal organ (a hearing membrane on the side of the first abdominal segment): it is anterior to the spiracle rather than posterior to it, as in all other erebids.

Quick key to the Noctuoidea families included in this book

Begin by looking at the venation on the forewing (FW); if it is trifine, it belongs to the family Notodontidae. If it is quadrifine, then examine the venation on the hind wing (HW).

If the venation on the hind wing is trifine, then the individual belongs to the family Noctuidae. If it is quadrifine, then examine the retinaculum on the ventral forewing (VFW).

If the retinaculum is long and bar-shaped, then the individual belongs to the family Nolidae. If the retinaculum is "normal," then the individual belongs to Erebidae, though note the exception case for members of the subfamily Arctiinae (described above).

1. FW venation trifine (Fig. 16B, p. 21)	Notodontidae
1. FW venation quadrifine (Fig. 16C, p. 21)	2
2. HW venation trifine (Fig. 16B, p. 21)	Noctuidae
2. HW venation quadrifine (Fig. 16C, p. 21)	3
3. VFW with a bar-shaped (long) retinaculum	Nolidae
3. VFW with a normal retinaculum	Erebidae (except Arctiinae)

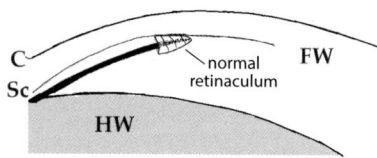
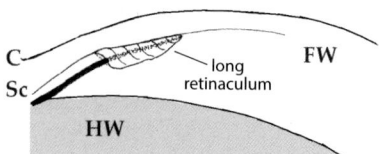

Epicorsia oedipodalis (Pyralidae: Pyraustinae)
Fiddlewood Leaf-roller

0–1300 m

FWL: 14 mm. Description: DFW and DHW are cream-colored with disperse very faint brown marks. In foreleg, femur and proximal half of tibia are brown and swollen, while distal end of tibia is white. **Similar species:** *Hahncappsia conisphora* has all-white legs, lacking swollen brown portion of foreleg. **Habitat:** Disturbed habitats such as cities and suburban areas on both Pacific and Caribbean slopes where the host plant is used as an ornamental. Also occurs in cattle ranches, pastures, and other open areas as well as primary forest in Braulio Carrillo and Rincón de la Vieja National Parks. **Distribution:** USA to Costa Rica, Caribbean islands. Introduced in Hawaii. **Seasonality:** Present all year. **Natural history:** Not much is known about this species in Costa Rica, but because of its voracious larvae and adaptability to disturbed habitats, it has the potential to become a serious pest of ornamental plants. **Early stages:** The young caterpillar skeletonizes leaf surfaces, and as it grows, rolls up leaves, knitting several together in a profuse silk web, within which the larva shelters inside one rolled-up leaf. Mature larva is pale green with a longitudinal white dorsal band and four dorsolateral black spots in each segment from T_2 to A_8, a white plate (scutellum) with many small black dots on T_1, and disperse long, rigid, thin, clear setae covering the body; head capsule is smooth and orange. Pupa is formed inside leaf shelter; it is long, cylindrical, and orange, with proboscis and antennal pad lying over the wing pad ventrally, reaching segment A_6. **Host plant:** *Citharexylum donnell-smithii* (Verbenaceae).

Fifth instar larva of *Epicorsia oedipodalis.*

Pupa of *Epicorsia oedipodalis.*

Adult *Epicorsia oedipodalis.*

Epicorsia oedipodalis pinned.
Dorsal view.

Mapeta xanthomelas (Pyralidae: Pyralinae)

0–1900 m

FWL: 15–21 mm. **Description:** DFW is bright orange with white and black transverse parallel lines from costal margin of apical area to most of its distal margin. DHW is also orange, with apical and sub-apical area black. **Similar species:** Black apical area on DHW distinguishes *M. xanthomelas* from *Phostria mapetalis* and *Mesene margaretta* (Riodinidae), on which the area is orange. *Oricia homalochroa* (Notodontidae) shares black apex, but black also extends all along distal margin. *Pyromorpha radialis* (Zygaenidae) also has black apex, but black area then narrows and widens again along distal margin, differing from single area of black in apical area of *M. xanthomelas*. **Habitat:** This moth is common on both Pacific and Caribbean slopes in all forest habitats and disturbed places such as gardens and parks where host plant is growing. In Central Valley it can be seen on sunny days in fast, fluttering flight. Other localities include preserved areas such as Monteverde, Coto Brus, Tapantí National Park, and volcanoes of Guanacaste. **Distribution:** Mexico to Venezuela and Peru, Jamaica. **Seasonality:** Present all year. **Natural history:** A bright-colored moth, *M. xanthomelas* is part of an intricate mimetic ring that includes day- and night-flying species of many families; *M. xanthomelas* is one of the unpalatable models. The caterpillars of this species sequester toxins from the host plant and retain them as adults, specifically females, for protection from predators. The similar species *P. radialis* is also considered to be unpalatable to predators and therefore a co-model sharing a Müllerian mimicry complex with *M. xanthomelas*, while the other similar species mentioned are part of a Batesian mimicry ring. **Early stages:** The caterpillars

Fifth instar larva of *Mapeta xanthomelas*.

Cocoon of *Mapeta xanthomelas*.

Adult *Mapeta xanthomelas*.

Mapeta xanthomelas pinned.
Dorsal view.

build shelters by putting two or three leaves together with silk and live inside them in groups of five to fifteen individuals. Larva is shiny black with long white setae arranged in disperse tufts and many small white dots on each body segment; head capsule is black with few white marks and setae. Pupa is black, cylindrical, and smooth, and is formed inside a light yellowish silk cocoon. **Host plants:** *Aristolochia anguicida*, *A. pilosa*, *A. maxima*, *A. grandiflora*, and other *Aristolochia* species (Aristolochiaceae).

Lygropia erythrobathrum (Crambidae: Spilomelinae)

0–650 m

FWL: 9–12 mm. **Description:** DFW is glossy brown with a yellowish-white post-medial band from costal margin to vein Cu$_1$ and a white bar in basal area from inside discal cell to 2A. DHW has same coloration but with dark red, or occasionally yellow, in basal area. **Similar species:** None. **Habitat:** Coastlines with mangrove habitats and surrounding areas, on both slopes but more common in Pacific localities such as Playa Naranjo, Sierpe, and Osa Peninsula. Occasionally found inland. **Distribution:** Costa Rica and Panama. **Seasonality:** Present all year. **Natural history:** This species seems to be somewhat variable in color pattern, as some yellowish individuals appear within populations. That is consistent with the original description by H. G. Dyar (1914), who noted that some of the type specimens were yellow in basal areas of DHW and DFW where others were dark red. The species is one of the few capable of living in extremely salty habitats such as mangroves, and it is the major leaf eater of its host plant, a species

Fifth instar larva of *Lygropia erythrobathrum*. AV

Cocoon of *Lygropia erythrobathrum*. AV

Adult *Lygropia erythrobathrum*. AV

Lygropia erythrobathrum pinned. Dorsal view.

of mangrove. **Early stages:** Larvae are solitary feeders and when young are leaf skeletonizers. They build a shelter to live inside by putting two leaves together. Mature larva is green with few disperse long white setae; head capsule is smooth and green. Pupation takes place inside a silk cocoon within the larval shelter. **Host plant:** *Avicennia germinans* (Acanthaceae).

Sufetula anania (Crambidae: Lathrotelinae)
Pineapple-root Caterpillar

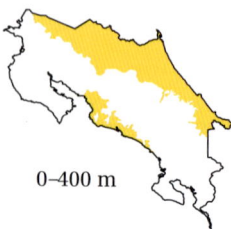

0–400 m

FWL: 5 mm. **Description:** DFW is white and gray with a conspicuous black dot in the distal portion of the discal cell; costal margin is black and white, and two thin, parallel black zigzag lines run transversely in sub-marginal area from costal margin near apex to anal margin. DHW has a small black spot in distal end of discal cell, and from medial area to distal margin is white splashed with gray. Both wings are shortened at level of M veins, producing a wavy distal margin. **Similar species:** No similar species occurs on pineapple. **Habitat:** Currently only known to occur in pineapple crops. **Distribution:** Costa Rica. **Seasonality:** Present all year. **Natural history:** *S. anania* is a recently described species and represents the first record of a crambid moth feeding on plants of the family Bromeliaceae. It has become an important pest of pineapple crops, and researchers have only just turned their attention to it. **Early stages:** Eggs are white, flattened, with reticulate surface, and are laid on organic material close to host plants. Once

Fifth instar larva of *Sufetula anania*. JDO

Pupae of *Sufetula anania*.

Pair of *Sufetula anania*. JDO

Sufetula anania pinned.
Dorsal view. JDO

hatched, the larvae move to the pineapple, where they feed from the tender roots the plant produces at ground level. Larva is long, thin, whitish, and covered in disperse brownish setae; each segment has six small brown warts, and T$_1$ has a dorsal prothoracic plate; head capsule is brownish. Pupa is formed at base of host plant inside a white cocoon covered in small dirt and frass particles; it is cylindrical, smooth, and whitish. **Host plant:** *Ananas comosus* (Bromeliaceae).

Amorbia depicta (Tortricidae: Tortricinae)

0–1300 m

FWL: 9–14 mm. **Description:** A small brown moth with long, forward-pointing labial palps and an elongate, acute apex to DFW (typical of tortricid moths). DFW is brown with some lighter brown areas; DHW is solid light brown. Costal vein of FW is curved in basal half. **Similar species:** *Amorbia cacao* has darker DHW and different light and dark brown color pattern on DFW. *A. osmotris* has more whitish DHW than *A. depicta*. Also, *A. depicta* has a small ocellus (simple eye) on each side of head at base of antenna in posterior position. E. Phillips (pers. comm.) proposes that *A. depicta* is not a member of *Amorbia* since it has some characters found in no other species of the genus. **Habitat:** Found on both slopes in all habitats except the northern Pacific dry forest. It is common in gardens, parks, and backyards; as well as in primary forest inside national parks such as Rincón de la Vieja and Braulio Carrillo. **Distribution:** Mexico to Panama. **Seasonality:** Present all year. **Natural history:** Nothing is known about the biology

Fifth instar larva of *Amorbia depicta*.

Pupa of *Amorbia depicta*.

Adult *Amorbia depicta*.

Amorbia depicta pinned.
Dorsal view.

of this species except that it visits lights at night and probably feeds on rotting fruits or tree sap. **Early stages:** The larva lives inside a shelter it builds by sticking two leaves together with silk. Larva is light green and somewhat translucent; it has a large dorsal prothoracic plate that is green with anterior side white; head capsule is very hard and black with mouth area white. Pupa is formed inside larval shelter and is reddish-brown with a row of small dorsal spikes on each abdominal segment. **Host plants:** *Annona reticulata* (Annonaceae), *Psidium guineense* (Myrtaceae), *Piper* (Piperaceae).

Euglyphis maria (Lasiocampidae: Poecilocampinae)

0–1600 m

FWL: 18–31 mm. **Description:** A finely mottled brown moth. DFW has some very faint, darker zigzag lines in sub-marginal distal margin. DHW is pale solid brown (not mottled) in anal half and has a darker zigzag line in sub-marginal distal margin. **Similar species:** Very similar to many other species of *Euglyphis*, especially *E. barda*, *E. capillata*, *E. phyllis*, and *E. sobrina*. *E. barda* has short, darker brown lines in sub-apical and medial areas of DFW close to distal margin, which are not present in *E. maria*. *E. capillata* has a darker, broken brown longitudinal line from costal margin (close to apex) to anal margin of DFW, absent in *E. maria*. *E. phyllis* is very similar to *E. maria* in adult stage, barely distinguishable by the sub-marginal zigzag lines on DFW, which are clearly broken in *E. phyllis*; but caterpillars and molecular characteristics differ. *E. sobrina* has a darker basal area and lighter brown markings in post-medial

Fifth instar larva of *Euglyphis maria*.

Pupa of *Euglyphis maria*.

♂

Euglyphis maria pinned.
Dorsal view.

Adult *Euglyphis maria*.

area of DFW, while *E. maria* has the same brown coloration on all of DFW surface. **Habitat:** Lives in many habitats (always evergreen), including secondary growth and abandoned plantations as well as inside primary forest of both Pacific and Caribbean slopes. Common in Rincón de la Vieja, Orosí Volcano, and Corcovado National Park. **Distribution:** Mexico to Brazil and Peru. **Seasonality:** Present all year but more common during rainy season. **Natural history:** This species is not very well known. It probably has the potential to become a pest of crops such as avocado and guava as its habitats become reduced by deforestation. Adults are attracted to lights at night. Males can fly several kilometers in search of females. The caterpillars are attacked by a parasitoid wasp (*Parapanteles mariae*, Braconidae). **Early stages:** Eggs are white and green, semi-oval, with smooth surface, and are laid on twigs and leaves in very organized rows in groups of up to fifty. Early instars are gregarious; they rest alongside each other and when moving from one leaf to another, form a line and travel in a procession. When mature, caterpillars become solitary, and each builds a tubular shelter with a host plant leaf. They are somewhat variable in coloration and pattern but in general terms, larvae have a velvet-black dorsal area with a transverse whitish-yellow band on each segment and a pair of pink tufts of setae anterior to each of those bands; lateral areas are yellow with a ventrolateral layer of long, soft pink setae surrounding the larva; head capsule is grayish-pink and smooth. In some individuals, the dorsal yellow bands are divided or reduced to only two spots and the yellow lateral band very reduced. Pupa is black and somewhat covered by a soft layer of a gray waxy powder; it is formed inside the larval shelter, in a silk cocoon. **Host plants:** *Mespilodaphne veraguensis*, *Persea americana* (Lauraceae), *Psidium friedrichsthalianum*, *P. guajava* (Myrtaceae).

Neopreptos marathusa (Eupterotidae: Eupterotinae)
Costa Rican Lappet Moth

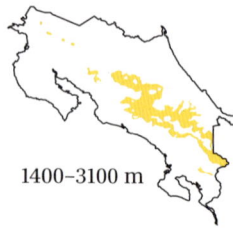

1400–3100 m

FWL: 26–39 mm. **Description:** *Female:* Grayish-brown with a continuous whitish longitudinal line from DFW costal margin at sub-apex to DHW anal margin (sometimes faint); inside discal cell of DFW is a white spot (sometimes absent). *Male:* Similar to female but browner and with more rounded wing shape. **Similar species:** None. **Habitat:** Montane habitats on both Pacific and Caribbean slopes; it is associated with well-preserved forests and sometimes flies to lights on house porches. Localities include Monteverde, Zurquí, Vara Blanca, Cerro de la Muerte, and Talamanca mountain range. **Distribution:** Costa Rica.
Seasonality: Present all year. **Natural history:** Although this species is common in many forest habitats, not much is known about it. It is likely adults do not live for long, since they do not feed at all. It is more often found as a caterpillar because of its bright colors. **Early stages:** One can spot the caterpillar of this species while walking on montane trails. Its bright colors betray it instantaneously on leaves and branches. Larva is entirely covered in a dense layer of white setae arranged in lateral and dorsal tufts, with a variable quantity of orange tufts on dorsum. The cocoon is built between leaves and cracks on trees with the larva's own body setae, giving it an orange coloration. **Host plants:** *Philodendron* (Araceae), *Gunnera insignis* (Gunneraceae).

Fifth instar larva of *Neopreptos marathusa*.

Cocoon of *Neopreptos marathusa*.

Neopreptos marathusa pinned.
Dorsal view.

Adult female *Neopreptos marathusa*.

Caio championi (Saturniidae: Arsenurinae)

0–1600 m

FWL: 70–90 mm. Description: DFW is brown with basal half lighter than distal half, a dark basal line from costal margin to anal margin, a dark comma-like figure at distal end of discal cell, and a series of wavy dark lines and shaded areas running longitudinally, distal to medial area. FW apex is elongate (more marked in males). DHW has the same patterns as DFW but without the comma-like shape. HW distal margin has a tail (vein M_2), very prominent in male and less so in female. VWs are almost solid brown. **Similar species:** Immediately recognized from all other species by comma-like figure on DFW; *Arsenura sylla* has a straight dark bar where *C. championi* has curved comma-like figure. **Habitat:** Associated with forest and rural habitats on both Pacific and Caribbean slopes. This species is most frequently seen at lights in small villages close to forest patches in localities such as Guanacaste, Garabito, Monteverde, Pérez Zeledón, San Vito in Coto Brus, and Osa Peninsula. **Distribution:** Mexico to Colombia and Venezuela. **Seasonality:** Present all year but much more common during rainy season. **Natural history:** This large moth appears seasonally in northern Pacific dry forest and more constantly in the rainy southern Pacific. In Guanacaste it has two generations per year, one of which passes the dry season in pupal stage, undergoing diapause for up to 150 days. On dark nights, it is common to find individuals inside

Fifth instar larva of *Caio championi.*

Pupa of *Caio championi.*

Adult *Caio championi.*

Caio championi pinned. Dorsal view.

houses and on porches. Adults are fast and powerful fliers, with an erratic flight pattern that makes it difficult for bats to catch them. They are reportedly a common prey of birds such as *Trogon elegans* and *T. melanocephala*, which feed their chicks the larvae (Miller et al. 2006). Larvae are attacked by parasitoid wasps (*Cotesia*, Braconidae) and a fly (*Winthemia subpicea*, Tachinidae). **Early stages:** Eggs are light green with a dark green dorsolateral ring, oval with smooth surface, and are laid singly or in small groups on underside of host plant's leaves. Mature larva is green, covered in many disperse white spots, with a purple dorsal band with white borders from segment A_1 to A_{10}; a dorsal hump in segment T_3 bearing a transverse purple and white dorsal band; diagonal black, red, and white bands followed by a ventral black spot in A_2, A_4, and A_6; and a black lateral slash on A_{10} accompanied by a white dorsal line that reaches A_7. Head capsule is smooth, with reddish-brown bands. Pupa forms while buried in the ground; it is black, cylindrical, and has a spike-like prolongation of cremaster. **Host plants:** *Pachira quinata*, *Ceiba pentandra* (Malvaceae).

Titaea tamerlan nobilis (Saturniidae: Arsenurinae)

0–1600 m

FWL: 70–80 mm. **Description:** Wings are different tones of light brown in both sexes. *Female:* FW has an irregular distal margin. DFW has two or three small (but variably sized) translucent areas in distal end of discal cell, a darker brown line running through post-medial area from costal margin to anal margin, and distal sub-marginal and marginal area slightly darker than rest of wing. DHW is also brown, with a darker post-medial line following the distal margin, which is also irregular. It has a small tail (M_3 vein) on HW. *Male:* Similar to female, but FW apex is very elongate, and HW has a prominent tail (veins M_2–M_3). **Similar species:** *Dysdaemonia boreas* is grayish and has a black line on DHW from tornus to base of elongate veins M_2–M_3, while *T. tamerlan* is light brown (though some males can be slightly grayish) and lacks line on DHW. Other members of the subfamily Arsenurinae do not have translucent areas in FW. **Habitat:** Preserved habitats with secondary and primary forest in rural areas on both Pacific and Caribbean slopes. May be seen perched inside buildings, cabins, or restaurants with lights that

Fifth instar larva of *Titaea tamerlan nobilis*.

Pupa of *Titaea tamerlan nobilis*.

Adult male *Titaea tamerlan nobilis*.

Titaea tamerlan nobilis pinned. Dorsal view.

Titaea tamerlan nobilis pinned. Dorsal view.

attract them at night. Common in Rincón de la Vieja and Braulio Carrillo National Parks. **Distribution:** Mexico to Peru and Venezuela. **Seasonality:** Present all year. **Natural history:** This species is seldom seen in nature unless attracted to lights, possibly because of its excellent camouflage. In Guanacaste, there are two generations per year (Janzen 1982). The first adults are seen during May–June; they lay eggs, and the next generation of adults emerges in August–September. It is probable that this species spends the dry season of Guanacaste in pupal stage, buried in the ground. **Early stages:** Eggs are light green, semispherical and somewhat flattened, and are laid on any part of host plant. First-instar larva is white with black dots and stripes, and has two long, black, two-forked fleshy projections arising dorsolaterally from segment T_3 and another on dorsal area of A_8. Third-instar larva is lime-green with two brown, curved lateral patches on A_1–A_3 and A_6–A_7; the fleshy projections persist but are no longer forked, and the anterior ones (T_3) are lime-green with black markings while the posterior one (A_8) is all black; and there are two small dorsolateral spikes on T_1; head capsule is brown with white front. Fifth-instar larva is variable: it can be all green and covered in many small orange spots; dorsally it has a longitudinal white line; spiracles are cream-colored; segment T_3 is swollen with two dorsolateral humps. Other forms additionally have variably sized, rounded brown blotches on segments A_2–A_3 and A_6–A_7, and sometimes smaller similar blotches are dispersed around the body. Pupa is shiny black and rounded; it is formed buried in soil near host plant. **Host plants:** *Paquira acuatica, Pachira quinata, Ceiba pentandra, Goethalsia meiantha* (Malvaceae).

Syssphinx molina (Saturniidae: Ceratocampinae)

0–1800 m

FWL: 31–55 mm. **Description:** *Female:* Much larger than male, with rounded wings. DFW is beige with many small brown stains dispersed all over surface; a brown line runs through medial area from costal to anal margin, and another faint line does the same in basal area. DHW is beige distally and pink in basal half with a large black dot at distal end of discal cell; a brown medial line runs from costal to anal margin. VFW is similar to DFW but has a black dot at distal end of discal cell. VHW is same as DHW but without the black dot. *Male:* Much smaller than female. DWs are similar to female's, but DFW has fewer brown markings and has an undulating distal margin, and DHW has more pink area. **Similar species:** Distinguished from all other species, including the similar *Syssphinx cola*, by prominent black dot at distal end of DHW discal cell. **Habitat:** This is a common species that can be seen on both Pacific and Caribbean slopes in many habitats, from primary forest to urban areas. Sporadic individuals may be seen resting close to house or porch lights or streetlights. **Distribution:** Mexico to Argentina. **Seasonality:** Present all year. **Natural history:** Males are strong fliers and may travel long distances in search of females. Adults seem to have aposematic coloration on the HW, which they quietly expose when alarmed, but it

Fifth instar larva of *Syssphinx molina*.

Pupa of *Syssphinx molina*.

Adult *Syssphinx molina*.

Syssphinx molina
pinned. Dorsal view.

is not known if they are toxic. Miller et al. (2006) state that the caterpillar's manner of feeding, by cutting leaves in large pieces, prevents it from absorbing the toxic secondary compounds of its host plants. The caterpillars are attacked by a parasitoid wasp (*Thyreodon santarosae*, Ichneumonidae), which itself is attacked by a hyperparasitoid wasp (*Neotheronia tacubaya*, Ichneumonidae) when *S. molina* enters the pupal stage. **Early stages:** Eggs are pale green, semispherical, and flattened, and are laid singly on the host plant's branches and leaves. Young larva is green with many setae all over its body and two long, black, fleshy tentacles armed with short setae on segment T_2, another two on T_3, and another, shorter dorsal tentacle on A_8. In fourth instar, segments A_7–A_9 are swollen, and there is a small spike on A_8, but these characteristics are lost in the fifth instar. Fifth-instar larva is green, covered in many small cream and blue dots, and has two small, red dorsolateral spikes each on T_2 and on T_3. Both fourth and fifth instars have a curved, bluish and yellow lateral line from segment A_9 to base of spikes on T_3. Head capsule is smooth and green with a whitish longitudinal line. Pupa is buried in the ground in a rudimentary pupal case; it is shiny black, covered in many small spikes, and has two large warts on dorsal side of T_3, cream-colored beanlike dots on each eye, and a forked cremaster. **Host plants:** *Inga punctata, Albizia adinocephala, Samanea saman, Calliandra calothyrsus* (Fabaceae).

Adeloneivaia centrojason (Saturniidae: Ceratocampinae)

0–1800 m

FWL: 35–61 mm. **Description:** *Female:* DFW is orange with a transverse black line from apex to medial area at anal margin, and another, longitudinal black line from basal area of costal margin to post-basal area of anal margin; DHW is pinkish with a faint, thin brown line from anal margin to apex. *Male:* Much smaller than female but similar color patterns on both DFW and DHW; basal and post-medial areas outside of the black lines have a variable amount of gray scales on an orange background. Some individuals (mostly males) have one or two small silverish spots at distal end of discal cell. **Similar species:** On DHW, female *Adeloneivaia boisduvalii* has much darker orange coloration, stained with multiple brown dotsa; female *A. isara* has brown DWs. Male *A. boisduvalii* has a thick, diffuse brown line on DHW that generally does not cross the whole wing, while on *A. centrojason* the line is thin and crosses from one margin to the other. Male *A. isara* has brown DWs and never has a silver-white spot on DFW. **Seasonality:** Present all year but more common during rainy season. **Habitat:** Associated with primary- and

Fifth instar larva of *Adeloneivaia centrojason.*

Pupa of *Adeloneivaia centrojason.*

Adult *Adeloneivaia centrojason.*

Adeloneivaia centrojason pinned. Dorsal view.

secondary-forest habitats on both Pacific and Caribbean slopes. It may be seen at lights on sides of forested roads, at mountain hotels, and in biological reserves. **Distribution:** Guatemala to Brazil. **Natural history:** This robust species is a strong flier, and males sometimes fly several kilometers in a single night in search of females. It is uncertain how the bright orange coloration helps this species; perhaps adults rest in forest litter where decaying orange leaves are abundant. The caterpillars are attacked by a parasitoid wasp (*Meteorus congregatus*, Braconidae). **Early stages:** Eggs are light green, semispherical, and are laid singly or in pairs on any part of the host plant's branches. Young larva is dark brown with six long, rigid fleshy prolongations on thorax and a dorsal one on segment A_8, and rings of short scoli on each segment. Mature larva is green with bluish spiracles; segment T_1 has six short, fleshy spikes; T_2 and T_3 each have four long, dorsolateral projections with many small spikes; segments A_1–A_7 each have four long, green dorsolateral fleshy prolongations; A_8 has two small dorsolateral horns and one curved, much larger dorsal horn covered in small spikes; A_2, A_4, and A_6 each have a cream-colored diagonal bar on lateral area with a silver, spike-like tip. Head capsule is smooth and pale green with a lighter longitudinal line on each side. Pupa is buried in ground litter inside a rudimentary pupal case; it is shiny black with a forked cremaster, and short spikes all over thorax, wing pads, and head areas; each abdominal segment has two rings of short spikes. **Host plant:** *Inga samanensis* (Fabaceae).

Hylesia continua (Saturniidae: Hemileucinae)

0–1700 m

FWL: 19–31 mm. **Description:** A small brown moth with grayish DWs interspersed with light brown longitudinal bands. VWs are similar but with diffuse color pattern. **Similar species:** Distinguished from *Hylesia dalina* by gray patch in DFW apex, which in *H. continua* is short but in *H. dalina* is longer, extending into sub-apical area. *H. invidiosa* has barely discernible medial longitudinal brown band on DFW, while in *H. continua* the band is very clear. Male *H. rubrifrons* has a brown band inside discal cell of DFW that does not touch the longitudinal medial band, as it does in *H. continua*. Female *H. rubrifrons* has post-medial longitudinal grayish band that does not touch the apical gray patch, while in *H. continua* it does. **Habitat:** Associated with disturbed habitats on both Pacific and Caribbean slopes. It is easier to find in larval stage in gardens and backyards, but adults are sometimes seen flying around lights or as occasional visitors inside houses. **Distribution:** Mexico to Venezuela and Ecuador. **Seasonality:** Present all year. **Natural history:** In Central Valley, this species is well known for two reasons: the caterpillar is a sporadic pest of guava trees and other ornamental plants, and it stings. Because it is a common species, the risk of contact is high, and people are frequently stung when they accidentally touch the caterpillar. Larvae are attacked by parasitoid wasps (*Hylesiicida*, Ichneumonidae; *Parapanteles tessares*, Braconidae) and flies (*Leptostylum*, Tachinidae). **Early stages:** Eggs are white, oval, with smooth surface, and are laid in large groups and covered in a dense layer of the adult female's scales. Mature larva is variable, from dark brown to black, and has two longitudinal white lines running dorsally and another two running laterally; each segment has six large, reddish fleshy appendages with stinging spines, the four dorsolateral ones in T_1 longer than those on other segments and with a white section close to the tip, and one pair is projected forward. Head capsule is smooth, black with a white band in the center and another at the sides right above stemmata. Pupa is formed inside a silk cocoon generally made between leaves of any plant or in natural or artificial crevices. **Host plants:** *Calathea crotalifera* (Marantacaeae), *Psidium guajava*, *P. friedrichsthalianum* (Myrtaceae), *Hampea appendiculata* (Malvaceae), and many species of the families Cannabaceae, Euphorbiaceae, Lauraceae, and Fabaceae.

Fifth instar larva of *Hylesia continua.*

Pupa of *Hylesia continua.*

Adult *Hylesia continua.*

Hylesia continua pinned.
Dorsal view.

Automeris tridens (Saturniidae: Hemileucinae)

0–1700 m

FWL: 35–55 mm. **Description:** Male is smaller and yellow; female is larger and orange-brown, although some females are more yellowish, resembling males. DFW has a curved longitudinal yellowish line in post-medial area, and a similar but wavy line in post-basal area; at distal end of discal cell is a patch slightly darker than background color. DHW has a prominent eyespot in medial area, formed of rings (from outside to inside) of yellow, black, brown, and black again, with an irregular white pupil; basal area is orange; in post-medial area a black line follows shape of distal margin; and sub-marginal area is brown, contrasting with a different tone of brown in distal margin. VFW has a slightly curved, dark longitudinal line in post-medial area, and a black eyespot with brown and a white pupil in distal part of discal cell. VHW is same color DHW, with a post-medial darker line from apex to anal margin, and a white spot

Fifth instar larva of *Automeris tridens*.

Pupa of *Automeris tridens*.

Adult *Automeris tridens*.

in distal part of discal cell. **Similar species:** *Automeris banus* and *A. zurobara* do not have an external yellow ring on DHW eyespot, and in *A. banus* the post-medial dark line on DFW extends to costal margin, while it curves to apex in *A. tridens*. (The name *A. rubrescens* is widely use in literature, but it is a synonym for *A. tridens*, representing a lowland form.) **Habitat:** Secondary habitats on both Pacific and Caribbean slopes. Easier to see as caterpillar, though adults may be seen resting during the day on house walls or floors, and sometimes at night they fly around lights or come into houses. **Distribution:** Mexico to Peru. **Seasonality:** Present all year. **Natural history:** Adults do not feed. Males fly at midnight in search of females, which spend most of the time perched in vegetation. Eggs are attacked by parasitoid wasps (*Telenomus*, Scelionidae). **Early stages:** Eggs are white, oval, with smooth surface, and are laid singly or in small groups on any part of host plant's leaves. Mature larva is fully covered in dense, rigid, fleshy green appendages with stinging spines; it has a longitudinal whitish-yellow lateral line; head capsule is green and smooth. Pupa is formed inside a dense silk cocoon generally placed in a hole or a shelter made by putting two or three leaves together. **Host plants:** *Gmelina arborea* (Lamiaceae), *Cordia alliodora* (Boraginaceae), *Erythrina poeppigiana* (Fabaceae), *Dypsis lutescens* (Arecaceae), *Guazuma ulmifolia* (Malvaceae), *Coffea arabica* (Rubiaceae).

Copaxa rufinans rufstralica (Saturniidae: Saturniinae)

700–2000 m

FWL: 48–61 mm. **Description:** FW has an acute point on apex. DFW has an almost straight brown and white line from anal margin to apical area, and at distal end of discal cell a small orange eyespot with a small translucent area in the center. DHW has a brown line from costal margin to basal anal margin, an orange eyespot in same position as in DFW, followed by a longitudinal, faint wavy brown line in medial area, and a post-medial grayish area that becomes wider as it reaches the anal margin. VWs are almost all brown. **Similar species:** *Copaxa rufinans rufstralica* is somewhat variable, and some adult individuals are indistinguishable from *C. moinieri* (p. 220); in most cases, *C. rufinans rufstralica* has a more strongly marked area of grayish shadows surrounding the dark lines on both DWs; the post-medial grayish area on DHW widens toward anal margin; and the translucent area in middle of DFW is slightly larger than in *C. moinieri*, and *C. curvilinea* has a strongly curved diagonal brown line on DFW; in *C. rufinans rufstralica* the line is straighter. **Habitat:** Secondary and primary wet and cloud forest habitats at mid-elevations along all mountain systems of Costa Rica, on both slopes. Localities include Coronado, Monteverde, and Tapantí National Park. Adults are seen at night in lights or perched on walls of houses. **Distribution:** Mexico to Ecuador. **Seasonality:** Present all year. **Natural history:**

Fifth instar larva of *Copaxa rufinans rufstralica*.

Cocoon of *Copaxa rufinans rufstralica*.

This species is common and widespread, probably owing to its capacity to eat numerous plant species of multiple genera. Adults have an erratic flight and are often preyed on by bats; larvae are hunted by many bird species. **Early stages:** Early stages are virtually identical to those of *C. moinieri*, but there is a substantial difference in cocoon building. *C. rufinans rufstralica* spins a rigid inner cocoon with many holes in it, which is then covered by another softer and more irregular cocoon as an outside layer. *C. moinieri* spins only a single rigid cocoon, usually protected by putting two or three leaves together. **Host plants:** Many species of *Persea, Ocotea, Nectandra, Cinnamomum* (Lauraceae).

Adult *Copaxa rufinans rufstralica.*

Copaxa rufinans rufstralica pinned. Dorsal view.

Copaxa moinieri (Saturniidae: Saturniinae)

0–1200 m

FWL: 45–55 mm. **Description:** Variably colored, from brown to gray. Sexes are very similar, but female may show a little more contrast between gray and brown areas. FW has an acute point on apex. DFW has an almost straight brown and white line from anal margin to apical area, a small orange eyespot with a small translucent area in center at distal end of discal cell, and a grayish shadow along distal margin. DHW has a brown line from costal margin to basal anal margin; an orange eyespot in same position as on DFW, followed by a longitudinal, faint brown wavy line in medial area; and a faint post-medial brown line from costal to anal margin. VWs are almost all brown. **Similar species:** *Copaxa curvilinea* has strongly curved brown and white line from anal margin to apical area of DFW; this line is straighter in *C. moinieri. C. rufinans rufstralica* (p. 218) is very similar and also variable, but generally has more marked grayish shadows surrounding the dark lines on DWs; and the translucent area in middle of DFW is slightly larger than in *C. moinieri.* **Habitat:** Secondary- and primary-forest habitats, and urban areas close to riparian forest or forest patches, on both Pacific and Caribbean slopes. Localities include dry forest of Guanacaste, Ciudad Colón, and lower parts of Central Valley. Generally, adults are seen at night in lights or perched on walls of houses. **Distribution:** Mexico to Peru. **Seasonality:** Present only during rainy season. **Natural history:** This species is found in forest habitats since its host plants

Fifth instar larva of *Copaxa moinieri.*

Pupa of *Copaxa moinieri.*

Adult male *Copaxa moinieri.*

Copaxa moinieri pinned. Dorsal view.

are not common outside them. Males are active fliers at 6:00 p.m. and again at midnight. In the dry forest, the pupa undergoes diapause during the dry season, and adults emerge with the beginning of rainy season. Because saturniid adults do not feed, their life span is only 1–2 weeks. Larvae are attacked by parasitoid flies (*Patelloa xanthura*, *Leschenaultia leucopheys*, *Leptostylum*, Tachinidae) and a wasp (*Enicospilus bozai*, Ichneumonidae). **Early stages:** Eggs are brown with a yellow transverse ring, ovoid, and are laid in groups of up to thirty on underside of host plant's leaves. Larvae rest and feed together until third instar, when they become solitary. Mature larva is green with a faint yellowish lateral line beneath the row of black spiracles, and six blue spots per segment forming rows along its body, which is also covered in yellowish setae and many disperse yellow spots; each segment has a dorsal hump, giving it an accordion-like shape; prolegs are black, and head capsule is shiny black. Pupa is formed inside a brownish-orange silk cocoon, which is made amidst leaves and stalks. **Host plants:** *Mespilodaphne veraguensis*, *Ocotea cernua*, *O. insularis*, *Cinnamomum triplinerve*, *Nectandra belizensis*, *N. smithii*, *Persea americana* (Lauraceae).

Rothschildia lebeau (Saturniidae: Saturniinae)
Lebeau Four-windows

0–1500 m

FWL: 51–73 mm. **Description:** Variably colored from grayish to orange or reddish-brown. DFW has a post-basal white line from costal margin to basal anal margin that almost touches a large transparent area in middle of wing; distal to transparent area, another white line runs longitudinally from costal to anal margin in medial area; apical area has a large grayish patch at costal margin followed by three small black blotches in sub-marginal area; there is another grayish shadow from vein M₃ to anal sub-margin; distal margin is light brown with a thin, wavy black longitudinal line. DHW has a black and white line from anal margin of post-basal area to costal margin, but it curves without touching the margin and continues, parallel to its first half, to anal margin in post-medial area; this line encloses a large transparent area in middle of wing; a sub-marginal line of small black dots parallels distal margin, and a grayish post-medial shadow extends from M₃ to 2A. VWs are similar to DWs, but pattern is less detailed. **Similar species:** *Rothschildia triloba* (p. 223) and *R. orizaba* are larger moths, and the transparent area of their wings is smaller and arrow-shaped with acute angles, in contrast to the larger, rounded transparent area on *R. lebeau*. *R. erycina* is distinguished by a prominent light brown oval with yellowish margins in apical area of DFW, very distinct from rest of wing. **Habitat:** Forest, disturbed areas, and towns on both Pacific and Caribbean slopes. Occasionally seen at daytime perched on walls or plants in backyards

Fifth instar larva of *Rothschildia lebeau*.   Cocoon of *Rothschildia lebeau*.

Adult *Rothschildia lebeau.*

Rothschildia lebeau pinned. Dorsal view.

and parks. Localities include Liberia, Puntarenas, Osa Peninsula, Limón, San Carlos, San José, and Central Valley. **Distribution:** USA to Ecuador. **Seasonality:** Present all year but more common during rainy season. **Natural history:** This species is common in many habitat types where its numerous and varied host plants are present. Adult males fly long distances at night searching for females by following their pheromones. Males live around 10 days, but females can live up to 1 month (R. Vasquez, pers. comm.). The color variability of adults has been studied by D. Janzen (1984), who found that the color is determined mostly by temperature during pupal development, and that pupae undergoing diapause of 4–5 months in the dry season of Guanacaste produced mostly light orange adults, while pupae formed during the second half of rainy season produced dark chocolate adults. He argues that the early rainy-season vegetation is lighter-colored and less dense; therefore, lighter-colored adults are better protected, while darker adults are better protected when much more dark green vegetation is present later in rainy season. Larvae are attacked by parasitoid flies (*Patelloa xanthura*, *Lespesia aletiae*, *Winthemia picea*, *Chetogena scutellaris*, *Leptostylum*, *Belvosia*, *Blepharipa*, Tachinidae) and wasps (*Enicospilus lebophagus*, Ichneumoniidae; *Cotesia*, Braconidae). **Early stages:** Eggs are pale yellowish-green, oval, with smooth surface, and are laid in small groups on underside of host plant's leaves. Many dozens of eggs are usually dispersed around the same plant. Larvae rest on underside of leaves and display gregarious behavior until third instar, when they disperse and become solitary. Mature larva is green with purple and white transverse rings in folds between segments from A_3 to A_7, a white lateral line from A_8 to A_{10}, and a green round patch on sides of prolegs on A_{10}; some short white setae are dispersed and in tufts on dorsal area of abdomen and much more densely on dorsal area of thorax. Head capsule is green with black stemmata and black sutures surrounding front and mouth area. Pupa is formed inside a strong, rounded silk cocoon that hangs from a branch or leaf. The adult moth leaves the cocoon through the softer region at the top (the point it hangs from). At times, some individuals undergo several months of diapause in pupal stage. **Host plants:** *Zanthoxylum setulosum* (Rutaceae), *Spondias mombim*, *S. purpurea* (Anacardiaceae), *Casearia nitida* (Salicaceae), *Exostema mexicanum* (Rubiaceae), *Prunus annularis*, *P. persica* (Rosaceae).

Rothschildia triloba (Saturniidae: Saturniinae)
Triloba Four-windows

0–1300 m

FWL: 72–90 mm. **Description:** A reddish-brown moth. DFW has a wavy, longitudinal white and black medial band from costal to anal margin, and a strongly curved white and black line in basal area from costal to anal margin; between both lines, in the middle of the wing, is a prominent clear triangular area; distally, there is a wavy, dark sub-marginal line. DHW has a strongly curved black and white line, which starts in anal margin, nears costal margin, and returns to anal margin, enclosing a large medial clear triangular area; there is a sub-marginal longitudinal row of small brownish dots on distal margin. **Similar species:** *Rothschildia triloba* is very similar to *R. orizaba*, of which it used to be considered a subspecies. The first clue to identification is habitat: *R. triloba* is a low- and mid-elevation species, and *R. orizaba* is a high-elevation species. In general terms, *R. triloba* is orange-brown, while *R. orizaba* is red-brown. In *R. triloba*, the medial white and black line on DFW touches the costal margin closer to the medial area, leaving a wide space between it and the apical margin; in *R. orizaba*, the line touches the costal margin closer to the apical margin, leaving a considerably smaller space. Male *R. orizaba* also has a pronounced

Fifth instar larva of *Rothschildia triloba*.

Cocoon of *Rothschildia triloba*.

Adult male *Rothschildia triloba*.

Rothschildia triloba pinned.
Dorsal view.

♂

and more acute apical margin to DFW. *R. lebeau* (p. 221) is differentiated by the clear areas on the wings, which are larger and more rounded than in *R. triloba*; *R. lebeau* is also a little smaller than *R. triloba* and often not reddish. **Habitat:** Primary and well-preserved secondary rainforest, in low and mid-elevations of Caribbean slope and from Carara south to Osa Peninsula on Pacific side. One can see rare individuals perched close to lights at night. **Distribution:** Mexico to Panama. **Seasonality:** Present all year. **Natural history:** This is a rare species that is often found in low population densities, relative to *R. lebeau* and *R. orizaba*. As with all members of the family, it does not feed as an adult; female adult likely makes only very short movements, staying close to its host plants, while males fly long distances following females' pheromones. It is thought that the clear area on each wing is a camouflage adaptation, enhancing its resemblance to a decomposing leaf that has been eaten in some areas by insects; this serves to disrupt a predator's search image of a moth shape. Larvae are attacked by parasitoid flies (*Lespesia*, Tachinidae). **Early stages:** Eggs are white, oval, smooth-surfaced, and are laid in groups on many parts of host plant. Mature larva is green in dorsal half with rings of small orange scoli with short black setae, and a black V stain on dorsal side of segment A_{10}; ventral half is dark brownish-green with abundant long, soft white setae; head capsule is green. Pupa is formed inside a dense silk cocoon hanging from a branch and often hidden amid a few dry leaves attached to it. **Host plants:** *Zanthoxylum caribaeum* (Rutaceae), *Sapium glandulosum* (Euphorbiaceae), *Tapirira mexicana* (Anacardiaceae), *Prunus annularis* (Rosaceae), *Coffea arabica* (Rubiaceae), and many others.

Manduca sexta (Sphingidae: Sphinginae)
Tobacco Hornworm Moth

0–1700 m

FWL: 48–55 mm. **Description:** DFW is brown with grayish and black margins. DHW is also brown but with three longitudinal gray bands from costal to anal margin. VWs are dull gray. Abdomen is brown with six large dorsolateral yellow patches. **Similar species:** *Manduca occulta* has similar amounts of black and white in distal margin (fringe) of DFW, while *M. sexta* has a very small amount of white compared with black. *M. pellenia* has much more color contrast in DFW, with many more whitish markings on apical area. *M. dilucida* has smaller yellow patches on abdomen and has diffuse gray lines on DHW. **Habitat:** All habitats from primary rainforest to dry forest to urban areas on both Pacific and Caribbean slopes. Sometimes seen perched in suburban areas or on tree trunks close to the base. **Distribution:** Canada to Chile. **Seasonality:** Present all year. **Natural history:** This species is famous for the damage its caterpillars cause to tobacco, tomato, peppers, potatoes, and other popular crops. Adult feeds from flowers of tobacco (its host plant) among other plants, and seems to lay larger amounts of eggs on leaves of good-quality nectar plants. *M. sexta* is probably an important pollinator of pitahaya, or dragon fruit (*Selenicereus undatus*), since its proboscis, at 108 mm long, is similar in length to the flowers, from which it is often seen feeding at night. Eggs are attacked by parasitoid wasps (*Telenomus connectans*, *T. monilicornis*, Scelionidae; *Trichogramma minutum*, Trichogrammatidae); larvae are attacked by parasitoid wasps (*Cotesia americanus*, *Apanteles throracius*, Braconidae) and a fly (*Sturmia distincta*, Tachinidae). **Early stages:** Eggs are green, semispherical, with smooth surface, and are laid singly. Mature larva is green with seven diagonal white lines on sides; on top of each white line are small black lines and a variable amount of yellow in a diffuse patch; segment A_8 bears a short, curved brown dorsal horn; head capsule is green. Pupa is buried in the ground; it is brown, cylindrical, and has a prominent curved proboscis case projecting out of head area. **Host plants:** *Nicotiana tabacum*, *Datura*, *Solanum tuberosum*, *S. torvum* (Solanaceae).

Fifth instar larva of *Manduca sexta*.

Pupa of *Manduca sexta*.

Adult *Manduca sexta*.

♂

Manduca sexta pinned. Dorsal view.

Manduca barnesi (Sphingidae: Sphinginae)

0–1300 m

FWL: 45–65 mm. **Description:** DFW is grayish with a variegated pattern of lines and dark brown shadows; it is often darker in medial area from costal margin almost to vein Cu_2. DHW is dark brown with gray markings at tornus. Abdomen is gray with black markings in lateral area. **Similar species:** This species and the similar *Manduca schausi* and *M. florestan* are very variable in color pattern, and it is almost impossible to tell them apart based only on adult coloration. It is necessary to check male genitalia (Fig. 28) or molecular information, or to examine early stages to ensure accurate identification. Fifth-instar larvae are very different among the three species. (In the author's opinion, many works have misidentified this species in the past by basing identification only on adults, leading to an even more confused taxonomic scenario. When future local and regional identifications are based on the definitive aspects mentioned, the morphological boundaries among the species may become clearer.) **Habitat:** This species has been recorded in Costa Rica only from the northern and central Pacific slope. Because of its crypsis, it is very difficult to see during the daytime; it may be found on walls in towns during daytime or around lights at night. **Distribution:** Mexico to Venezuela. **Seasonality:** Present all year but more common during rainy season. **Natural history:** Little is known about the natural history of this moth; its proboscis indicates that it probably feeds from long tubular flowers, as do other *Manduca* species. It rests on tree trunks at any height, from a few centimeters above the

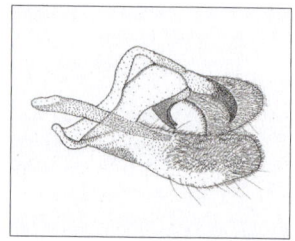

Fig. 28. Male genitalia of *Manduca barnesi.*

Fifth instar larva of *Manduca barnesi.*

Pupa of *Manduca barnesi.*

Adult *Manduca barnesi.*

Manduca barnesi pinned. Dorsal view.

ground to the canopy, where its lichenlike wing pattern protects it from daytime predators. **Early stages:** Mature larva is green with seven thin, yellowish diagonal lines on sides of segments A_1–A_8, the lattermost leading to a long, straight dorsal horn armed with many tiny spikes; many small yellowish warts are dispersed over most of body; spiracles are black with an internal white line; head capsule is smooth and green, crossed by a transverse pale green line (variable in intensity) on each side. Pupa is formed in the ground, buried with leaf litter; it is brown, cylindrical, and has a prominent external proboscis case that is short, curved, and swollen at the tip. **Host plant:** *Stachytarpheta mutabilis* (Verbenaceae).

Manduca albiplaga (Sphingidae: Sphinginae)
White-plagued Sphinx

0–1600 m

FWL: 64–78 mm. **Description:** DFW is brown in basal area, with a large white patch in post-basal area, a larger brown irregular patch filled with some black lines in medial area, a white sub-apical area, and distal margin at tornus is brown; there is a white spot at distal end of discal cell. DHW is mainly brown with some white in tornus. Dorsally, the head is white, thorax is brown, and abdomen has three yellow patches on sides in anterior segments and white segment A_1. Ventrally, the head is dark brown, and thorax and abdomen are white. Labial palps are white. **Similar species:** None. **Seasonality:** Present all year. **Habitat:** Occurs on both slopes but seems to be absent in lowlands of northern Pacific dry forest. It is a regular inhabitant of secondary and primary forest as well as cattle ranches and disturbed habitats near riparian areas. Localities include Osa Peninsula, all Caribbean flatlands, and foothills of all Costa Rican mountains. **Distribution:** Mexico to Brazil. **Natural history:** This species is not very common and seldom seen in

Fifth instar larva of *Manduca albiplaga*.

Pupa of *Manduca albiplaga*.

Adult *Manduca albiplaga*. HZ

Manduca albiplaga pinned.
Dorsal view.

nature. It feeds from flowers of families Solanaceae and Cactaceae. Larvae are attacked by parasitoid flies (*Leschenaultia*, Tachinidae) and wasps (*Cryptophion manueli*, Ichneumonidae; *Euplectrus testaceipes*, Eulophidae); larvae of *E. testaceipes* emerge by the dozens from the caterpillar and pupate on its body. **Early stages:** Young larvae are whitish with an orange head and a long black horn on segment A_8. Mature larvae rest on underside of host plant's leaves. In mature larva, segments T_1–T_3 are green with black legs; A_1–A_7 are lime-green in dorsolateral area, divided by a diagonal black line from bluish-white lateroventral half; segment A_8 is green with a purple horn that meets a dorsal band with white borders that continues to A_2; A_9 and A_{10} are green with many black dots; head capsule is dark green with a white band on each side. Pupa is black, rounded, and elongate, with short, curved external proboscis case. **Host plants:** *Annona rensoniana* (Annonaceae), *Aegiphila laevis* (Lamiaceae).

Eumorpha anchemolus (Sphingidae: Macroglossinae)

0–1800 m

FWL: 50–65 mm. **Description:** A large, robust moth with a dorsal pattern of light and dark brown and gray areas. DFW has a transverse whitish line with a thick brown shadow from mid-costal margin to distal margin. DHW is dark brown with some whitish patches in basal and tornus areas. VWs are mostly brown. HW is much smaller than FW. Abdomen has transverse white lines and gray patches on each segment. **Similar species:** The dark brown shadow distal to white line on DFW is wide and evenly shaped in *Eumorpha obliquus*, while in *E. anchemolus* it is irregularly shaped, with thinner areas; in the same wing, a dark brown area in anal margin right at side of tornus is triangular with clear edges in *E. anchemolus* but in *E. obliquus* is more rounded and diffuse. A dark brown inverted triangle with its base at costal margin

Fifth instar larva of *Eumorpha anchemolus*.

Pupa of *Eumorpha anchemolus*.

Adult *Eumorpha anchemolus*.

Eumorpha anchemolus pinned. Dorsal view.

of DFW close to apical area has a flat tip in *E. anchemolus*, and an acute and prolonged tip in *E. triangulum*. **Habitat:** Associated with pristine and disturbed habitats on both Pacific and Caribbean slopes. Adults are generally seen perched on walls in urban areas but very seldom seen in wild habitats due to their crypsis. Never common. **Distribution:** Mexico to Argentina. **Seasonality:** Present all year but more relatively more common during rainy season. **Natural history:** This species is a powerful flier and is known to feed early at night, probably from a wide variety of flowers, if it follows the pattern of other *Eumorpha* species. Larvae are attacked by a parasitoid fly (*Drino rhoeo*, Tachinidae). **Early stages:** Young larvae are very colorful with a bluish longitudinal dorsal band and longitudinal yellow and green lateral bands; segment A_8 bears a long black dorsal whip. Mature larva is green with elongate light green lateral patches from A_5 to A_7 (sometimes also A_4) and smaller spots of the same color in lateral areas of A_1 and A_2; anterior half of body is covered in many disperse small black dots; head capsule is smooth and light green. Early in the larva's last instar, A_8 bears a very small, black curved whip, which disappears later in instar, and a few days later, the larva turns dark brown. Pupa is cylindrical and brown; it is formed in the ground. **Host plants:** *Doliocarpus multiflorus* (Dilleniaceae), *Cissus verticillata* (Vitaceae).

Eumorpha labruscae (Sphingidae: Macroglossinae)
Gaudy Sphinx

0–1400 m

FWL: 46–58 mm. **Description:** DFW is green with a darker inverted triangle from costal to anal margin in medial area. DHW is black with a blue spot in middle of wing followed by a blue band, which becomes yellow as it approaches distal margin, which is all yellow. Abdomen and thorax are green. **Similar species:** No similar species in Costa Rica. **Habitat:** Associated with many habitats, including primary and secondary forest, on both Pacific and Caribbean slopes. May be seen at lights or perched on walls at night in Central Valley and suburban areas. **Distribution:** USA to Argentina. **Seasonality:** Present all year.

Natural history: This widespread but uncommon species is a sporadic visitor to Central Valley houses and buildings. Its presence in highly urbanized areas could be attributed to its capacity to fly several kilometers during a single night. Also, its host plants are common in abandoned fields and riversides in San José, so reproduction can take place there, at least in low densities. Both sexes feed from a wide variety of flowers including *Impatiens walleriana* and *Hamelia patens*. Larvae are attacked by parasitoid wasps (*Cotesia*, Braconidae) and flies (*Drino*, Tachinidae). **Early stages:** Mature larva has an irregular pattern of dark brown in dorsal area and light mottled brown in lateral areas; head capsule is smooth and dark brown. Segments T_1 to A_1 are swollen, and A_1 has a large eyespot patch with a white pupil on each side; when larva is alarmed, these segments retract and become larger, producing a convincing

Fifth instar larva of *Eumorpha labruscae*.

Pupa of *Eumorpha labruscae*.

Adult *Eumorpha labruscae*.

♂

Eumorpha labruscae pinned. Dorsal view.

snake-face impression from all angles. Younger instars have a long black dorsal whip on A_8, but mature caterpillar has a rounded, black shiny plate inside a large circular patch. This structure appears to be wet, and when alarmed, the larva moves it quickly in and out, perhaps to distract predators. Pupation takes place between leaves on the ground or when buried; pupa is shiny, reddish-brown, and cylindrical. **Host plant:** *Cissus biformifolia* (Vitaceae).

Madoryx oiclus oiclus (Sphingidae: Macroglossinae)

0–1300 m

FWL: 32–45 mm. **Description:** A robust gray and brown moth. FW has a very irregular, toothed distal margin. DFW has an irregular brown basal patch with a white spot in costal area, a diffuse brown medial area with two conspicuous silver-white blotches toward costal margin, and a straight, dark brown line from apical margin to post-medial anal margin. DHW is gray with basal and apical areas dark brown. VWs are grayish and not very patterned. Thorax and abdomen are robust and gray; abdomen has a dark brown patch on each side in penultimate segment. **Similar species:** In *Madoryx bubastus*, the FW distal margin is less toothed, and a small sub-marginal dark line, which begins in same spot in DFW apical area as the large transverse line, is straight and produces a triangular form with the larger line, while in *M. oiclus oiclus*, this smaller line is curved and ends abruptly on distal margin. **Habitat:** Associated with primary- and secondary-forest habitats or nearby crops and rural communities on both Pacific and Caribbean slopes. Not always noticed because of its cryptic coloration. **Distribution:** Mexico to Brazil. **Seasonality:** Present all year but more common during rainy season. **Natural history:** Both early stages and adults of this species are well camouflaged, and adults probably spend their lives in treetops. That is why encounters with this widespread species, found almost everywhere in low- and mid-elevation locations, are rare. **Early stages:** Eggs are green, spherical, smooth, and are glued to bark of host plants. Mature larva is grayish-brown with variable variegated brownish lines all along its body; segment T_3 has a large black dorsolateral eyespot with a little blue on each side; head capsule is grayish; and the prolegs on segment A_{10} are well developed, triangular, and project backward. A short spike-like projection on A_8 is not always obvious, as the caterpillar keeps it against its body most of the time. Mature larvae rest among tree branches up in the canopy, behavior more typical of moths of Geometridae or Erebidae than of Sphingidae. Even more astonishing, the pupa is formed inside a silk cocoon attached to the host plant's trunk or branches, sometimes with dry leaves stuck to it. Such behavior is very rare in this family, in which most species pupate in the ground. *M. oiclus oiclus* spends its whole life cycle in the canopy, never descending to the ground. The pupa is cylindrical, brown, and has three yellow rings on A_4, A_5, and A_6. **Host plants:** *Crescentia alata*, *Tecoma*, *Handroanthus* (Bignoniaceae).

Fifth instar larva of *Madoryx oiclus oiclus*. CB

Pupa of *Madoryx oiclus oiclus*.

Madoryx oiclus oiclus
pinned. Dorsal view.

Adult *Madoryx oiclus oiclus*.

Aellopos titan titan (Sphingidae: Macroglossinae)
Titan Hummingbird Moth

0–3400 m

FWL: 26–31 mm. Description: DFW is dark brown with a longitudinal straight white line from costal to anal margin in medial area, and a post-medial line of semitranslucent patches from costal margin to vein Cu_1. DHW dark brown with a thin cream patch on costal margin and white tornus. VWs are brown, largely unpatterned. Body is brown with a conspicuous white transverse band crossing middle of abdomen. On each side of abdomen tip are modified mobile scales, which are wide and rounded, and are used as a rudder to help in flight maneuvers. **Similar species:** In *Aellopos fadus*, the post-medial line of semitranslucent patches on DFW becomes a double line from M_1 to M_3 and continuous all the way to tornus. In *A. clavipes*, the line does not reach costal margin or tornus, and DHW lacks white anal margin and tornus. **Habitat:** This is a widespread and fairly common species that can be seen in all habitats within its elevation range, although it requires the relatively pristine forest conditions necessary for its host plants. It can be seen in suburban areas of Central Valley in gardens and riversides on both Pacific and Caribbean slopes, but is more common in rural areas such as Osa Peninsula, Guanacaste coastline, Limón, San Carlos, and Upala. **Distribution:** USA to Argentina. **Seasonality:** Present all year on the Caribbean slope but only during rainy season on the northern Pacific slope. **Natural history:** This species flies exactly like a hummingbird, moving fast from one flower to another, buzzing and hovering in the air while feeding. Owing to this behavior and its large body, short and fast-beating wings, and the white abdominal band, this moth greatly resembles and is often confused with hummingbirds such

Fifth instar larva of *Aellopos titan titan*.

Pupa of *Aellopos titan titan*.

Adult *Aellopos titan titan*. JCn

Aellopos titan titan pinned. Dorsal view.

as *Lophornis helenae*, *L. adorabilis*, and *Discosura conversii*. Researchers have discussed whether this species actually mimics the hummingbirds; it is more common and widespread than they are, but according to mimicry theory, the model must be more commonly found than the mimic (suggestions that the hummingbirds copy the moth have not been well received). Escaping mimicry is another option to consider (Pinheiro 1996); in this case, both organisms are model and mimic simultaneously. The moth body is somewhat conical, and the enormous quantity of loose scales on its abdomen makes it hard to hold, either in the paw of a mammal or the beak of a bird. The effort to successfully catch and eat this moth is likely too much for most predators and not worth the effort; and many insect predators will not attempt to catch a hummingbird. Larvae are attacked by parasitoid flies (*Chetogena*, *Belvosia*, Tachinidae) and wasps (*Cryptophion*, Ichneumonidae). Both sexes feed from flowers such as *Stachytarpheta mutabilis*, *Lantana camara*, *Hamelia patens*, *Duranta erecta*, and many species of Asteraceae. They can be seen feeding at any time of day, but their activity peaks in early morning and late afternoon. **Early stages:** Eggs are green, somewhat ovoid, with smooth surface, and are laid on upper side of host plant's young leaves. Mature larva is variable in coloration, from light brown to green, with seven parallel, diagonal lateral white and brown lines; though the six anterior lines can be almost absent in some individuals, the most posterior white line is always present, beginning in segment A_6 proleg and continuing to base of large, curved black horn on A_8; head capsule is green. Pupa is cylindrical, dark brown, and is formed in the ground. **Host plants:** *Chomelia spinosa*, *Genipa americana*, *Randia aculeata*, *Alibertia edulis* (Rubiaceae).

Pachylia ficus (Sphingidae: Macroglossinae)
Fig Sphinx

0–2100 m

FWL: 47-67 mm. Description: DFW is greenish-brown with a small brown ring in medial area at distal end of discal cell, and a light brown inverted bowl-shaped figure in costal margin of apical area. DHW has a yellowish-brown basal area, followed by a black longitudinal band from costal to anal margin, then a yellowish-brown band, and finally a black distal margin, with a bright white spot in tornus. VWs are dull brown. **Similar species:** *Pachylia syces* has two inverted "bowls" on DFW, the second one in medial area of costal margin. **Habitat:** Present in most habitats on both Pacific and Caribbean slopes, *P. ficus* is a common moth that can be found perched on walls and fences. Encounters with this species are more common in early stages, as the huge caterpillars are associated with trees that are common in the country, and when they abandon the host tree to pupate, they may be seen on sidewalks or in backyards. **Distribution:** USA to Uruguay, Cocos Island. **Seasonality:** Present all year but more common during rainy season. **Natural history:** This is a huge and hardy moth that has adapted to living in urban areas as well as rural habitats all around the country. As its host plants are among the most common Costa Rican trees and also very diverse, this species has access to abundant oviposition places. It is the commonest sphinx moth on Cocos Island, far out in the Pacific Ocean. Orchids may be among the flowers this moth visits at night, and it is suspected to be a pollinator of the ghost orchid, *Dendrophylax lindenii.* **Early stages:** Eggs are light green, rounded, with smooth surface, and are laid on underside of host plant's leaves. Fourth-instar larva is gray with black markings that make it cryptic on bark or branches. Fifth-instar larva is green with a diagonal yellowish line on side of each body segment, and

Fifth instar larva of *Pachylia ficus.*

Pupa of P*achylia ficus.*

Adult *Pachylia ficus.*

Pachylia ficus pinned.
Dorsal view.

a longitudinal yellowish dorsolateral line enclosing a yellowish-green dorsal area; head capsule is smooth and green with a lateral yellowish line that continues from dorsolateral line on body. Pupa is cylindrical, reddish-brown, and is formed buried in the ground or among soil litter. **Host plants:** *Ficus elastica, F. benjamina, F. aurea, Maclura tinctoria* (Moraceae).

Xylophanes tersa (Sphingidae: Macroglossinae)
Tersa Sphinx

0–2500 m

FWL: 37 mm. **Description:** Robust orange body. DFW is mostly grayish-brown with faint orange marks in post-medial area close to costal margin. DHW is black with a post-medial longitudinal row of yellow spots parallel to distal margin, which is also yellow, as is the tornus. When resting, HWs are hidden by FWs. VWs are dull brown. **Similar species:** *Xylophanes titana* has more black/dark coloration but is best distinguished by two longitudinal yellow lines on dorsal side of thorax and abdomen that are absent in *X. tersa*. *X. crotonis* has a longitudinal line in medial area of DFW from apex to anal margin that separates a distal darker tone from a lighter proximal tone; *X. tersa* lacks the different tones, and its medial line is parallel to costal margin, while in *X. crotonis* the line angles to anal margin. **Habitat:** Common on both Pacific and Caribbean slopes in all habitats; can be seen at night on walls or porches or flying fast

Light and dark form of fifth instar larvae of *Xylophanes tersa*.

Pupa of *Xylophanes tersa*.

Xylophanes tersa pinned. Dorsal view.

Adult *Xylophanes tersa*.

around lights. **Distribution:** Canada to Argentina, Cocos Island. **Seasonality:** Present all year on the Caribbean slope and in the southern Pacific, but only during rainy season in the dry northern Pacific area. **Natural history:** The adult moth feeds from the nectar of *Cestrum nocturnum, Brugmansia,* and other species. This species has the ability to detect far-off sources of infrared radiation with its antennal spines and parts of its eye structure; it uses this ability to find host plants and mates, detecting other *X. tersa* by body temperature, which reaches 37–38 °C when flying (Callahan 1965). It is also found, and thought to have a stable population, on Cocos Island. **Early stages:** Eggs are green, spherical, with smooth surface, and are laid singly on underside of host plant's leaves. Mature larva is polychromatic, with some individuals brown and others light green. Abdominal segment A_1 is strongly swollen, and thoracic segments are reduced and concealed inside it; A_1 has two large, colorful dorsolateral eye-spots; slightly smaller dorsolateral circular spots follow from A_2 to A_7, framing a fine-checkered dorsal pattern; a conspicuous straight, black dorsal horn arises in segment A_8. Brown form has a visible whitish dorsolateral band running through the eyespots from A_8 horn to head capsule, as well as a thin black dorsal line. Pupation takes place buried in the ground inside a weak silk cocoon. Pupa is yellowish-brown with a longitudinal black line, black patches surrounding spiracles, and profuse dark brown markings on wing pads. **Host plants:** *Eumachia microdon, Hamelia patens, Psychotria exilis, Pentas lanceolata,* and others (Rubiaceae).

Hemeroplanes triptolemus (Sphingidae: Macroglossinae)

0–1500 m

FWL: 32–40 mm. Description: DFW has a complex pattern of many tones of gray and brown, a horizontal "femur-shaped" silver-white mark right in middle of wing, and a bright white anal margin in medial area. DHW is brown with gray basal area and a gray sub-marginal band that begins in tornus and becomes blurred as it approaches costal margin. FWs and HWs have very serrate, irregular margins. Body is very thick, and gray and brown with thin orange dorsal rings on some of the anterior abdominal segments, hidden when segments are contracted. **Similar species:** *Hemeroplanes ornatus* has a short "femur-shaped" white mark on DFW, while on *H. triptolemus* the mark is long. **Habitat:** Occurs on both Pacific and Caribbean slopes from primary forest to semi-disturbed habitats. It is a rare species, and adults are

Fifth instar larva of *Hemeroplanes triptolemus* in threatening posture.

Fifth instar larva of *Hemeroplanes triptolemus* in threatening posture.

Pupa of *Hemeroplanes triptolemus.*

Hemeroplanes triptolemus pinned. Dorsal view.

Adult *Hemeroplanes triptolemus.*

very seldom seen at lights, though larvae may be seen at riversides in forest edges at understory level. **Distribution:** Mexico to Brazil. **Seasonality:** Present during rainy season. **Natural history:** Even though it occurs in many habitats, this rare moth is always difficult to find. Not much is known about its adult biology, but it certainly feeds from flowers, probably at canopy level. **Early stages:** Young caterpillars are green and very cryptic. Fifth-instar larva is gray dorsally, the color of bark, and dark brownish-green ventrally, the same color as that of many snakes. It is cryptic when resting, resembling a thin dry stick, but if touched or disturbed, it suddenly twists its body and separates the anterior half from the perch at the same time it contracts the head inside the thoracic segments, which widen, exposing large black circles with white "pupils" inside, resembling snake eyes, on sides of segment T$_3$. The sides of the thorax are cream-colored with white lines, now forming the snake's mouth scales, and the swollen thorax acquires the typical triangular head shape. In this alarm position, the prolegs retract, becoming invisible, and the true caterpillar legs, now on the upper side of the false snake head, retract and become part of the cryptic pattern. This remarkable posture and the larva's color pattern make it perfectly resemble a snake; certainly any predator would think twice before approaching such a creature. This is a case of extreme mimicry, where the model belongs to another phylum. Pupation takes place in the ground inside a weak cocoon of leaf litter; pupa is brown, cylindrical, and smooth. **Host plants:** *Mesechites trifidus, Mandevilla subsagittata, M. hirsuta, Prestonia quinquangularis, Fischeria panamensis* (Apocynaceae), *Tassadia obovata* (Asclepiadaceae).

Urania fulgens fulgens (Uraniidae: Uraniinae)
Green Swallowtail Day-flying Moth

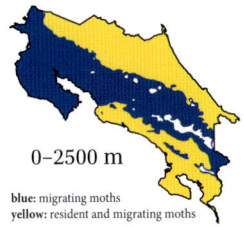

0–2500 m

blue: migrating moths
yellow: resident and migrating moths

FWL: 35–45 mm. Description: DFW is black with longitudinal parallel green bands on basal half. DHW is also black but with post-medial green markings in distal area, faint green lines in basal area, and a long, white-tipped tail (vein M_3). VWs similar to DWs but bluish instead of green. **Similar species:** None. **Habitat:** Resident individuals can be seen in forest habitats of the central and southern Pacific slope, such as Carara and Corcovado National Parks; and in the Caribbean slope in Tortuguero, Cahuita, and Braulio Carrillo National Parks, from sea level to 600 m. Migrating individuals may be seen in any part of the country and in any habitat. **Distribution:** USA to Venezuela, Colombia, and Ecuador. **Seasonality:** Present all year where it is resident; migrates during February–April and July–September. **Natural history:** Although this is a conspicuous, fascinating, and relatively common species, much is still unknown about it. Resident populations live inside primary and secondary forest; males set up territories 2–4 m above the ground, usually in semi-open areas where they perch on the upper side of leaves toward the tip, with the head pointing downward and the wings wide open. Female lays eggs at canopy level in clusters of up to 150 eggs. N. G. Smith (1983, 1992) proposed that extensive defoliation by the larvae stimulates the host plant to increase toxic compounds in its leaves, making them less suitable for the caterpillars and forcing them to migrate as adults to other host plant patches where the toxicity level is not as high. J. Cordoba-Alfaro (pers. comm.), however, based on field observations in Corcovado National Park, proposes that the migration is triggered by the absence of host plant leaves; five or more years of consecutive generations of the moth results in the gradual destruction of the source of leaves. Either hypothesis (or a combination of the two) ends in a mass migration, which varies in distance from 1000 to 2500 km (Dudley & DeVries 1990, N. G. Smith 1992). During migration, the moths can be seen flying straight ahead, 1–2 m above the ground in pastures or open areas, or high at canopy level in forested habitats. Their mean speed is 15 km/h (DeVries & Dudley 1990), and they spend some hours a day feeding from flowers such as *Acnistus arborescens*, *Terminalia catappa*, *Lantana camara*, *Stachytarpheta*, *Inga*, *Comaclinium montana*, *Havardia*, and *Croton*. The flower nectar provides these insects with energy to accomplish their migration, and the current decrease in flowers in fields, forests, and rural habitats may be one of the causes, together with deforestation and incidental killing by vehicles on highways, of the current decrease of the moth population. Massive migrations of millions of individuals occur in some years (in a mean of every 5.6 years, but varying from 3 to 9 years apart; Smith 1992). In Costa Rica, the moths migrate every year in a southeasterly direction in July–September and in northwesterly direction in February–April. The adults are unpalatable to predators, due to the sequestration of the host plant's toxins, but dragonflies prey on them (pers. obs.), and birds such as *Tachycineta albilinea* and *Quiscalus mexicanus* are important predators. J. Cordoba-Alfaro (pers. comm.) has observed spider monkeys

Fifth instar larva of *Urania fulgens fulgens*. NR

Pupa of *Urania fulgens fulgens*.

Adult *Urania fulgens fulgens.*

(*Ateles geoffroyi*) hunting mature larvae by shaking vines and collecting the hanging larvae, and notes that eating them seems to produce a psychotropic effect on the monkeys. Large aggregations of fresh, recently hatched males are commonly seen, apparently drinking from wet mud and sand during hot mornings (puddling); what they are searching for is uncertain. **Early stages:** Eggs are yellow and laid in large clusters on any part of host plant's leaves at canopy level. The larvae feed gregariously when young and disperse when mature. Fifth-instar larva is black with a fragmented yellow dorsal longitudinal band and another less conspicuous band at lateral position; few, thin medium-size setae occur along sides of body, but segments T_2, T_3, and A_7–A_{10} bear long, black, white-tipped spatulate setae in dorsolateral area; head capsule is red with black spots, from which black setae arise, in lateral and frontal areas. Cocoon is orange and netlike; pupa is cylindrical and orange with many black lines on wing pads and leg area, and many disperse black dots on abdomen. J. Cordoba-Alfaro (pers. comm.) reports that the mature larvae descend to the forest floor by "bungee jumping" with a silk string, and pupation takes place in ground litter; this process occurs synchronously as does the daytime emergence of adult moths from the pupae. **Host plant:** *Omphalea diandra* (Euphorbiaceae).

Himeromima aulis (Geometridae: Ennominae)

0–2600 m

FWL: 18–28 mm. **Description:** DFW is pale green with two faint, parallel lines of darker green from costal to anal margin in basal and medial areas, small sub-marginal grayish spots on veins R_5–M_1 and M_2–Cu_2, and many scattered small pale or white spots. DHW is mostly pale green with a weak continuation of the darker green lines from DFW. FW has an acute tip on apical area and a slight extension of vein M_3. HW has a small, peaked tail (vein M_3). **Similar species:** Superficially similar to many species; *Euclysia columbipennis* can be distinguished by its darker green parallel lines and grayish spots on

Fifth instar larva of *Himeromima aulis*.

Pupa of *Himeromima aulis*.

Himeromima aulis pinned. Dorsal view.

Adult *Himeromima aulis*.

DFW. **Habitat:** Found from primary to highly urbanized habitats on both Pacific and Caribbean slopes. In Central Valley, it is not unusual to see it at house and building lights. **Distribution:** Mexico to Colombia. **Seasonality:** Present all year. **Natural history:** This species is very common and can become a serious pest of its host tree, a common species, known as *uruca*, that is often grown as an ornamental in parks and streets in towns and neighborhoods. Adults feed from rotting fruits at night and are difficult to observe during daylight due to their effective crypsis, in contrast to the caterpillars, which possess a bright coloration that suggests they are unpalatable. **Early stages:** The caterpillars of this species are generally found in great numbers; when a tree is attacked it is likely to lose all its leaves to the voracious caterpillars. Larva is long, with a longitudinal black dorsal band and white sides with three or four thin black longitudinal lines and a thicker one bordering the yellowish-green ventral coloration; prolegs are in segments A_6 and A_{10}; head capsule is smooth and red. When the larvae are alarmed, they drop off the tree by hanging from a silk string; sometimes there are so many that the tree ends up covered in silk. To pupate, they drop off the tree in the same way and build a cocoon with leaf litter and soil. Pupa is cylindrical and reddish-brown. **Host plant:** *Trichilia havanensis* (Meliaceae).

Nematocampa completa (Geometridae: Ennominae)
Filament-bearing Caterpillar Moth

0–1300 m

FWL: 11–14 mm. Description: DFW has yellow basal and medial areas interrupted by many thin gray lines; post-medial area is brownish-gray, and apical area has some yellow blotches from costal margin to vein Cu_1. DHW has yellow basal area, but medial area to distal margin is grayish-brown, sometimes with small black dots in distal margin close to tornus. Ventral pattern is very similar to dorsal. **Similar species:** In *Nematocampa arenosa*, the DFW apical yellow patch follows distal margin in shape, while in *N. completa*, the distal border of the yellow patch gradually separates from distal margin. *N. reticulata* has basal area cream-colored instead of yellow. **Habitat:** Occurs in all habitats from primary forest to urban areas on both Pacific and Caribbean slopes. This small moth is attracted to streetlights and sometimes enters houses and buildings. In natural places its cryptic pattern makes it more difficult to see. **Distribution:** Mexico to Argentina. **Seasonality:** Present all year. **Natural history:** Not much is known about this species, but its wide distribution and capacity to persist in many habitats is probably due to its having many potential host plants. Adults feed from rotting fruits at night. **Early stages:** Larva is variegated brown, with orange on dorsal side of thorax; two dorsal white spots on segment A_2 and, just posterior to them, two long, retractile fleshy appendages; two shorter dorsal appendages on A_3; and a dark brown dorsal disk on A_8–A_9; head capsule is variegated brown. The four dorsal appendages look like tendrils (probably for camouflage) and can be as long as half of the abdomen length. Pupa is formed on the ground inside a few leaves held together by a few silk strings; pupa is orange and tear-shaped. **Host plants:** *Stachytarpheta mutabilis* (Verbenaceae), *Heterocondylus vitalbae* (Asteraceae), *Piper aduncum* (Piperaceae).

Fifth instar larva of *Nematocampa completa*.

Pupa of *Nematocampa completa*.

Adult *Nematocampa completa*.

Nematocampa completa pinned.
Dorsal view.

Oxydia vesulia (Geometridae: Ennominae)
Spurge Spanworm Moth

0–1300 m

FWL: 28–40 mm. **Description:** A brown moth with very variable wing patterns. DFW is brown with an oblique black post-medial line from apex to anal margin; remainder of wing is variegated with darker bands and spots, though in some cases may be almost solid brown. FW apex ends in an acute point. DHW is brown with a continuation of black post-medial line from costal to anal margin, but this line can be very faint or even absent; a dark brown to black blotch sits below costal margin just distal to beginning of black line; variable brown markings may be present or absent. **Similar species:** In *Oxydia bilinea*, black line on DFW begins in costal margin closer to discal cell, not in apical tip as it does in *O. vesulia*. *O. sociata* lacks a dark, wavy, diffuse, zigzag line in sub-marginal area of DHW, which is present in *O. vesulia* and becomes a black line just distal to the dark brown blotch. **Habitat:** Associated with all habitats on both Pacific and Caribbean slopes. Sometimes comes inside houses, mostly in rural areas with few artificial lights. **Distribution:** USA to Argentina. **Seasonality:** Present all year. **Natural history:** *O. vesulia* has been reported as a sporadic pest of avocado and citrus trees and a defoliator of eucalyptus. It would not be surprising if this species became a more recurrent pest of many different crops, since it seems to have the capacity to feed on many plant families. Adults feed from rotting fruits at night. Caterpillars are attacked by parasitoid wasps (Microgastrinae, Braconidae). **Early stages:** Eggs are green, rounded, and laid in lines on host plant's branches. Although they look like typical inchworms, larvae can become very large. They rest on the margin or petiole of leaves they are eating, and when resting are attached to the substrate by only the last abdominal segments, holding the body at an angle so that it resembles a wooden twig. The thorax segments become shrunken, with

Fifth instar larva of *Oxydia vesulia*.

Pupa of *Oxydia vesulia*.

Adult *Oxydia vesulia*.

Oxydia vesulia pinned. Dorsal view.

the head capsule flattened against them, resembling a small branch with its end chopped off. Larva is smooth and brown with yellow spiracles that may mimic fungi or lichens on the wood. Pupation takes place on the ground, with the cylindrical, brown pupa hidden in leaf litter but not wrapped in a cocoon. **Host plants:** *Solanum betaceum* (Solanaceae), *Eucalyptus* (Myrtaceae), *Persea americana* (Lauraceae), *Citrus* (Rutaceae).

Simena luctifera (Geometridae: Ennominae)

500–1600 m

FWL: 21–28 mm. **Description:** DFW is grayish-blue with a wide white band from medial area of costal margin to anal margin almost at tornus. DHW is grayish-blue and sometimes has a thin white distal margin. Head and prothorax are bright orange. **Similar species:** None. **Seasonality:** Present all year but more common during rainy season. **Habitat:** Primary and secondary forest as well as riparian forest on both Pacific and Caribbean slopes. It may be seen flying slowly in the understory at light gaps or forest edges. Localities include Rincón de la Vieja, Arenal Volcano, Braulio Carrillo, and La Cangreja National Parks.
It is common in Central Valley along riparian forest and in weedy abandoned terrain. **Distribution:** Mexico to Panama. **Natural history:** This is a day-flying moth; its fluttery and slow flight together with its bold coloration suggest that it is unpalatable to predators. When flying it resembles *Heliconius cydno* species (p. 146), so there is a possibility that this is a case of Batesian mimicry, at least in places where the two species are found together. However, the caterpillar is also bright-colored, so it may have chemical protection from its host plant. Adults rest on underside of leaves with wings wide open. **Early stages:** Mature larva is long and black with four longitudinal white lines along sides, and wide orange transverse rings on segments A_2–A_5; head capsule is smooth and black with irregular white lines. To pupate, larvae drop off the host plant on a silk string and inside ground litter form the long, smooth, reddish-brown pupa. **Host plant:** *Megaskepasma erythrochlamys* (Acanthaceae).

Fifth instar larva of *Simena luctifera*.

Pupa of *Simena luctifera*.

Adult *Simena luctifera*.

Simena luctifera pinned. Dorsal view.

Melanchroia chephise (Geometridae: Ennominae)
White-tipped Black

0–1400 m

FWL: 13–17 mm. **Description:** DFW is bluish-black with apical area white, and a faint white line on top of all veins. DHW is all bluish-black, also with faint white lines following venation. VWs are similar to DWs. Dorsal side of thorax is orange; ventral side of abdomen is orange on male and black on female. **Similar species:** *Melanis cephise* (Riodinidae) has a conspicuous red spot in basal area of each wing on dorsal side. **Habitat:** Gardens, fields, parks, and other open areas. **Distribution:** USA to Brazil, Caribbean islands. **Seasonality:** Present all year. **Natural history:** This small day-flying moth flutters close to the ground during the hottest hours of the day, stopping on plants to walk around and then taking flight again. Its slow flight and bold coloration are characteristics of an unpalatable species. Adults feed from nectar of *Ixora* and many species of Asteraceae. **Early stages:** Mature larva is black with yellow lateral and dorsal patches on every segment, and very few, disperse thin setae; head capsule is smooth and orange. Pupa is formed in the ground; it is dark brown and cylindrical. **Host plants:** *Phyllanthus anisolobus, P. niruri, P. amarus, P. acidus* (Phyllanthaceae).

Fifth instar larva of *Melanchroia chephise.*

Adult *Melanchroia chephise.*

Pupa of *Melanchroia chephise.* JMR

Melanchroia chephise pinned. Dorsal view.

Disphragisella baracoana (Notodontidae: Heterocampinae)

0–1500 m

FWL: 13–22 mm. **Description:** *Female:* DFW is brown with numerous irregular darker lines and stains, and a whitish lichenlike mark in post-medial area touching only a bit of costal margin and apex. DHW is grayish-brown except for a short, transverse brown-and-white band on costal margin and sub-margin close to apex. *Male:* DFW is grayish-brown with numerous, irregular darker lines and stains. DHW is similar to female's. **Similar species:** Female *Disphragis edwardsi* has a roughly C-shaped whitish patch with a black shadow on DFW, while in *Disphragisella baracoana* the patch is more rounded and lacks black shadow. Male *D. baracoana* is distinguished by a sub-marginal brown zigzag band on DFW from costal margin to vein Cu_1, which is absent in *D. edwardsi.* **Habitat:** Disturbed habitats such as gardens and parks in suburban ecosystems; also in secondary forest or riverine forest. Present on both Pacific

Fifth instar larva of *Disphragisella baracoana.*

Pupa of *Disphragisella baracoana.*

Adult *Disphragisella baracoana.*

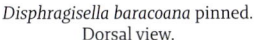

Disphragisella baracoana pinned.
Dorsal view.

Disphragisella baracoana pinned.
Dorsal view.

and Caribbean slopes. Although this moth is not rare, it is difficult to see as adult owing to its effective crypsis. Adults occasionally enter houses and buildings in rural areas such as Guanacaste, Limón, and Garabito. **Distribution:** Costa Rica, Panama, Cuba. **Seasonality:** Present only during rainy season. **Natural history:** Adults feed from rotting fruits and decaying matter. They are good fliers and skilled at escaping predators such as bats and night birds. Caterpillars are attacked by parasitoid flies (*Lespesia*, Tachinidae). **Early stages:** Eggs are green and laid in any part of host plant's leaves. Caterpillars are green and smooth; early instars have a pair of long, black dorsolateral spines on all abdominal segments except A_7 and A_{10}, and a forked pair on thoracic segment T_1. Mature larva loses all the black spines; it is green with a yellowish-white dorsal band that becomes forked in A_7, and has purplish dorsal rhomboid blotches, variable in intensity, on A_1–A_5 and A_8; head capsule is smooth, green, and has longitudinal black lines from top of head to mouthparts with some brownish markings in inner area. Pupa is formed in ground litter; it is cylindrical, shiny black, and smooth. **Host plants:** *Sapium glandulosum*, *Hura crepitans*, and others (Euphorbiaceae).

Dicentria rustica (Notodontidae: Heterocampinae)

0–1500 m

FWL: 15–21 mm. **Description:** DFW is grayish-brown with a single small brown bar surrounded by light gray distinguishable at distal end of discal cell. DHW is pale gray in basal half and a little darker in distal half, with a small semi-eyespot at tornus. **Similar species:** *Dicentria vallima* and *D. missilis* lack the small semi-eyespot on DHW. **Habitat:** Urban areas, secondary forest, and riverine forest patches as well as parks and gardens in towns. Present on both Pacific and Caribbean slopes. This species is rarely seen as adult unless it enters a house and flies vigorously around lights. It is a sporadic visitor of backyards around the Central Valley. **Distribution:** Mexico to Panama. **Seasonality:** Present all year but more common during rainy season. **Natural history:** This species is poorly known, and the adult is rarely seen in nature, probably because of its crypsis. It is easier to find the caterpillars eating on host plants. Adults are fruit feeders but will take advantage of any juicy decaying matter they find. **Early stages:** The caterpillars of this species are a great example of disruptive coloration accompanied by an unexpected body shape. Mature larva is green with a large, fleshy, two-forked dorsal protuberance on segment A_1, followed by two smaller dorsolateral projections on A_5, and two similar but smaller ones on A_8; a few long black setae are arranged in a transverse ring on each body segment; thorax is green in lateral areas and has a brown dorsal band from head to A_1 appendage. Abdomen from A_1 to A_8 has a brown and cream reticulate pattern on sides, and dorsal area is cream from A_1 to A_4, and brown from A_4 to A_{10}, except for A_6 and A_7, which have two dorsal cream spots inside the brown band. Head capsule is light green with two longitudinal thin black lines on frontal area on each side. **Host plants:** *Rosa* (Rosaceae).

Fifth instar larva of *Dicentria rustica*.

Pupa of *Dicentria rustica*.

Adult *Dicentria rustica*.

Dicentria rustica pinned.
Dorsal view.

Didugua argentilinea (Notodontidae: Nystaleinae)

0–1700 m

FWL: 14–18 mm. **Description:** DFW is brown with thin darker lines following the veins, and two elongate silverish-white blotches of similar shape, one in medial area and one in post-medial area, both along sub-costal margin; in distal sub-marginal area are two transverse parallel dark lines. DHW is dull brown. **Similar species:** *Didugua asymetrica* has a series of distal sub-marginal parallel lines on DFW in the following order, from basal to distal: a thin light line, a thin dark line, a thick light line, and a thin dark line. *D. argentilinea* has: a thin light line, a thin dark line, a thick light line, a thin dark line, a thin light line, and a thick dark line. *D. leona* has only one slightly curved, dark brown line in distal sub-margin, from vein M_3 to Cu_2. In *D. beckeri* (facing page) the distal sub-marginal line is wavy and broken, and distal margin of DFW is dark. **Habitat:** Secondary and primary forest on both Pacific and Caribbean slopes. It is more often found in larval stage since adults are very cryptic, but adults attracted by lights occasionally enter houses and buildings. Localities include Rincón de la Vieja, Monteverde, and Tapantí, La Cangreja, and Corcovado National Parks. **Distribution:** USA to Panama. **Seasonality:** Present all year but more common during rainy season. **Natural history:** Not much is known about this species; it is a powerful flier and feeds from rotting fruits at night. Larvae are attacked by parasitoid flies (*Lespesia*, *Hyphantrophaga virilis*, Tachinidae). **Early stages:** Caterpillars are gregarious, resting in groups of

Fifth instar larva of *Didugua argentilinea*.

Pupa of *Didugua argentilinea*.

Adult *Didugua argentilinea*.

Didugua argentilinea pinned.
Dorsal view.

up to fifty individuals on underside of leaves when young and on stalks when mature. Larva is shiny black with disperse short, rigid setae; each body segment has one yellow and a few white transverse rings; segment T_1 is red in anterior half; A_8 has a small dorsal red hump; A_{10} is red; and true legs are red; head capsule is shiny bright red with few setae. Pupa is formed inside a silk cocoon, which can be constructed in any secure place; pupa is smooth, cylindrical, and reddish-brown. **Host plant:** *Serjania atrolineata* (Sapindaceae).

Didugua beckeri (Notodontidae: Nystaleinae)

400–1800 m

FWL: 15-19 mm. **Description:** DFW is brown with two elongate silver blotches in sub-costal region and some transverse dark brown lines in medial area. DHW is brown. **Similar species:** *Didugua asymetrica*, *D. leona*, and *D. argentilinea*; for details, see *D. argentilinea* (opposite page). **Habitat:** Secondary and primary forest habitats along a mid-elevation mountain belt on both Pacific and Caribbean slopes. Localities include Monteverde, San Vito, and Rincón de la Vieja, Braulio Carrillo, and Tapantí National Parks. **Distribution:** Mexico and Costa Rica. **Seasonality:** Present all year. **Natural history:** Since this is a recently discovered species, not much is known about its biology. The specimen illustrated in this book was collected as an egg by entomologist Laura Longoria in Tequecholapa reserve in Córdoba, Mexico; it

Fifth instar larva of *Didugua beckeri*.

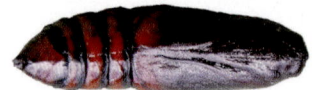

Pupa of *Didugua beckeri*.

Didugua beckeri pinned. Dorsal view.

Adult *Didugua beckeri*.

represents its first report for Mexico, and no other specimen is known between Mexico and Costa Rica at this time. **Early stages:** Mature larva is black and has two large lateral yellow patches per segment, a longitudinal dorsal band that is white with black borders, followed by a yellow longitudinal band, and a black dorsolateral area with irregular yellow and white lines and reddish patches; segment A_8 has a red-tipped dorsal hump; legs and prolegs are reddish; head capsule is orange, smooth, and shiny. Pupation occurs inside a silk cocoon; pupa is cylindrical, brown, and smooth. **Host plant:** *Serjania atrolineata* (Sapindaceae).

Xylodonta terrena (Notodontidae: Nystaleinae)

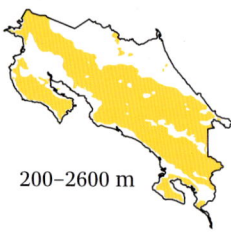

200–2600 m

FWL: 19–23 mm. **Description:** A gray moth with a complex pattern of zigzag lines and different tones on DFW. In DHW, pattern is reduced to dull gray with whitish distal margins. **Similar species:** *Xylodonta markvanputteni, X. patrickgoodwilliei, X. andrewrusselli, X. xylinata, X. robertodelgadoi, X. billhaberi* and *X. guarana.* A revision of the Costa Rican status of this species was conducted by Chacón et al. (2017). Based on genitalia and molecular evidence, they recognized six new cryptic species in addition to two very similar existing ones, which were then redescribed. For a definitive identification of *X. terrena*, it is necessary to analyze genitalia and perhaps molecular evidence. However, there are some superficial nondefinitive characters to look for. In Costa Rican species, females of *X. patrickgoodwilliei, X. andrewrusselli, X. billhaberi, X. guarana*, and *X. terrena* have a lighter gray semicircular patch in basal area of DFW; it is absent in males. Among females, *X. patrickgoodwilliei* and *X. andrewrusselli* are darker gray than *X. terrena*; *X. billhaberi* has a curved, transverse post-medial line caused by a change of tone much more marked than in *X. terrena* (in which the line is faint and change of tone slight), and *X. guarana* has that same line but does not have different tones. Among males, *X. markvanputteni*, *X. patrickgoodwilliei*, *X. andrewrusselli*, and *X. billhaberi* have larger light brown patches and bands on DFW than *X. terrena*. *X. robertodelgadoi* does not have a straight black line in basal area of DFW; *X. terrena* does. *X. xylinata* can be distinguished only by genitalia and molecular analysis. **Habitat:** Primary and secondary forest on both Pacific and Caribbean slopes. It usually approaches lights of houses and buildings; otherwise, it is hard to find due to its crypsis. Localities include Rincón de la Vieja, Braulio Carrillo, Irazú Volcano, and Turrialba Volcano National Parks. **Distribution:** Mexico to Costa Rica. **Seasonality:** Present all year. **Natural history:** Not much is known about this species' habits or behavior. Caterpillars are attacked by parasitoid flies (*Winthemia*, Tachinidae). **Early stages:** Mature larva is smooth and shiny, with various longitudinal bands: the light yellow dorsal band is followed by some thin green dorsolateral lines, then a lateral white line with black markings, followed by another yellow band; ventral color is green; segment A_8 has a short, fleshy dorsal protuberance with a purple patch at the base; head capsule is light green, and taller than segment T_1. The caterpillar lives alone, inside a shelter built by putting two leaves together; a couple of days before pupation, its body and head turn grayish-purple, and only small fragments of the lateral lines persist. Pupa forms inside larval shelter; pupa is black, cylindrical, and shiny. **Host plants:** *Erythrina lanceolata*, *Lonchocarpus oliganthus*, *Pterocarpus officinalis* (Fabaceae).

Fifth instar larva of *Xylodonta terrena*.

Pupa of *Xylodonta terrena*.

Xylodonta terrena pinned. Dorsal view.

Adult *Xylodonta terrena*.

Josia frigida (Notodontidae: Dioptinae)
Frigida False Tiger Moth

0–1600 m

FWL: 13–19 mm. **Description:** DFW is black with a yellow transverse band from basal area nearly to (but not touching) distal margin, and a bluish iridescence on distal margins. DHW is also black with a yellow transverse band from basal area nearly to distal margin, and a yellow anal margin all the way to tornus. **Similar species:** *Lyces cruciata* lacks yellow anal margin extending to tornus of DHW, which is present in *J. frigida*. *J. gigantea* is distinguishable first by its larger size (FWL: >20 mm). And *J. frigida* is distinguished from all other Dioptinae species by vein M_1 of FW, which arises distally from fork of R_2; in other species, M_1 separates at same point as radial veins, at distal end of discal cell. *Josiomorpha penetrata* (Erebidae) is distinguishable by yellow band on DHW, which runs right along costal margin, not in center of wing as in *J. frigida*. **Habitat:** All habitats from rainforest to croplands on both Pacific and Caribbean slopes and around towns in Central Valley. Flies fast in open areas such as gardens and along streets in the morning. **Distribution:** Mexico to Panama. **Seasonality:** Present all year. **Natural history:** The bright coloration of this species is related to the caterpillar's capacity to sequester toxic secondary compounds from its host plant, making both larva and adults unpalatable to predators. In consequence, various species of the same tribe (Josiini) have developed a similar pattern, both in coloration and behavior, producing a Müllerian mimetic ring including *J. gigantea* and *Lyces cruciata*. *Mesenopsis melanochlora*

Fifth instar larva of *Josia frigida*.

Pupa of *Josia frigida*.

Josia frigida pinned. Dorsal view.

Adult *Josia frigida*.

(Rodinidae) has very similar pattern and behavior, but it is not known if this represents Müllerian or Batesian mimicry. Adults of *J. frigida* fly in a straight line with fluttery wingbeats; they feed from flower nectar and decaying organic matter. **Early stages:** Fifth-instar larva is black and yellow striped; segments T$_1$ and A$_9$ are white, and segments T$_2$, A$_1$, A$_3$, A$_5$, and A$_8$ have thicker black transverse bands, the abdominal bands with a white mark inside the band; head capsule is shiny, black, and smooth. Pupa is long, cylindrical, and reddish with transverse black lines in abdominal area and black lines along legs, antennae, proboscis, and wing-pad edges. **Host plant:** *Passiflora capsularis* (Passifloraceae).

Getta tica (Notodontidae: Dioptinae)

500–1400 m

FWL: 19–23 mm. **Description:** DFW is black with a transverse yellowish-orange band from costal margin of medial area to tornus; basal half of wing has a bluish iridescence. DHW is dark metallic bluish. Female and male are very similar: female has more light blue iridescence, and male has a convex costal margin on HW. **Similar species:** *Erbessa lindigii* and *E. albilinea* have orange in costal margin and apex of DHW and VHW, which *G. tica* never has. **Habitat:** Pristine, very humid forest habitats; localities include Braulio Carrillo National Park and rivers on the Caribbean slope of La Amistad International Park and Tilarán mountain range. **Distribution:** Costa Rica and Panama. **Seasonality:** Present all year. **Natural history:** This day-flying moth is probably unpalatable to predators, given its bright colors and host plant

Fifth instar larva of *Getta tica*.

Pupa of *Getta tica*.

♀

Adult *Getta tica*.

Getta tica pinned.
Dorsal view.

♂

♀

Getta tica pinned.
Dorsal view.

biochemistry. It seems to be very sensitive to habitat destruction, as its host plant is highly specialized for life on very humid cliffs on sides of rocky rivers. Not much is known about the biology of this moth, probably because of its very recent discovery and its restricted distribution. **Early stages:** The bright-colored caterpillars rest on the edges of the host plant's leaves right where they have been eating. Mature larva is smooth, shiny, and yellow, black, and white; segments T_1 and A_{10} are mostly white, A_8 is black with a white blotch in dorsal area, and remaining segments are yellow with irregular black markings mainly in dorsal area; head capsule is smooth and shiny, with upper half black and mouth area white. Pupa is formed buried in forest-floor leaf litter; it is cylindrical, orange, and has transverse black bands on abdomen and on edges of wing pads, legs, antennae, and proboscis. **Host plant:** *Passiflora tica* (Passifloraceae).

Spodoptera dolichos (Noctuidae: Noctuinae)
Sweet-potato Armyworm Moth

0–2500 m

FWL: 12–22 mm. **Description:** DFW has a complex pattern of gray, brown, and black lines. DHW is whitish-silver with a thin brown line sub-marginal to distal margin. **Similar species:** Distinguished from *Spodoptera cosmioides* by white curved line on DFW at sub-marginal position of distal margin: in *S. dolichos*, the line is heavily curved, begins at vein M_2, and breaks in Cu_2; in *S. cosmioides*, the line is less curved and runs to vein 2A without breaking. **Habitat:** Associated with all habitats from primary forest to croplands and urban areas on both Pacific and Caribbean slopes. Can be found inside houses or at porch lights. Localities include Rincón de la Vieja, Upala, San Carlos, Alajuela, and Central Valley. **Distribution:** USA to Argentina, Caribbean islands. **Seasonality:** Present all year. **Natural history:** Adults feed on rotting fruits and tree sap. This species is considered an important pest of several crops. Its impact as a pest is poorly known, but it has been recorded feeding on more than ninety different host plants of several families. Eggs are attacked by a parasitoid wasp (*Telenomus remus*, Platygastridae). **Early stages:** Eggs are pale green, spherical, with smooth surface, and are laid on underside of leaves in masses of up to 100. Caterpillars are gregarious until third instar, when they disperse and become solitary. Mature larva is grayish-brown with two longitudinal thin, brown dorsolateral lines with white marks; and black marks that are square in T_2 and become triangles pointing toward dorsal area as they approach segment A_8; head capsule is brown and smooth. Pupa is formed in a capsule made with soil on the surface on the ground; pupa is cylindrical, smooth, and reddish-brown. **Host plants:** *Phaseolus vulgaris*, *Senna* (Fabaceae), *Solanum tuberosum* (Solanaceae), *Ipomoea batatas* (Convolvulaceae), *Gossypium hirsutum* (Malvaceae).

Fifth instar larva of *Spodoptera dolichos*.

Pupa of *Spodoptera dolichos*.

Spodoptera dolichos pinned.
Dorsal view.

Adult male *Spodoptera dolichos*.

Spodoptera latifascia (Noctuidae: Noctuinae)
Velvet Armyworm Moth

0–1400 m

FWL: 17–22 mm. Description: *Female:* DFW pattern is a mix of gray, brown, and whitish lines; DHW is translucent white. *Male:* DFW has a complex pattern of brown, black, and gray lines; DHW is silver-white with a brown sub-marginal line. **Similar species:** *Spodoptera ornithogalli* has wide brown distal fringes with very thin, equidistant white lines on DFW; *S. latifascia* has a thin brown margin and thicker white lines. Also, in sub-marginal area of distal margin of DFW, *S. ornithogalli* has a thin, curved white line from vein M_2 to Cu_2 that follows a black patch; this character is absent in *S. latifascia*. Female *S. latifascia* is distinguishable from *S. dolichos* (opposite page) by a sub-marginal whitish line on distal margin of DFW, which runs uninterrupted from Cu_2 to R_5 in *S. latifascia*, but in *S. dolichos* goes from Cu_2 to M_2, where it becomes a large grayish blotch. **Habitat:** Disturbed habitats including gardens, backyards, and croplands, in most areas of the country. Common in Central Valley, Guanacaste, Pérez Zeledón, and San Carlos. **Distribution:** Canada to Argentina, Caribbean islands. **Seasonality:** Present all year. **Natural history:** Common and widespread, this species is an important pest of many crops. It can fly long distances, is attracted by lights, and is a common intruder of houses and buildings. Adults feed from rotting fruits and tree sap. Eggs are attacked by a micro-wasp (*Trichogramma fasciatum*,

Fifth instar larva of *Spodoptera latifascia*.

Pupa of *Spodoptera latifascia*.

Adult male *Spodoptera latifascia*.

Spodoptera latifascia
pinned. Dorsal view.

Spodoptera latifascia
pinned. Dorsal view.

Trichogrammatidae); larvae are attacked by parasitoid wasps (*Chelonus antillarum*, Braconidae; *Euplectrus plathypenae*, Eulophidae) and flies (*Winthemia*, *Archytas analis*, *A. piliventris*, Tachinidae). **Early stages:** Eggs are pale green with smooth surface, and are laid on underside of host plant's leaves in masses of up to 150 eggs, covered by female's abdomen scales. Caterpillars are gregarious when young but become solitary when mature. Fourth-instar larva has a longitudinal dark brown dorsolateral band, which is lost in fifth instar. Mature larva is grayish-brown with a longitudinal dorsolateral line that is brown with small white and orange dots; as the line approaches segment A$_8$, dark brown marks appear and increase size in each segment, becoming triangular; head capsule is smooth and brown. To pupate, the larva makes a cell semi-buried in the ground; pupa is bright brown and cylindrical. **Host plants:** *Solanum lycopersicum*, *Capsicum annuum* (Solanaceae), *Phaseolus vulgaris* (Fabaceae), *Zea mays* (Poaceae), and many others.

Diphthera festiva (Nolidae: Diphtherinae)
Hieroglyphic Moth

0–1600 m

FWL: 16–23 mm. **Description:** DFW is lemon-yellow with metallic-blue, white, and gray curved lines on basal and medial areas; post-medial area has three parallel longitudinal rows of sub-marginal dark gray spots. DHW is dark gray with thin yellowish distal margin. **Similar species:** None. **Habitat:** All habitats on both Pacific and Caribbean slopes. Though rarely seen, it is locally common. Sometimes found perched on walls and sidewalks in areas such as Liberia, Playa del Coco, Grecia, Turrialba, and Osa Peninsula. **Distribution:** USA to Brazil. **Seasonality:** Present all year but more abundant during rainy season. **Natural history:** The bright coloration of this species draws attention despite its nocturnal habits. The coloring is likely a warning sign to predators that the moth is distasteful; Becker & Miller (2002) observed birds catching the caterpillars and immediately rejecting them. Adults feed from flower nectar and extrafloral secretions. Eggs are always laid in large clusters on many plant species; larvae are thus sporadically found in very high numbers and are considered a pest of some garden plants and crops. **Early stages:** Eggs are pale green, spherical, with smooth surface, and are laid on underside of leaves in clusters of up to 150. Young larvae feed gregariously by scraping the leaves. Mature larva becomes solitary; it is transversely striped in white and black, and head capsule is bright orange. Pupa is brown and cylindrical; it is formed inside a cocoon built with pieces of host plant's leaves. **Host plants:** *Waltheria indica, W. glomerata, Helicteres guazumifolia* (Malvaceae).

Fifth instar larva of *Diphthera festiva.*

Pupa of *Diphthera festiva.*

Adult *Diphthera festiva.*

Dorsal view of pinned
Diphthera festiva.

Cecharismena anartoides (Erebidae: Phytometrinae)

0–1400 m

FWL: 12–14 mm. **Description:** DFW is brown with some faint black lines and dots. Distal margin of FW is elongated at M₃ vein. DHW is light yellow with brown distal margin. **Similar species:** The yellow DHW distinguishes *Cecharismena anartoides* from similar species such as *C. jalapena*. **Habitat:** All habitats from primary forest to abandoned fields on both Pacific and Caribbean slopes. Generally seen as sporadic visitors of house lights. **Distribution:** USA to Venezuela. **Seasonality:** Present all year. **Natural history:** Adults are active and strong fliers that generally stay close to the ground. They feed on rotting fruits and tree sap. **Early stages:** Mature larva is pale green with a single, longitudinal dark green and white lateral line; body is covered in disperse thin, almost transparent setae; prolegs are present only on segments A₅, A₆, and A₁₀; head capsule is also pale green. Larva lives solitarily inside a roll shelter made from leaf edge. Pupa is formed inside larval shelter; it is brown, elongate, and cylindrical. **Host plant:** *Tragia volubilis* (Euphorbiaceae).

Dorsal view of pinned *Cecharismena anartoides.*

Fifth instar larva of *Cecharismena anartoides.*

Pupa of *Cecharismena anartoides.*

Adult *Cecharismena anartoides.*

Gonodonta nitidimacula (Erebidae: Ophiderinae)
Nitidimacula Fruit-piercing Moth

200–1600 m

FWL: 17–19 mm. Description: A robust moth with FW tornus excavated at anal margin, forming an acute distal peak. DFW is brown with a wavy and irregular pattern of orange-brown bands and lines of different tones. DHW is brown with a rounded yellowish patch in medial area from costal margin to vein Cu$_1$. VFW is brown with orange costal margin; VHW is brown with orange basal to medial area. **Similar species:** Can be distinguished from *Gonodonta fulvangula* by color pattern of DFW: *G. nitidimacula* has a dark brown line in sub-marginal area that is wavy, thin, does not run parallel to distal margin, and generates triangular peaks pointing to basal area; in *G. fulvangula*, that line is not clearly present, and instead a straight band of different tones runs parallel to distal margin. **Habitat:** Secondary forest and disturbed rural areas associated with regenerating forest patches, on both Pacific and Caribbean slopes. Rare individuals are seen perched on walls or floors in rural and urban areas or on ground litter in forest habitats. Localities include Liberia, Ciudad Colón, San Isidro de Heredia, and Upala. **Distribution:** USA to Ecuador, Caribbean islands. **Seasonality:** Present all year. **Natural history:** This moth is a strong flier, and it is possible that it performs nighttime migrations from one slope to the other across the central mountain ranges (Miller et al. 2007). Adults are an important pest on many fruit plantations, feeding on ripe fruits still on plants by piercing the skin with the proboscis, as well as rotting fruits. Corro (2018) reports *Gonodonta incurva* is an important flower pollinator,

Fifth instar larva of *Gonodonta nitidimacula*.

Pupa of *Gonodonta nitidimacula*.

Adult *Gonodonta nitidimacula*.

Gonodonta nitidimacula pinned. Dorsal view.

especially of orchids, which raises the possibility that *G. nitidimacula* also exploits that resource. **Early stages:** As with all *Gonodonta* species, the mature caterpillar rests in a particular twisted position. Prolegs are in segments A_4–A_6 and A_{10}. Mature larva is grayish with profuse small, thin black lines all over the body, a white transverse band on dorsal side of segment A_2, and one black dorsolateral blotch each on A_3 and A_4; dorsal side of T_1 is black in anterior half with a small dorsolateral white line; a dorsal hump is present on A_8; head capsule is black and smooth. As with other *Gonodonta* species, the caterpillar seems to be polychromatic and very variable in coloration and pattern; it is not known whether this is attributable to intraspecific variation, geographic forms, or larvae of different species with identical adults. Pupa is formed inside a cocoon built with leaf pieces. Pupa is shiny, brown, and cylindrical. **Host plants:** *Piper* (Piperaceae).

Sarsina purpurascens (Erebidae: Limantriinae)

0–1400 m

FWL: 12–20 mm. **Description:** A small, robust moth. DFW has a veined pattern of darker and lighter tones with a purplish gloss; it has three longitudinal dark brown lines from costal to anal margin: the basal one is curved, the medial one is slightly curved, and the distal one is wavy. DHW is solid brownish-gray. **Similar species:** *Sarsina festiva* has a black spot in medial area of DFW at distal end of discal cell, which is absent in *S. purpurascens*; also, DHW is yellowish in *S. festiva* and brownish-gray in *S. purpurascens*. *S. electa* differs in DFW color, which is greenish, not purplish as in *S. purpurascens*. **Habitat:** Most habitats, from forest to suburban areas, on both Pacific and Caribbean slopes. Generally seen perched on house walls or in porch lights. **Distribution:** USA to Brazil. **Seasonality:** Present all year but more common

Fifth instar larva of *Sarsina purpurascens*.

Pupa of *Sarsina purpurascens*.

♀

Sarsina purpurascens pinned.
Dorsal view.

Adult *Sarsina purpurascens*.

during rainy season. **Natural history:** This species is considered a pest of plantations of the guava fruits guayaba (*Psidium guajava*) and cas (*P. friedrichsthalianum*), as the caterpillars damage the plants' leaves. Its capacity to adapt to most habitats is probably due to the abundance of its host plants in disturbed places. Caterpillars are attacked by parasitoid flies (*Leschenaultia*, Tachinidae) and wasps (*Glyptapanteles*, Braconidae). **Early stages:** Eggs are white, rounded, with smooth surface, and are laid in large clusters of up to sixty or seventy eggs. The caterpillars are white in early stages and very hairy in all stages. Mature larva is colored in different tones of brown, with a longitudinal brown double line running along all segments; segment T_3 has a black band in dorsal position, and A_4 and A_7 have brown dorsal markings; each body segment is armed with many warts bearing long setae that may cause skin irritation if touched. Head capsule is variable, from green to brown or veined. Larvae rest on underside of leaves or in branches with the body flattened against the substrate, where they are very well camouflaged. Pupa is suspended in a thin silk web with the head facing upward; except for the last three segments, which are brown, the pupal abdomen is grayish, as is dorsal side of thorax; abdominal segments have light tufts of setae on sides; wing pads and dorsal part of head are dark brown; ventral side of thorax, leg pads, and head are light gray and rounded and flattened ventrally. **Host plants:** *Psidium guajava*, *Psidium friedrichsthalianum*, *Eugenia acapulcensis* (Myrtaceae).

Ardonea morio (Erebidae: Arctiinae)

400–1800 m

FWL: 15-20 mm. **Description:** A small, dark blue moth with a black head, thorax, and abdomen. **Similar species:** *Opharus procroides* is very similar but has red lateral markings on abdomen while *A. morio* does not. *Inopsis scylla* has an orange head, and *A. morio* does not. *Gardinia magnifica* is much larger, with a FWL of 25 mm in the smallest individuals. *Cloesia digna* is smaller, with a FWL of 15 mm in the largest specimens, and its wings are thinner and more elongate. **Habitat:** Disturbed habitats on both Pacific and Caribbean slopes. Usually seen as rare individuals hidden in low bush vegetation. **Distribution:** Mexico to Venezuela. **Seasonality:** Present all year. **Natural history:** Not much is known about this moth, which may be seen flying during the day in open areas and secondary forest (pers. obs.), and also along roads and gardens in suburban areas of Central Valley. **Early stages:** Mature larva is black, covered in long setae distributed in five tufts on each body segment; each segment also has small dorsal and dorsolateral orange patches. Caterpillars feed and rest in groups of about 20 individuals, then disperse in smaller groups to pupate. The cocoon is made of grayish silk but has a thin layer of black setae on top. **Host plants:** Liverworts (hepatics), fungi, lichens, and algae on tree bark.

Fifth instar larva of *Ardonea morio.*

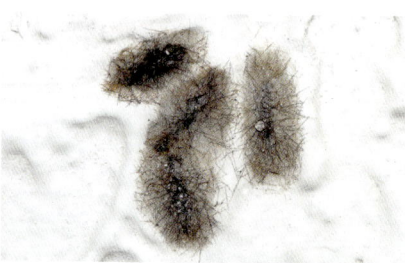

Cocoon of *Ardonea morio.*

Adult *Ardonea morio.*

Ardonea morio pinned.
Dorsal view.

Eudesmia menea (Erebidae: Arctiinae)
Lunar Eudesmia Moth

0–1400 m

FWL: 12–22 mm. **Description:** A small moth with an orange head, thorax, and abdomen. DFW is black with a longitudinal orange band in medial area from costal to anal margin, and a sub-marginal orange band that curves to follow distal margin shape. DHW has orange basal area and black distal half. **Similar species:** None. **Habitat:** Occurs on Pacific slope in secondary and primary forest, where it flies at low altitude in forest understory. **Distribution:** USA to Brazil. **Seasonality:** Present all year but more common during rainy season. **Natural history:** This diurnal moth is usually seen fluttering slowly among herbaceous plants during the hot hours of the day. Its bright colors and slow flight suggest that at some point in its development, it probably has the ability to produce secondary compounds that make it unpalatable to predators. If that is so, it does not obtain these toxins from its larval host plant, as many other lepidopterans do, since the caterpillars of this species feed from liverworts (hepatics), lichens, and algae associated with fungi that grow on the bark of trees (Chacón & Montero 2007). There are reports of larvae of this species feeding from leaves of the species noted as host plants below (Sermeño-Chicas & Parada-Berríos 2013, Sermeño-Chicas et al. 2014); further study is needed to find out whether these are exceptional cases. The caterpillars have been found on many unrelated species of trees. **Early stages:** Mature larva is black with few whitish markings and entirely covered in not-too-dense, long, soft grayish setae. The cocoon in which the pupa forms is a rounded capsule made of the larval setae. Pupa is rounded and yellowish with conspicuous black dorsolateral dots

Fifth instar larva of *Eudesmia menea.*

Pupa of *Eudesmia menea.*

Adult *Eudesmia menea.*

Eudesmia menea pinned.
Dorsal view.

on abdominal segments and a few more on thorax. **Host plants:** *E. menea* belongs to a group of moths whose larvae typically feed on liverworts, algae, and very small lichens that grow on bark, regardless of the tree species. Some reports list *Pouteria sapota* (Sapotaceae) and *Brosium alicastrum* (Moraceae) as larval host plants.

Euchaetes mitis (Erebidae: Arctiinae)
Milkweed Tussock Moth

0–1400 m

FWL: 15–20 mm. **Description:** A small whitish moth with orange abdomen with a small black dot on dorsal area of each segment. DFW is pearly grayish-white; DHW is white. **Similar species:** *Euchaetes expressa* has a darker gray DFW. **Habitat:** Open areas, croplands, and backyards around Central Valley, Tilarán, and Guanacaste. Present on Pacific slope. **Distribution:** USA to Costa Rica. **Seasonality:** Present all year. **Natural history:** This moth is not often seen as an adult since it does not commonly fly to artificial light. It is easier to see in daytime resting on low vegetation in areas with abundant milkweed plants.

Early stages: The caterpillars are highly gregarious; they feed together from the same leaves of the host plant when young and from the plant tip and new shoots when mature. Larva is covered in a dense layer of soft black setae; segments A_1 and A_7 also bear four long white setae tufts, and A_2–A_7 have short yellow dorsal tufts of setae; head capsule is smooth and black. Pupa is formed inside a rounded, mostly black cocoon built with the larval setae; the pupa inside is rounded and brown. **Host plant:** *Asclepias curassavica* (Apocynaceae).

Fifth instar larva of *Euchaetes mitis*.

Adult *Euchaetes mitis*.

Pupa of *Euchaetes mitis*.

Euchaetes mitis pinned.
Dorsal view.

Opharus consimilis (Erebidae, Arctiinae)

0–1800 m

FWL: 20–25 mm. **Description:** A dark brown moth with bright blue dots on dorsal area of head and thorax, orange marks on ventral side of thorax, orange rectangular marks on abdomen, and a yellow proboscis. FWs and HWs are solid dark brown. **Similar species:** *Opharus procroides* is very similar, but ventral area of its thorax is black, while in *O. consimilis* it is orange. **Habitat:** Occurs on both slopes in primary and secondary forest as well as riparian forest. Recorded in rain- and cloud-forest habitats in Rincón de la Vieja and Braulio Carrillo National Parks; less often in lowlands such as Tortuguero and Corcovado National Parks; and only a few records in dry forest, such as that in Santa Rosa National Park. **Distribution:** Mexico to Colombia. **Seasonality:** Present all year but more common during rainy season. **Natural history:** Not much is known about this moth; adults probably feed on flower nectar of plants such as *Cestrum racemosum* and *C. nocturnum*. Its proboscis is very short, and it likely accesses the nectar through holes at the base of flowers previously made by beetles of family Curculionidae (J. Flores, pers. comm.). **Early stages:** Eggs are laid in small groups. Young caterpillars are black and covered in long setae. Mature caterpillars, in groups of up to 30 individuals, rest on underside of leaves, mostly closer to the ground. Mature larva is white with many black dots all over the body, those on the sides merge together to form a black lateral line; segments A$_1$ and A$_8$ are all black; body is covered in many disperse, long black and white setae; head capsule is shiny black. Pupa is shiny black and cylindrical, formed inside a cocoon made of silk mixed with larval setae. Generally, this moth pupates in groups on bark of plants close to the host plant. **Host plants:** *Urera caracasana, U. lianoides, U. simplex, U. elata, Myriocarpa bifurca, M. longipes* (Urticaceae).

Fifth instar larvae of *Opharus consimilis*.

Pupa of *Opharus consimilis*.

Adult *Opharus consimilis*.

Opharus consimilis pinned.
Dorsal view.

Amastus aconia rumina (Erebidae, Arctiinae)

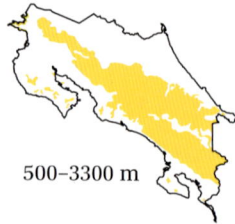

500–3300 m

FWL: 34–40 mm. **Description:** FWs and HWs are different tones of gray with black veins. Thorax is gray with four orange longitudinal lines; abdomen is orange. **Similar species:** *Amastus rothschildi* has a brown abdomen, lacks orange lines on thorax, and lacks distinctive black wing veins. **Habitat:** Occurs on both Pacific and Caribbean slopes, in most habitats except cities. Individuals are commonly seen around street lights and house lights. **Distribution:** Costa Rica and Panama. **Seasonality:** Present all year. **Natural history:** Not much is known about the biology of this common and widely distributed species. It is uncertain whether it is palatable to predators, but its habit of staying still when threatened and exposing the bright orange coloration of its abdomen is possible evidence of unpalatability. **Early stages:** Caterpillars are covered in a dense layer of rigid, stinging setae. Mature larva is black-bodied with a yellowish transverse dorsal band on each abdominal segment; each thoracic and abdominal segment has row of twelve warts bearing tufts of long, rigid red, orange, and black setae; head capsule is smooth, shiny, and red-wine colored. Pupa is reddish-brown, smooth, and cylindrical; it is formed inside a cocoon composed of densely knit silk and larval setae. The cocoon is generally placed on a tree trunk or branch close to the host plant. **Host plant:** *Inga longispica* (Fabaceae).

Fifth instar larva of *Amastus aconia rumina*.

Pupa of *Amastus aconia rumina*.

Amastus aconia rumina pinned. Dorsal view.

Adult *Amastus aconia rumina*.

Ammalo helops (Erebidae: Arctiinae)

0–1600 m

FWL: 25–42 mm. **Description:** A medium-size moth with orange head and thorax, and an orange abdomen with transverse black bands. DFW is brown with a variable number of faint pale orange and pink stains, and a more conspicuous orange patch in basal area, matching color of thorax and head. DHW is light brown. **Similar species:** None. **Habitat:** Secondary forest, gardens, and parks in towns and disturbed habitats. Present on both slopes. Sometimes visits porch and house lights in suburban areas of Central Valley, where there is a healthy population. **Distribution:** Mexico to Uruguay and Bolivia; Cuba, Haiti, Trinidad, and Grenada. **Seasonality:** Present all year. **Natural history:** Adults are rarely seen in nature, probably because of their great crypsis. They feed from any flower or extrafloral nectary of which their proboscis can reach the nectar or sweet secretions. **Early stages:** This species is more commonly found in larval form, as its caterpillars gather in very large groups of up to 200 individuals, which rest at the base of the host plants' trunks. They feed at night and rest gregariously during the day, side-by-side, forming a black-velvet-like surface on tree trunks and branches or inside tree holes. Mature larva is covered in long, dense, semi-rigid black setae; head capsule is bright orange. Pupation takes place in smaller groups at ground level or in low-altitude shelters; it forms a black cocoon with its own larval setae. **Host plants:** *Ficus aurea*, *F. benjamina*, other *Ficus* species (Moraceae).

Fifth instar larvae of *Ammalo helops.*

Pupa of *Ammalo helops.*

Ammalo helops pinned. Dorsal view.

Adult *Ammalo helops.*

Halysidota schausi schausi (Erebidae, Arctiinae)
Schaus's Tussock Moth

0–2500 m

FWL: 25–28 mm. **Description:** DFW is light whitish-orange with multiple bars and blotches with dark brown outlines; blotches in costal area are orange. VFW coloration is similar but paler. HWs are semitranslucent. Thorax is light orange with bluish lines; dorsal side of abdomen is bright orange. Legs are orange with white transverse bands with black borders. **Similar species:** Presence of a longitudinal black line in center of tegula (a dorsolateral structure of segment T_1 akin to a "shoulder pad") distinguishes this species from the very similar *Halysidota orientalis*, in which the line is absent or very reduced.

Habitat: Occurs on both slopes, in primary- and secondary-forest habitats. It is easiest to find at lights of houses and forest stations at night or in the morning. **Distribution:** USA to Colombia. **Seasonality:** Present all year. **Natural history:** Not much is known about this species; its nighttime activity makes opportunistic observations uncommon. There is some evidence that adults search for sources of certain chemicals (pyrrolizidine alkaloids), such as *Eupatorium* flowers or decaying vegetative matter, to feed from and obtain important chemical defenses. This is probably done by males, which then use the chemical as a mating cost to gain success to females (Krasnoff & Dussourd 1989). This might explain why a species with a polyphagous caterpillar, such as this one, becomes unpalatable to predators, which is indicated by its behavior of staying still and exposing its bright orange abdomen when threatened. Otherwise, the adults probably feed on nectar of Solanaceae flowers. **Early stages:** Eggs are laid in large groups of up to forty, and the young caterpillars are gregarious as well. Fourth- and fifth-instar larvae disperse and become solitary. Mature larva is covered in a dense layer of long, rigid

Fifth instar larva of *Halysidota schausi schausi*.

Pupa of *Halysidota schausi schausi*.

Adult *Halysidota schausi schausi*.

Halysidota schausi schausi
pinned. Dorsal view.

tufts of setae; it has a black longitudinal dorsal line from A_3 to A_{10} and is grayish laterally; segments A_1 and A_2 are completely black; T_2 and T_3 bear four tufts of long, white setae from a red base; A_8 bears two long, white, dorsolateral tufts of setae; head capsule is smooth and shiny black with white mouth area. Pupa is cylindrical, smooth, and brownish-red; it is formed inside a dense cocoon of silk and mature larval setae. The cocoon is usually found in the vegetation surrounding its host plant. **Host plant:** *Celtis iguanaea* (Cannabaceae).

Phaloesia saucia (Erebidae: Arctiinae)
Saucy Beauty

0–300 m

FWL: 18–26 mm. **Description:** DFW is dark brown with a row of white sub-apical dots; basal area of costal margin is red, followed distally by brown and then two small white lines; there is another red spot inside discal cell and distal to it another small white spot. DHW is completely dark brown with a thin white marginal line. VWs coloration is same as DWs. Prothorax is red dorsally. **Similar species:** None. **Habitat:** In Costa Rica, this moth has been found only in the tropical dry forest of Guanacaste. **Distribution:** USA to Venezuela. **Seasonality:** Present during rainy season. **Natural history:** Not much is known about this rare moth in Costa Rica. It is diurnal, flying quickly amid understory vegetation from 10:00 a.m. to early afternoon when the sun shines. It flies along trails, rivers, and forest edges, always associated with forest patches. Adults feed on flowers of *Eupatorium* (Asteraceae) and probably obtain some alkaloids from them, which could explain its bold coloration. **Early stages:** Eggs are green,

Fifth instar larva of *Phaloesia saucia*.

Cocoon of *Phaloesia saucia*.

Adult *Phaloesia saucia*.

Phaloesia saucia pinned.
Dorsal view.

semispherical, with smooth surface, and are laid in groups on any part of host plant. Young caterpillars are gregarious but disperse along the plant as they grow. Mature larva is shiny and black; all body segments have irregular yellow dorsal and dorsolateral markings and eight large, bluish-black warts armed with straight, rigid black setae; segments T_1–T_3, A_1, and A_8 also have some longer white setae; head capsule is shiny, black, and smooth with white mouth area. Pupa is black, cylindrical, and is formed inside a cocoon built with branches or dry leaves and silk; the silk cover is not very profuse. **Host plant:** *Tournefortia volubilis* (Boraginaceae).

Cosmosoma bogotensis (Erebidae: Arctiinae)
Bogotensis Wasp Moth

800–1500 m

FWL: 14-16 mm. **Description:** A small, colorful moth. DFW is black with basal, medial, and post-medial transparent areas; basal area is bright metallic blue with a conspicuous white dot. VHW is mostly transparent, with a thick black distal margin and veins. Head and thorax also have bright metallic blue and white markings; anterior half of abdomen is bright red, with white, and metallic blue in the center; posterior half is black with dorsal, dorsolateral, and lateral rows of white spots. **Similar species:** Distinguished from *Cosmosoma orathidia* and nearly all other species by white tarsi on second and third pair of legs. *C. regia* shares that characteristic but can be differentiated by the basal translucent area on DHW, which is large and touches the medial translucent area, while in *C. bogotensis* it is reduced to a small basal spot. **Habitat:** This species is represented in Costa Rican collections by only two specimens, one from Ciudad Colón and the other from Heredia, both in the central mountain ranges at mid-elevation on the

Fifth instar larva of *Cosmosoma bogotensis*.

Pupa of *Cosmosoma bogotensis*.

Adult *Cosmosoma bogotensis*.

Cosmosoma bogotensis pinned.
Dorsal view.

Pacific slope. Generally seen at roadsides, rivers, and light gaps. **Distribution:** Costa Rica to Colombia. **Seasonality:** Present during rainy season. **Natural history:** Not much is known about this beautiful moth; rare individuals have been seen flying actively on sunny mornings within low vegetation. As with the other *Cosmosoma* species, it mimics a wasp in appearance and behavior, which protects it from birds and lizards. Its bright coloration may be indicative of unpalatability, which might be tied to its host plant family, which is known to produce strong secondary metabolites. **Early stages:** The solitary caterpillar rests on underside of host plant's leaves. Larva has black and white transverse bands along its abdomen on segment A_1–A_7, and thorax and A_8–A_{10} are light brown; it is covered in a dense layer of long, soft, brown setae, which arise in tufts from warts; segments A_1 and A_7 have lateral tufts of yellowish setae; head capsule is smooth and orange. Pupa is suspended in the middle of a weblike capsule built with brown silk; pupa is cylindrical, whitish, with black lines highlighting the wing veins, legs, and proboscis, and black transverse lines on each abdominal segment. **Host plant:** An unidentified member of Asteraceae, probably *Tithonia*.

Cacostatia sapphira (Erebidae: Arctiinae)

0–1300 m

FWL: 16–20 mm. **Description:** A small black moth with very strong dark blue iridescence on wings. FW has a semitranslucent transverse band in post-medial area from vein Sc to Cu$_2$, and a translucent area in basal position of wing. HW has a translucent area from basal to post-medial area almost reaching distal margin. **Similar species:** Distinguished by division of basal translucent area on FW into only two parts: a large one from discal cell to 2A vein, and a small one inside discal cell. In the similar species *Agyrta dux*, *A. vitrea*, *Uranophora flaviceps*, and *Metastatia pyrrhorhoea*, the translucent area is divided into more parts, or it does not enter the discal cell. **Habitat:** Primary and secondary forest along trails and forest understory; occasionally flies out to open areas adjacent to forest edges. Localities include San Carlos, San Vito, Rincón de la Vieja, and San Pedro of Montes de Oca, in preserved forest patches. **Distribution:** Mexico to Panama. **Seasonality:** Present all year on the Caribbean slope but only during rainy season on the Pacific slope. **Natural history:** This day-flying moth is active in the hotter hours of morning. It flies very quickly and prefers dark areas inside forest. Adult rests on underside of leaves, where it stays still when alarmed. Caterpillars are parasitized by flies (*Pseudochaeta*, Tachinidae) and wasps (*Diradops zootita*, Ichneumonidae). Miller et al. (2007) state that this species is unpalatable to birds, probably from sequestering alkaloids from its host plants; but they also state that other, palatable species are more common in the same areas, and the complex does not fit the model-mimic hypotheses. Adding to this unsolved conundrum, this species' extremely fast and powerful flight (pers. obs.) is a trait more often seen in edible species; but in an escaping mimicry scenario, its fast flight might provide cover for similar species that fly more slowly and are edible. **Early stages:** Adults prefer to oviposit on young saplings in the understory. Larva is black with white ventrolateral area, and a white dorsal patch on segments T_3, A_8, and A_9, and has six tufts of long, black setae on most body segments; head capsule is shiny and bright orange. Mature caterpillars rest in small groups on underside of host plant leaves. They pupate in small groups or alone, generally on dry leaves, branches, or tree trunks. Pupa is formed inside a sparse cocoon made of silk mixed with larval setae; pupa is cylindrical, shiny, and black. **Host plants:** *Stemmadenia alfari, S. litoralis, S. donnell-smithii, S. robinsonii, S. obovata, Tabernaemontana alba* (Apocynaceae).

Fifth instar larva of *Cacostatia sapphira*.

Pupa of *Cacostatia sapphira*.

Adult *Cacostatia sapphira*.

Cacostatia sapphira pinned.
Dorsal view.

Cyanopepla arrogans (Erebidae: Arctiinae)
Arrogans Metalmark Moth

400–2200 m

FWL: 17–20 mm. Description: DFW is black with an orange post-medial transverse band from sub-costal area to vein Cu$_1$, another orange spot inside discal cell, and a third one in post-basal area; basal area is iridescent greenish-blue. DHW is black with distal margin and sub-margin orange, and basal and post-basal areas iridescent greenish-blue. **Similar species:** None. **Habitat:** Occurs on both Pacific and Caribbean slopes in all habitats, generally seen as a fast, day-flying moth in open fields or inside forest trails on sunny days. Localities include downtown areas such as San José, Heredia, and Cartago; as well as rural areas such as Orosí, San Ramón in Alajuela, Monteverde, and Coto Brus. **Distribution:** Mexico to Venezuela and Ecuador. **Seasonality:** Present all year. **Natural history:** This colorful

Fifth instar larva of *Cyanopepla arrogans.*

Cocoon of *Cyanopepla arrogans.*

Adult *Cyanopepla arrogans.*

Cyanopepla arrogans pinned.
Dorsal view.

diurnal moth has a fast and agile flight; it can be found in many habitats since its ecological requirements occur commonly in most places within its elevation range. Adults feed from nectar of *Tithonia*, *Melampodium*, and various other Asteraceae flowers. Males set up territories 2–3 m above the ground, where they chase each other in the morning. Females lay eggs at midday and in early afternoon on leaves close to the ground in open areas. **Early stages:** Eggs are pale green, spherical, with smooth surface, and are laid on underside of leaves in clusters of up to thirty or forty eggs. Young caterpillars stay together at first but disperse after a few days. Mature larva has a white body and is densely covered in long, soft white setae, with a few scattered black spatulate setae on thorax, segment A_1, and last abdominal segments; head capsule is bright orange and smooth. It rests on underside of leaves close to the ground. Cocoon is pale yellow, built with larval setae, including the black spatulate setae; it is oval and generally formed close to or on its host plant. **Host plants:** *Ixophorus unisetus*, *Cynodon nlemfuensis*, and other grasses (Poaceae).

Bertholdia yashoquintela (Erebidae: Arctiinae)

300–1900 m

FWL: 16–20 mm. **Description:** DFW is gray with a translucent post-medial area from costal margin to vein Cu_1, one or two small orange spots in basal area, and orange spots alternating with gray along distal margin. DHW is translucent gray. Abdomen is orange in dorsal area and white in ventral area. **Similar species:** Other species of *Bertholdia* are smaller and darker in coloration. *B. flammea* is distinguished by a conspicuous yellow distal margin and a yellow patch in medial area close to anal margin on DFW. **Habitat:** Occurs on both Pacific and Caribbean slopes in dry to wet, primary, secondary, and disturbed forest habitats. **Distribution:** Mexico to Panama. **Seasonality:** Present all year. **Natural history:** This species is common in many habitats, but because it is nocturnal and small, it is not often perceived unless light traps are used. In general, it is easier to find it as larva than as adult. **Early stages:** Caterpillars feed alone, resting on underside of leaves. Mature larva is black with segments T_3, A_2, A_6, and A_8 orange; it is covered entirely in long black setae that arise in six tufts from each segment; T_3 has two white, forward-pointing tufts; head capsule is smooth and dark red. Cocoon, grayish with some black setae, is formed in low vegetation; pupa is brown and rounded. **Host plants:** *Genipa americana* (Rubiaceae), *Solanum betaceum* (Solanaceae).

Fifth instar larva of *Bertholdia yashoquintela*.

Cocoon of *Bertholdia yashoquintela*.

Adult *Bertholdia yashoquintela*.

Bertholdia yashoquintela pinned.
Dorsal view.

Turuptiana oblicua (Erebidae: Arctiinae)
Oblique Tiger Moth

400–1800 m

FWL: 16–26 mm. **Description:** DFW is white with brown costal and anal margins, and thick brown band crossing wing from medial area of costal margin to tornus; distal margin may or may not have a thin brown margin. DHW is white with a little brown on tornus. **Similar species:** Distinguished by abdomen coloration, which is brown in anterior half and orange in posterior half, from *Turuptiana annesmithae*, which has brown dorsal marks on all abdominal segments. Also, in *T. oblicua*, tegula ("shoulder pad") is all white, while that of *T. annesmithae* has a brown border. *Robinsonia dewitzi* differs in having a very thick brown distal margin on DFW, while in *T. oblicua*, the brown margin is very thin if present. **Habitat:** Occurs in all habitats of both Pacific and Caribbean slopes but is more common in secondary forest and urban areas. It is present in avocado plantations, where it can become a pest. In Central Valley, adults sometimes perch on walls inside houses or on porches, attracted by lights; but it is often found in larval stage on avocado trees in backyards and parks. Localities include Guanacaste volcanoes, Palmares, and Turrialba. **Distribution:** Mexico to Colombia and Ecuador. **Seasonality:** Present all year. **Natural history:** This is a common moth, although not much is known about its habits or biology. Larva is attacked by a parasitoid fly (*Calolydella discalis*, Tachinidae). **Early stages:** Eggs are laid in large groups of up to fifty on underside of leaves. Young larvae are dark-colored, gregarious (in groups of varying sizes), and skeletonize leaves. In mature stages they disperse and feed on leaves in smaller groups. Mature larva is densely covered in long, rigid setae that are orange in anterior and posterior areas, and black in the center (segments A_3–A_6); from the orange areas, four tufts of longer black setae project forward or back, making it difficult to distinguish which end holds the head; head capsule is shiny, black, and smooth. Setae can sting if touched. Pupa is rounded, smooth, shiny, and black; it is formed inside a dense cocoon built with silk and larval setae, which can still sting. Generally, cocoon is formed in ground litter, on host tree, or on a wall. **Host plant:** *Persea americana* (Lauraceae).

Fifth instar larvae of *Turuptiana oblicua*.

Pupa of *Turuptiana oblicua*.

Adult *Turuptiana oblicua*. KW

Turuptiana oblicua pinned.
Dorsal view.

Utetheisa ornatrix ornatrix (Erebidae: Arctiinae)
Ornate Bella Moth

0–1500 m

FWL: 16–22 mm. Description: DFW is white with pink costal margin interrupted by three black spots; in distal area is a sub-marginal row of black spots, then a row of pink blotches, then a row of black marginal spots; a pair of black spots sits right outside discal cell at its distal end; other black spots occur in basal area. DHW is white with black in distal and apical areas, and rose-pink on costal margin and in apical area. Pink color on DFW is variable in tone and amount; some specimens are almost entirely rose-pink. Number of black spots on DFW and amount of black on DHW are also variable. The description provided here is for the Costa Rican phenotype; in other places, such as eastern USA, the color pattern varies. **Similar species:** None. **Habitat:** Occurs in disturbed and open areas of both Pacific and Caribbean slopes, in any place where host plants are growing. Localities include Guápiles, Ciudad Colón, Liberia, Pérez Zeledón, and Santa Rosa National Park. **Distribution:** Canada to Chile and Argentina. **Seasonality:** Present during rainy season. **Natural history:** This day-flying moth is often found patrolling open areas in sunshine around thickets of its host plants. The bright colors are a warning to predators that it tastes bad, resulting from toxins (pyrrolizidine alkaloids) that the larvae obtain when feeding on its host plants' seeds. Females receive an important quantity of this alkaloid from the male's spermatophore during copulation; females mate with several males to enhance their chemical protection, and even pass it to the eggs, offering them protection as well (Bezzerides & Eisner 2002). Larvae are attacked by parasitoid wasps (*Apanteles*, Braconidae) and flies (*Architas*, Tachinidae). **Early stages:** Eggs are yellowish, rounded, with smooth surface, and are laid in groups

Fifth instar larva of *Utetheisa ornatrix ornatrix*.

Pupa of *Utetheisa ornatrix ornatrix*.

Adult *Utetheisa ornatrix ornatrix*.

Utetheisa ornatrix ornatrix
pinned. Dorsal view.

on underside of leaves. Some days after emerging, the young caterpillars bore into their host plant's fruit pods and feed from the immature seeds. Mature larvae then leave the pods and feed some more from leaves. Mature larva is black with many yellow patches and various long, black and white setae on each segment; head capsule is bright red and smooth. The larva spins a silk cocoon in a crevice or in a shelter formed by putting together dry leaves. Pupa is cylindrical and black with some orange markings on ventral side of abdomen, dorsal side of thorax, and wing pads. **Host plants:** *Crotalaria retusa* and other *Crotalaria* species (Fabaceae).

Hypercompe icasia (Erebidae: Arctiinae)

0–1400 m

FWL: 22–35 mm. **Description:** *Female:* DFW is white with longitudinal rows of grayish-brown rings/cells, and parallel brown lines on costal margin. DHW is white with few brownish rings and blotches. Thorax is white with brown cells; abdomen is orange with transverse white lines. *Male:* DFW is white in basal half and translucent in distal half; costal margin has about five brown patches, and area from vein Cu$_2$ to anal margin is filled with brown rings. DHW is translucent white with a sub-marginal brown band in anal margin. Thorax similar to female's; abdomen is mostly black with a longitudinal faint dorsal orange line.

Similar species: Male *Hypercompe icasia* is distinguished by brown line parallel to anal margin of DHW from male *H. caudata*, which lacks the line; females are difficult to distinguish but, generally, the DFW rings are dark brown in *H. caudata* and light brown in *H. icasia*. *H. albescens* lacks extension to HW tornus, which is present in *H. icasia*. *H. perplexa* has brown-filled blotches in costal and sub-costal area of DFW, while in *H. icasia*, if any blotches are filled (most are not), they are only in costal, not sub-costal, position. Also, in some individuals of *H. perplexa*, brown band on DHW divides, with a diagonal branch that enters medial area, always absent from *H. icasia*. B. Espinoza (pers. comm.), who has studied this group's taxonomy for many years, considers the current *H. icasia* to represent a species complex, which might include up to six or seven species; further studies of molecular data, morphology, and ecology may result in splitting of the species. **Habitat:** Occurs in all habitats on both Pacific and Caribbean slopes. More often seen as caterpillars, but adults are sometimes found near lights or perched on the ground in neighborhoods or towns. Localities include towns such as Golfito, Limón, and Cartago, as well as Santa Rosa National Park. **Distribution:** Mexico to Colombia and Venezuela, Caribbean islands. **Seasonality:** Present all year. **Natural history:** Adults of this species are powerful fliers, but if surprised by a predator, they will fall to the ground and show their large and colorful abdomen; if the threat persists, orange drops of foul-smelling and -tasting liquid are secreted from each side of the thorax, deterring predators. Larvae are attacked by parasitoid wasps (*Protopelmus atrocaeruleus*, *Enicospilus glabratus*, Ichneumonidae) and flies (*Juriniopsis adusta*, *Carcelia flavirostris*,

Fifth instar larva of *Hypercompe icasia*.

Pupa of *Hypercompe icasia*.

Adult *Hypercompe icasia.*

Hypercompe icasia pinned.
Dorsal view.

C. orellana, Patelloa xanthura, Leschenaultia echinacea, Chetogena scutellaris, Tachinidae). **Early stages:** Eggs are small, whitish, and rounded, and are laid in clusters of hundreds in virtually any place. Young larvae are covered in rigid setae, which are black anteriorly and posteriorly, and orange in middle part of abdomen. Mature larva is covered dorsally and dorsolaterally in a dense layer of long, rigid, black setae arising from raised warts; ventrolateral setae and all those on segment T_1 are red; head capsule is shiny, black, and smooth. The setae are irritating to the skin of some people. Pupa is formed inside a dense cocoon built with silk; pupa is cylindrical, black, very smooth, and shiny. **Host plants:** *Handroanthus ochraceus* (Boraginaceae), *Emilia fosbergii* (Asteraceae), *Colocasia esculenta* (Araceae), *Guazuma ulmifolia* (Malvaceae), and many other species from many families.

APPENDIX: BUTTERFLY AND MOTH PLANTS OF COSTA RICA

Many butterfly species associate with specific plant species or groups of plants, and thus plants are useful in identifying lepidoptera species. Note that the species accounts mention only the plants' scientific names. This list includes not only the scientific names for plants but also family names and, when relevant, English and Spanish (in italics) common names, which should come in handy for the nonspecialist who is familiar with the common names for plants but knows very few scientific names. The Index of Plant Names (p. 308) indicates the page numbers on which the scientific names for plants occur.

Acnistus arborescens (Solanaceae); hollow heart, wild tree tobacco; *güitite*

Acroceras zizanioides (Poaceae); oat grass; *arrocillo*

Acrocomia aculeata (Arecaceae); coyol palm, grugru palm; *coyol*

Adelia triloba (Euphorbiaceae); copperleaf; *clavillo*

Aegiphila laevis (Lamiaceae)

Ageratum conyzoides (Asteraceae); billygoat-weed; *Santa Lucía*

Aiouea montana (Lauraceae); wild avocado; *aguacatillo*

Albizia adinocephala (Fabaceae); cream albizia; *gavilán, gavilancillo, gallinazo*

Albizia niopoides (Fabaceae); silk tree; *cenízaro macho, guanacaste blanco*

Alchornea costaricensis (Euphorbiaceae); *fósforo, ira*

Alchornea latifolia (Euphorbiaceae); achiotillo; *canelito, chapaneo*

Alibertia edulis (Rubiaceae); wild guava; *trompillo*

Allophylus racemosus (Sapindaceae); palo de caja; *cafecillo, cajita, esquitillo*

Alternanthera pubiflora (Amaranthaceae); alligatorweed; *sanguinaria*

Amyris pinnata (Rutaceae); torchwood; *manzana, manzano*

Anacardium excelsum (Anacardiaceae); wild cashew; *espavel*

Anacardium occidentale (Anacardiaceae); cashew nut; *marañón, anacardo*

Ananas comosus (Bromeliaceae); pineapple; *piña*

Andira inermis (Fabaceae); cabbagebark tree; *carne asada, almendro de río*

Annona cherimola (Annonaceae); cherimoya; *chirimolla, anona*

Annona holosericea (Annonaceae); little annona; *anonillo, anonillo de cerro*

Annona pruinosa (Annonaceae); *anonillo*

Annona purpurea (Annonaceae); annona; *zoncoya, soncoya*

Annona rensoniana (Annonaceae); *anonillo*

Annona reticulata (Annonaceae); red custard apple; *anona, anonillo*

Aristolochia anguicida (Aristolochiaceae); harlequin dutchman's pipe; *canastilla*

Aristolochia arborea (Aristolochiaceae); dutchman's pipe; *canastilla*

Aristolochia elegans (Aristolochiaceae); elegant dutchman's pipe; *canastilla*

Aristolochia grandiflora (Aristolochiaceae); pelican flower; *carraco, canastilla*

Aristolochia leuconeura (Aristolochiaceae); dutchman's pipe; *canastilla*

Aristolochia maxima (Aristolochiaceae); Florida dutchman's pipe; *cuajilote, carraquito, curare, canastilla*

Aristolochia pilosa (Aristolochiaceae); dutchman's pipe; *canastilla*

Aristolochia tonduzii (Aristolochiaceae); dutchman's pipe; *curarina, tripa de cusuco*

Arracacia xanthorrhiza (Apiaceae); Peruvian parsnip; *arracache*

Asclepias curassavica (Apocynaceae); milkweed; *vivorana, bailarina*

Asclepias oenotheroides (Apocynaceae); Texas milkweed; *leche de perro*

Asterogyne martiana (Arecaceae); *suita*

Avicennia germinans (Acanthaceae); black mangrove; *mangle negro, palo de sal*

Bacopa spp. (Plantaginaceae); water hyssop; *isopo de agua*

Bactris gasipaes (Arecaceae); peach palm; *pejibaye*

Bactris guineensis (Arecaceae); Guinea bactris; *huiscoyol, viscoyol, uvita de monte*

Bactris mayor (Arecaceae); *huiscoyol, viscoyol*

Bambusa vulgaris (Poaceae); common bamboo; *bambú común*

Bidens pilosa (Asteraceae); hairy beggar ticks; *moriseco, mozote*

Blechum pyramidatum (Acanthaceae); Browne's blechum; *sornia*

Brassica oleracea (Brassicaceae); cabbage; *col, repollo, brócoli, coliflor*

Brassica rapa (Brassicaceae); field mustard; *nabo*

Bromelia pinguin (Bromeliaceae); wild pineapple, pinguin; *piñuela*

Brosium alicastrum (Moraceae); breadnut; *ojoche, ojoche de fruta*

Brugmansia spp. (Solanaceae); angel's trumpets; *reina de la noche*

Byrsonima crassifolia (Malpighiaceae); maricao cimun, shoemaker's tree; *nance, nancite*

Caesalpinia pulcherrima (Fabaceae); dwarf poinsettia, pride of Barbados; *hoja de sen, clavelina*

Calathea crotalifera (Marantaceae); rattlesnake plant; *bijagua, platanilla*

Calathea inocephala (Marantaceae); pampano; *bijagua, platanilla*

Calathea latifolia (Marantaceae); pampano; *bijagua, platanilla*

Calathea lutea (Marantaceae); pampano; *bijagua, platanilla, hoja blanca*

Calathea macrosepala (Marantaceae); pampano; *platanilla*

Calathea marantifolia (Marantaceae); pampano; *bijagua, platanilla*

Calathea warsczewisczia (Marantaceae); pampano; *bijagua, platanilla*

Calycophyllum candidissimum (Rubiaceae); lancewood, degame; *madroño*

Canna indica (Cannaceae); Indian shot; *platanilla, periquitoya*

Capparidastrum discolor (Capparaceae); *granadilla de árbol*

Capsicum annuum (Solanaceae); bell pepper, chili pepper; *chile, chile dulce*

Cardiospermum grandiflorum (Sapindaceae); showy balloonvine; *chimbolillo*

Casearia nitida (Salicaceae); smooth honeytree; *yaya, cerito*

Casimiroa edulis (Rutaceae); Mexican apple, white sapote; *matasano, tapaculo*

Cassia biflora (Fabaceae); *saragundí*

Cassia emarginata (Fabaceae); *saragundí*

Cassia grandis (Fabaceae); pink shower; *carao*

Cecropia obtusifolia (Urticaceae); guarumo, trumpet tree; *guarumo colorado*

Cecropia peltata (Urticaceae); guarumo, trumpet tree; *guarumo*

Ceiba pentandra (Malvaceae); kapok; *ceiba*

Celtis iguanaea (Cannabaceae); iguana hackberry; *cagalera*

Centrosema macrocarpum (Fabaceae); butterfly pea; *gallinita*

Centrosema pubescens (Fabaceae); butterfly pea; *gallinita*

Centrosema sagittatum (Fabaceae); arrowleaf butterfly pea; *gallinita*

Cestrum glanduliferum (Solanaceae); *zorrillo*

Cestrum microcalyx (Solanaceae); *zorrillo*

Cestrum nocturnum (Solanaceae); night-blooming jessamine; *zorrillo*

Cestrum tomentosum (Solanaceae); *zorrillo, zorrillo blanco*

Chamaedorea costaricana (Arecaceae); pacaya palm; *pacaya*

Chamaedorea tepejilote (Arecaceae); pacaya palm; *pacaya*

Chimarrhis parviflora (Rubiaceae); *yema de huevo, pejibayito*

Chomelia spinosa (Rubiaceae); tom bush; *malacahüite, chocolatico*

Chromolaena odorata (Asteraceae); Christmas bush, jack in the bush; *charralera*

Chusquea spp. (Poaceae); *chusquea*

Cinnamomum brenesii (Lauraceae); cinnamon; *aguacatillo*

Cinnamomum hammelianum (Lauraceae); *aguacate de potrero*

Cinnamomum triplinerve (Lauraceae); Mexican cinnamon; *aguacatillo*

Cissus biformifolia (Vitaceae); *azulillo, yasú*

Cissus verticillata (Vitaceae); seasonvine; *bejuco ubí, uvilla*

Citharexylum donnell-smithii (Verbenaceae); fiddlewood tree; *dama, huelenoche*

Citrus × auriantum (Rutaceae); sour orange, Seville orange; *naranja dulce*

Citrus aurantiifolia (Rutaceae); key lime; *limón agrio*

Citrus medica (Rutaceae); citron; *cidro, toronja*

Citrus reticulata (Rutaceae); mandarin orange; *mandarina*

Cleome gynandra (Cleomaceae); spiderwisp

Cleome spinosa (Cleomaceae); spiny spider flower; *espuela de caballero*

Cleome viscosa (Cleomaceae); Asian spider flower

Clibadium surinamense (Asteraceae)

Cocos nucifera (Arecaceae); coconut palm; *coco, palo de pipa, cocotero*

Coffea arabica (Rubiaceae); coffee plant, Arabian coffee; *café*

Cojoba arborea (Fabaceae); wild tamarind; *lorito*

Colocasia esculenta (Araceae); coco yam; *ñampí, malanga*

Comaclinium montana (Asteraceae); *girasolillo*

Combretum farinosum (Combretaceae); orange flame vine; *papa miel*

Conostegia rufescens (Melastomataceae); Luquillo Mountain snailwood; *lengua de vaca, purra*

Cordia alliodora (Boraginaceae); Spanish elm; *laurel*

Cosmos bipinnatus (Asteraceae); garden cosmos; *cambrai*

Cosmos sulphureus (Asteraceae); garden cosmos; *cambrai*

Costus scaber (Costaceae); spiral ginger; *cañagria, caña agria*

Crateva tapia (Capparaceae); garlic pear tree; *ajillo, muñeco*

Crescentia alata (Bignoniaceae); morrito; *jícaro, tiquí*

Crotalaria retusa (Fabaceae); rattleweed; *quiebraplatos, chipilín*

Croton billbergianus (Euphorbiaceae)

Croton draco (Euphorbiaceae); dragon's blood; *targuá*

Croton niveus (Euphorbiaceae); *copalchí, colpachí*

Croton schiedeanus (Euphorbiaceae); *copalchí, colpachí*

Croton verapazensis (Euphorbiaceae); *targua*

Cupania glabra (Sapindaceae); Florida toadwood; *cascuá, tres huesos, huesillo*

Cyclospermum leptophyllum (Apiaceae); marsh parsley; *culantrillo, eneldillo, eneldo*

Cynodon nlemfuensis (Poaceae); African Bermuda grass; *estrella africana, pasto estrella*

Dalbergia retusa (Fabaceae); rosewood; *cocobolo, ñámbar*

Dalechampia scandens (Euphorbiaceae); spurgecreeper; *ortiguilla*

Dalechampia triphylla (Euphorbiaceae); dalechampia; *ortiguilla*

Dalechampia websteri (Euphorbiaceae); dalechampia; *ortiguilla*

Datura spp. (Solanaceae); *reina de la noche*

Delonix regia (Fabaceae); flamboyant tree; *malinche*

Dendropanax spp. (Araliaceae); *cacho de venado, fosforillo*

Dendrophthora costaricensis (Santalaceae); tree destroyer; *matapalo*

Dendrophylax lindenii (Orchidaceae); ghost orchid; *orquídea fantasma*

Desmodium distortum (Fabaceae); tick trefoil; *pega-pega*

Desmodium glabrum (Fabaceae); zarzabacoa dulce; *pega-pega*

Desmodium incanum (Fabaceae); zarzabacoa comun; *pega-pega*

Desmodium infractum (Fabaceae); tick trefoil; *pega-pega*

Desmodium nicaraguense (Fabaceae); tick trefoil; *engorda caballo, pega-pega*

Dichapetalum grayumii (Dichapetalaceae)

Dichapetalum morenoi
(Dichapetalaceae)

Dioclea malacocarpa (Fabaceae); *ojo de buey*

Diphysa americana (Fabaceae); *guachipelín*

Doliocarpus multiflorus (Dilleniaceae)

Drypetes standleyi (Putranjivaceae)

Duranta erecta (Verbenaceae); golden dewdrops; *pingo de oro, miguelito*

Dypsis lutescens (Arecaceae); Areca palm; *palmera múltiple*

Dyschoriste quadrangularis (Acanthaceae); snakeherb

Emilia fosbergii (Asteraceae); Florida tasselflower; *clavelillo*

Emilia sonchifolia (Asteraceae); lilac tasselflower; *clavelillo, pincel*

Epidendrum paniculatum (Orchidaceae); paniculate epidendrum; *orquídea estrella*

Erblichia odorata (Passifloraceae); butterfly tree; *flor de fuego, periquita*

Erythrina berteroana (Fabaceae); coral bean; *poró, poró de montaña, poró espinoso*

Erythrina costaricensis (Fabaceae); machete plant; *poró, poró cimarrón, poró colorado*

Erythrina gibbosa (Fabaceae); machete plant; *poró*

Erythrina lanceolata (Fabaceae); machete plant; *poró*

Erythrina poeppigiana (Fabaceae); mountain immortelle; *poró gigante, poró extranjero*

Esenbeckia berlandieri (Rutaceae); Berlandier's jopoy; *parra*

Eugenia acapulcensis (Myrtaceae); *guayabillo, murta*

Eugenia hypargyrea (Myrtaceae)

Eugenia monticola (Myrtaceae); birdcherry; *cacique*

Eugenia salamensis (Myrtaceae); *guayabillo, güisaro macho*

Eumachia microdon (Rubiaceae)

Eupatorium spp. (Asteraceae)

Euphorbia pulcherrima (Euphorbiaceae); poinsettia; *pastora*

Exostema mexicanum (Rubiaceae)

Ficus aurea (Moraceae); Florida strangler fig; *higuerón blanco, higuerón*

Ficus benjamina (Moraceae); weeping fig; *laurel de la India*

Ficus cahuitensis (Moraceae); fig tree; *higuerón, matapalo*

Ficus costarica (Moraceae); fig tree; *higuerón colorado*

Ficus cotinifolia (Moraceae); strangler fig; *higuerón*

Ficus crassinervia (Moraceae); fig tree; *higuerón*

Ficus crocata (Moraceae); fig tree; *matapalo, barrú, barú*

Ficus elastica (Moraceae); Indian rubberplant; *árbol de hule, hule*

Ficus obtusifolia (Moraceae); amate; *capulamate, higuerón*

Ficus pertusa (Moraceae); fig tree; *higuito, higuerón*

Ficus tinctoria (Moraceae); fig tree; *higuerón*

Fischeria panamensis (Apocynaceae)

Foeniculum vulgare (Apiaceae); sweet fennel; *hinojo*

Forchhammeria trifoliata (Capparaceae)

Fuchsia arborescens (Onagraceae); Andre's fuchsia; *achiotillo, candelilla*

Fuchsia paniculata (Onagraceae); shrubby fuchsia; *achiotillo*

Galactia striata (Fabaceae); Florida hammock milkpea

Genipa americana (Rubiaceae); jagua; *guatíl, tapaculo, jagua*

Geonoma ferruginea (Arecaceae); *caña de danta*

Gitara nicaraguensis (Euphorbiaceae); *barraquillo*

Gmelina arborea (Lamiaceae); gumhar, white teak; *melina*

Goeppertia villosa (Marantaceae); *bijagua, platanilla*

Goethalsia meiantha (Malvaceae); *guácimo blanco*

Gomphocarpus physocarpus (Apocynaceae); balloon milkweed, balloon plant; *chayote de aire*

Gonolobus edulis (Apocynaceae); cuayote; *cuayote*

Gossypium hirsutum (Malvaceae); upland cotton; *algodón*

Gouania lupuloides (Rhamnaceae); whiteroot

Graptophyllum pictum (Acanthaceae); caricature plant; *planta caricatura, unión de los casados*

Guadua angustifolia (Poaceae); bamboo; *bambu*

Guarea glabra ssp. *tuerckheimii* (Meliaceae); alligator wood; *cocorilla, cocora, pocora*

Guazuma ulmifolia (Malvaceae); bastard cedar; *guácimo*

Gunnera insignis (Gunneraceae); poorman's umbrella; *sombrilla de pobre*

Gurania spp. (Cucurbitaceae); *pata de danta*

Guzmania spp. (Bromeliaceae); tufted airplant; *bromelia*

Hamelia patens (Rubiaceae); scarlet bush; *coralillo*

Hampea appendiculata (Malvaceae); doll's eyes; *burío blanco, burío ratón, azajardillo*

Handroanthus ochraceus (Bignoniaceae); yellow ipê; *cortez amarillo*

Havardia spp. (Fabaceae); havardia

Hedychium coronarium (Zingiberaceae); butterfly ginger; *flor de San Juan, heliotropo*

Heliconia imbricata (Heliconiaceae); heliconia; *platanilla*

Heliconia irrasa (Heliconiaceae); wild plantain; *platanilla*

Heliconia latispatha (Heliconiaceae); red-yellow gyro; *caliguate, platanilla*

Heliconia longa (Heliconiaceae); heliconia; *platanilla*

Heliconia longiflora (Heliconiaceae); heliconia; *platanilla*

Heliconia mathiasiae (Heliconiaceae); heliconia; *platanilla*

Heliconia metallica (Heliconiaceae); shining bird of paradise; *platanilla*

Heliconia pogonantha (Heliconiaceae); heliconia; *platanilla*

Heliconia tortuosa (Heliconiaceae); red twist; *platanilla*

Heliconia umbrophila (Heliconiaceae); heliconia; *platanilla*

Heliconia vaginalis (Heliconiaceae); heliconia; *platanilla*

Heliconia wagneriana (Heliconiaceae); heliconia; *platanilla*

Helicteres guazumifolia (Malvaceae); *gúacimo torcido, rabo de chancho, tornillo*

Heliotropium spp. (Boraginaceae); heliotrope; *alacrancillo, cola de alacrán*

Heterocondylus vitalbae (Asteraceae)

Heteropterys laurifolia (Malpighiaceae); dragon withe; *bejuco de corral, bejuco real*

Hibiscus spp. (Malvaceae); hibiscus, rosemallow; *amapola, hibisco*

Hura crepitans (Euphorbiaceae); sandbox tree; *jabillo*

Hygrophila costata (Acanthaceae); water wisteria; *yerba de hicotea*

Hylaeanthe hoffmannii (Marantaceae)

Impatiens walleriana (Balsaminaceae); busy lizzy; *china*

Inga chocoensis (Fabaceae); inga; *cuajiniquíl, guaba, guabo*

Inga densiflora (Fabaceae); densely flowered inga; *guaba caite, guabo salado*

Inga longispica (Fabaceae); inga; *guabo, guaba, guabo ronron*

Inga oerstediana (Fabaceae); inga; *cuajiniquíl peludo, guaba chilillo*

Inga punctata (Fabaceae); inga; *guabo, cuajiniquil, cuajiniquil colorado*

Inga ruiziana (Fabaceae); inga; *guaba de río*

Inga samanensis (Fabaceae); inga; *cuajiniquíl, guaba, guabo*

Inga sapindoides (Fabaceae); inga; *guabo cuadrado*

Inga vera (Fabaceae); river koko; *guabo de río, cuajiniquíl*

Ipomoea batatas (Convolvulaceae); sweet potato; *camote*

Ixophorus unisetus (Poaceae); Mexican grass; *zacate de Honduras, zacate blanco*

Jatropha spp. (Euphorbiaceae); nettlespurge; *jatrofa*

Justicia spp. (Acanthaceae); water-willow; *camaron*

Lantana camara (Verbenaceae); lantana; *cinco negritos*

Lasiacis spp. (Poaceae); smallcane; *bambucillo*

Lasianthaea fruticosa (Asteraceae); *quitirrí*

Lepidium virginicum (Brassicaceae); Virginia pepperweed; *mastuerzo*

Licania arborea (Chrysobalanaceae); licania; *alcornoque, canilla de mula*

Lindernia spp. (Linderniaceae); false pimpernel

Lippia bracteosa (Verbenaceae); *orégano de monte*

Lonchocarpus macrophyllus (Fabaceae); lancepod; *chaperno*

Lonchocarpus oliganthus (Fabaceae); lancepod; *chaperno*

Lycianthes spp. (Solanaceae); lycianthes

Mabea occidentalis (Euphorbiaceae);

Machaerium biovulatum (Fabaceae); machaerium; *siete cueros, espino negro*

Machaerium floribundum (Fabaceae); machaerium; *uña de gato*

Machaerium kegelii (Fabaceae); machaerium; *uña de gato*

Machaerium salvadorense (Fabaceae); machaerium; *uña de gato*

Machaerium seemannii (Fabaceae); machaerium; *uña de gato*

Maclura tinctoria (Moraceae); fustic tree; *mora, palo de mora, brasil, morillo*

Maianthemum spp. (Asparagaceae)

Malvaviscus arboreus (Malvaceae); wax mallow; *amapola, amapolita*

Mandevilla hirsuta (Apocynaceae); rocktrumpet

Mandevilla subsagittata (Apocynaceae); rocktrumpet

Mangifera indica (Anacardiaceae); mango; *mango*

Marila spp. (Calophyllaceae)

Matelea spp. (Apocynaceae); milkvine

Megaskepasma erythrochlamys (Acanthaceae); Brazilian red-cloak; *pavón rojo, pavoncillo rojo*

Melampodium divaricatum (Asteraceae); *florecilla, comunismo*

Melanthera nivea (Asteraceae); snow squarestem; *paira, totolquelite*

Mesechites trifidus (Apocynaceae)

Mespilodaphne macrophylla (Lauraceae); sweetwood; *aguacate de mono, ira zopilote*

Mespilodaphne veraguensis (Lauraceae); sweetwood; *aguacatillo, canelillo, sigua canelo*

Miconia appendiculata (Melastomataceae); johnnyberry; *uña de gato*

Miconia calvescens (Melastomataceae); velvet tree; *uña de gato*

Miconia donaeana (Melastomataceae); johnnyberry; *oreja de mula*

Miconia elata (Melastomataceae); johnnyberry; *lengua de vaca*

Miconia impetiolaris (Melastomataceae); johnnyberry; *chirré, hoja de pasmo*

Miconia longifolia (Melastomataceae); johnnyberry; *uña de gato*

Miconia trinervia (Melastomataceae); johnnyberry; *uña de gato*

Mikania micrantha (Asteraceae); mile-a-minute

Mollinedia costaricensis (Monimiaceae); *limoncillo*

Mollinedia viridiflora (Monimiaceae); *curilla*

Morisonia americana (Capparaceae); ratapple; *tamalcahua*

Morus spp. (Moraceae); mulberry; *morera*

Mucuna monticola (Fabaceae); horse-eye bean; *ojo de buey*

Musa acuminata (Musaceae); edible banana; *banano, banano enano*

Musa sapientum (Musaceae); French plantain; *plátano*

Myriocarpa bifurca (Urticaceae); *ortiga*

Myriocarpa longipes (Urticaceae); cow itch; *estrella, ortiga*

Nasturtium officinale (Brassicaceae); watercress; *berro*

Nectandra belizensis (Lauraceae); sweetwood; *aguacatillo, quizarrá, ira, sigua*

Nectandra hihua (Lauraceae); shingle wood; *aguacatillo*

Nectandra salicifolia (Lauraceae); sweetwood; *quizarrá, sigua blanco, ira, quizarrá blanco*

Nectandra smithii (Lauraceae); sweetwood; *aguacatillo, quizarrá, ira, sigua*

Nectandra umbrosa (Lauraceae); sweetwood; *aguacatillo, quizarrá, ira, sigua*

Nicotiana tabacum (Solanaceae); cultivated tobbaco; *tabaco*

Ochroma pyramidale (Malvaceae); West Indian balsa; *balsa*

Ocotea atirrensis (Lauraceae); sweetwood; *quizarrá, tiquizarrá macho*

Ocotea cernua (Lauraceae); sweetwood; *aguacatillo, torito, quizarrá lorito*

Ocotea insularis (Lauraceae); sweetwood; *ira marañón, aguacatón, pocora*

Ocotea puberula (Lauraceae); sweetwood; *ira, quizarrá, sigua*

Odontonema tubiforme (Acanthaceae); firespike; *coral*

Omphalea diandra (Euphorbiaceae); Jamaican cobnut; *bejuco de sangre*

Oyedaea verbesinoides (Asteraceae); *margarita de monte*

Pachira quinata (Malvaceae); red ceiba, spiny cedar; *pochote*

Pachistachys lutea (Acanthaceae); golden shrimp-plant, lollipop plant; *camaroncillo, olotillo*

Palicourea spp. (Rubiaceae); cappel; *cafecillo*

Paquira acuatica (Malvaceae); Guiana chestnut, water chestnut; *poponjoche*

Passiflora adenopoda (Passifloraceae); velcro passionflower; *ococa, pococa, tococa*

Passiflora alata (Passifloraceae); fragrant granadilla; *granadilla colombiana*

Passiflora ambigua (Passifloraceae); passionflower; *granadilla, granadilla de monte*

Passiflora apetala (Passifloraceae); batwing passionflower; *calzoncillo*

Passiflora auriculata (Passifloraceae); passionflower; *flor de pasión*

Passiflora bicornis (Passifloraceae); wing-leaf passionflower; *flor de pasión*

Passiflora biflora (Passifloraceae); twin-flower passionflower; *calzoncillo, ñorbito*

Passiflora caerulea (Passifloraceae); bluecrown passionflower; *flor de pasión*

Passiflora capsularis (Passifloraceae); passionflower; *flor de pasión*

Passiflora coriacea (Passifloraceae); batleaved passionflower; *ala de murciélago*

Passiflora costaricensis (Passifloraceae); passionflower; *flor de pasión*

Passiflora edulis (Passifloraceae); maypop, purple granadilla; *maracuyá*

Passiflora foetida (Passifloraceae); stinking passionflower; *ñorbo, bombillo*

Passiflora lobata (Passifloraceae); passionflower; *flor de pasión*

Passiflora menispermifolia (Passifloraceae); passionflower; *flor de pasión*

Passiflora oerstedii (Passifloraceae); passionflower; *flor de pasión*

Passiflora pedata (Passifloraceae); passionflower; *flor de pasión*

Passiflora pittieri (Passifloraceae); passionflower; *flor de pasión*

Passiflora platyloba (Passifloraceae); acid granadilla; *calala*

Passiflora quadrangularis (Passifloraceae); giant granadilla, badea; *granadilla real*

Passiflora serratifolia (Passifloraceae); passionflower; *flor de pasión*

Passiflora suberosa (Passifloraceae); corky passionflower; *flor de pasión*

Passiflora talamancensis (Passifloraceae); passionflower; *flor de pasión*

Passiflora tica (Passifloraceae); passionflower; *flor de pasión*

Passiflora vitifolia (Passifloraceae); crimson passionflower; *flor de pasión, pastora de monte, pasionaria*

Paullinia costaricensis (Sapindaceae); *ojo de pájaro*

Pentaclethra macroloba (Fabaceae); oil bean tree; *gavilán*

Pentagonia donnell-smithii (Rubiaceae); *lengua de vaca, tabacón*

Pentas lanceolata (Rubiaceae); pentas; *pentas*

Persea americana (Lauraceae); avocado; *aguacate*

Persea caerulea (Lauraceae); wild avocado; *aguacatillo, irá café*

Persea povedae (Lauraceae); wild avocado; *aguacatillo*

Persea veraguasensis (Lauraceae); wild avocado; *aguacatillo*

Phaseolus lunatus (Fabaceae); sieva bean; *frijolillo, frijol lima*

Phaseolus vulgaris (Fabaceae); kidney bean; *frijol*

Philodendron herbaceum (Araceae); heart-leaved philodendron

Phoradendron quadrangulare (Santalaceae); mistletoe; *matapalo*

Phoradendron tonduzii (Santalaceae); mistletoe; *matapalo*

Phoradendron undulatum (Santalaceae); mistletoe; *matapalo*

Phyllanthus acidus (Phyllanthaceae); Tahitian gooseberry tree; *cimbilín, grosella*

Phyllanthus amarus (Phyllanthaceae); carry-me seed; *riñoncillo*

Phyllanthus anisolobus (Phyllanthaceae); leaflower

Phyllanthus niruri (Phyllanthaceae); gale of the wind; *tamarindillo, riñoncillo*

Pilocarpus racemosus (Rutaceae); *talcacao*

Piper aduncum (Piperaceae); spiked pepper; *cordoncillo*

Piper amalago (Piperaceae); pepper elder; *alcotán*

Piper arboreum (Piperaceae); candle bush

Piper auritum (Piperaceae); Veracruz pepper, anise piper; *anisillo, hoja de estrella, alcotán, hinojillo*

Piper colonense (Piperaceae); candle bush

Piper jacquemontianum (Piperaceae); Caracas pepper

Piper lanceifolium (Piperaceae); candle bush

Piper linearifolium (Piperaceae); candle bush; *cordoncillo*

Piper marginatum (Piperaceae); marigold pepper; *anisillo, hoja de estrella*

Piper multiplinervium (Piperaceae); candle bush; *cordoncillo*

Piper peltatum (Piperaceae); Santa María plant; *baquiña, canfut, hoja de estrella*

Piper phytolaccifolium (Piperaceae); candle bush

Piper psilorhachis (Piperaceae); *cordoncillo*

Piper reticulatum (Piperaceae); aneisi wiwiri; *cordoncillo*

Piper sancti-felicis (Piperaceae); candle bush

Piper trigonum (Piperaceae); candle bush

Piper tuberculatum (Piperaceae); candle bush; *cigarrillo*

Piper umbricola (Piperaceae); candle bush; *cordoncillo*

Pithecellobium spp. (Fabaceae); blackbead; *michigüiste*

Platymiscium parviflorum (Fabaceae); *cristóbal, ñambar*

Pleiostachya pruinosa (Marantaceae); platanillo, prayer plant; *platanilla*

Pouteria sapota (Sapotaceae); mammee sapote; *zapote, zapote mamey*

Prestoea decurrens (Arecaceae);
palmitillo, palmito mantequilla

Prestonia longifolia (Apocynaceae);
prestonia

Prestonia portobellensis (Apocynaceae);
prestonia

Prestonia quinquangularis
(Apocynaceae); prestonia

Prunus annularis (Rosaceae); wild
apricot; *duraznillo*

Prunus persica (Rosaceae); peach;
durazno, melocotón

Pseuderanthemum cuspidatum
(Acanthaceae)

Psidium friedrichsthalianum
(Myrtaceae); wild guava; *cas*

Psidium guajava (Myrtaceae); guava;
guayaba

Psidium guineense (Myrtaceae);
Brazilian guava; *guísaro*

Psiguria spp. (Cucurbitaceae)

Psychotria exilis (Rubiaceae)

Pterocarpus hayesii (Fabaceae);
sangrillo, sangregado, manteco

Pterocarpus officinalis (Fabaceae);
dragon's blood tree; *sangrillo,
targuayugo, cuajada amarilla*

Pterocarpus rohrii (Fabaceae);
bloodwood, Mexican pterocarpus;
paleta, sangregado

Quadrella cynophalloflora
(Capparaceae); bay-leaved caper-tree,
falseteeth

Randia aculeata (Rubiaceae); white
indigo berry; *espino, espino blanco,
horquetilla*

Randia armata (Rubiaceae); *horquetilla,
mostrenco, palo de cruz*

Randia echinocarpa (Rubiaceae);
shacua; *papache picudo, papache*

Randia monantha (Rubiaceae)

Raphanus raphanistrum ssp. *sativus*
(Brassicaceae); cultivated radish; *rábano*

Rhipidocladum pittieri (Poaceae);
carrizo

Rhizophora mangle (Rhizophoraceae);
red mangrove; *mangle colorado*

Rosa spp. (Rosaceae); rose; *rosa*

Roupala montana (Proteaceae); *danto,
carne asada, ratón*

Ruellia inundata (Acanthaceae); slender
ruellia; *hierba del cabro*

Ruellia simplex (Acanthaceae); Britton's
wild petunia; *petunia mexicana*

Ruellia terminalis (Acanthaceae); wild
petunia

Saccharum officinarum (Poaceae);
sugarcane; *caña de azúcar*

Samanea saman (Fabaceae); raintree;
cenízaro

Sapium glandulosum (Euphorbiaceae);
gum tree; *yos*

Sapranthus palanga (Annonaceae);
palanco, turrú

Selenicereus undatus (Cactaceae);
nightblooming cactus; *pitahaya*

Senecio spp. (Asteraceae)

Senna alata (Fabaceae); emperor's
candlesticks

Senna atomaria (Fabaceae); flor de San
José; *vainillo*

Senna bicapsularis (Fabaceae);
Christmasbush; *abejon*

Senna hayesiana (Fabaceae)

Senna hirsuta (Fabaceae); woolly senna

Senna obtusifolia (Fabaceae); Java-bean;
dormilona

Senna occidentalis (Fabaceae); septicweed; *pico de pájaro, pisabed*

Senna pallida (Fabaceae); twin-flowered cassia; *abejón*

Senna papilosa (Fabaceae); *candelillo, vainillo*

Senna reticulata (Fabaceae); *saragundí, sorocontil*

Senna septemtrionalis (Fabaceae); arsenic bush; *candelillo, quiebraplato, vainillo*

Senna spectabilis (Fabaceae); casia amarilla; *candelillo, vainillo*

Serjania atrolineata (Sapindaceae)

Solanum aturense (Solanaceae); contenete, *tomatillo, uña de gato*

Solanum betaceum (Solanaceae); tree tomato; *tomate de palo, tomate de árbol*

Solanum hayesii (Solanaceae)

Solanum hazenii (Solanaceae)

Solanum jamaicense (Solanaceae); Jamaican nightshade

Solanum lanceifolium (Solanaceae); lanceleaf nightshade; *tomatillo*

Solanum lycopersicum (Solanaceae); garden tomato; *tomate*

Solanum rudepannum (Solanaceae); aubergine, eggplant; *berenjena cimarrona*

Solanum rugosum (Solanaceae); tabacon áspero; *zorro, lengua de vaca*

Solanum schlechtendalianum (Solanaceae)

Solanum torvum (Solanaceae); turkey berry; *berenjena silvestre, berenjena espinuda*

Solanum tuberosum (Solanaceae); Irish potato; *papa*

Solanum umbellatum (Solanaceae); nightshade; *zorrillo, bodoque*

Spananthe paniculata (Apiaceae); *carricillo*

Spondias mombim (Anacardiaceae); yellow mombin, hog plum; *jobo*

Spondias purpurea (Anacardiaceae); purple mombin; *jocote*

Stachytarpheta calderonii (Verbenaceae); *rabo de gato*

Stachytarpheta frantzii (Verbenaceae); purple poterweed; *rabo de gato*

Stachytarpheta mutabilis (Verbenaceae); changeable velvetberry; *rabo de gato*

Stauranthus perforatus (Rutaceae)

Stemmadenia alfari (Apocynaceae); milky way tree; *huevos de caballo*

Stemmadenia donnell-smithii (Apocynaceae)

Stemmadenia litoralis (Apocynaceae)

Stemmadenia obovata (Apocynaceae)

Stemmadenia robinsonii (Apocynaceae)

Stevia lucida (Asteraceae); *jarilla*

Struthanthus orbicularis (Loranthaceae); *matapalo*

Swietenia macrophylla (Meliaceae); American mahogany; *caoba*

Symphonia spp. (Clusiaceae); *botoncillo, cerillo*

Tapirira mexicana (Anacardiaceae); *cirrí, cirrí blanco, cirrí colorado*

Tassadia obovata (Asclepiadaceae)

Tecoma spp. (Bignoniaceae); yellow trumpet bush; *vainillo*

Terminalia amazonia (Combretaceae); white olive; *surá, guayabón*

Terminalia catappa (Combretaceae); tropical almond; *almendro de playa*

Tetrapterys discolor (Malpighiaceae)

Thalia geniculata (Marantaceae); bent alligator-flag; *platanilla*

Thouinidium decandrum (Sapindaceae); *escobillo, sardino, mata pulgas*

Tithonia spp. (Asteraceae); tithonia; *mirasol*

Tournefortia hirsutissima (Boraginaceae); chiggery grapes

Tournefortia volubilis (Boraginaceae); bastard rat-root; *cola de alacrán*

Toxosiphon lindenii (Rutaceae)

Tragia volubilis (Euphorbiaceae); fireman; *picapica, ortiga brava*

Trichilia havanensis (Meliaceae); *uruca*

Tropaeolum majus (Tropaeolaceae); nasturtium; *capuchina, mastuerzo*

Tropaeolum moritzianum (Tropaeolaceae)

Turnera ulmifolia (Passifloraceae); ramgoat dashalong; *dachalong, mariposa amarilla*

Uncaria spp. (Rubiaceae); uncaria; *uña de gato*

Urera caracasana (Urticaceae); flameberry; *ortiga*

Vachellia collinsii (Fabaceae); bullhorn acacia; *cornizuelo*

Vigna vexillata (Fabaceae); zombie pea

Waltheria glomerata (Malvaceae); sleepy morning; *escobilla blanca*

Waltheria indica (Malvaceae); uhaloa; *escobilla blanca*

Warszewiczia coccinea (Rubiaceae); pride of Trinidad; *pastora de monte, lengua de diablo*

Zamia fairchildiana (Zamiaceae); zamia; *zamia*

Zamia neurophyllidia (Zamiaceae); zamia; *zamia*

Zamia skinneri (Zamiaceae); zamia; *zamia, yuquilla*

Zanthoxylum americanum (Rutaceae); common prickly ash; *lagartillo*

Zanthoxylum caribaeum (Rutaceae); prickly yellow; *lagartillo*

Zanthoxylum melanostictum (Rutaceae); *lagartillo, lagarto colorado, arcabú*

Zanthoxylum setulosum (Rutaceae); *lagartillo*

Zea mays (Poaceae); corn; *maíz*

Zinnia elegans (Asteraceae); elegant zinnia; *San Rafael, zinnia*

Zygia longifolia (Fabaceae); sotacaballo

Zygia palmana (Fabaceae)

GLOSSARY

abdomen. The posterior region of the insect body, located behind the thorax.

alkaloid. Naturally occurring chemical compound that contains at least one nitrogen atom. Alkaloids are produced by many plants (and some animals) as a chemical defense; well-known examples include caffeine and nicotine.

androconia (singular: androconium). Specialized scent scales on adult male butterflies that produce pheromones used to attract females. May occur in an **androconial patch** on wings or body.

anterior. Toward the front or the head.

anthropogenic. Involving the influence of humans on nature.

aposematic. Having conspicuous coloration, usually with bright and contrasting colors, that transmits a message to other species, such as unpalatability.

basal. Near or at the base of something or closer to the body.

Batesian mimicry. Resemblance of a species, such as a palatable insect, to an unpalatable one to gain protection against its predators.

bioaccumulation. The accumulation of chemicals in an organism.

biodiversity. All the different kinds of life found in one area.

bioindicator. An organism that reveals the health of an ecosystem.

bioregion. A region defined by characteristics of the natural environment rather than by human-made divisions.

charral. Spanish word for an early successional habitat composed mainly of herbaceous plants adapted to intense sunlight.

chemoreceptor. A sense organ capable of detecting chemicals.

chorion. The external shell of a butterfly egg.

chrysalis. The pupa of a butterfly.

cloud forest. A type of forest that is almost always covered in clouds.

cocoon. Silk shelter built by moth larvae (and other insects) inside which the pupa is formed.

compound eye. An eye composed of many independent, tiny simple eyes crowded into one unit.

convergence. In evolution, the development of similar characteristics in unrelated organisms, such as wings in insects, birds, and bats.

cordillera. Mountain range or system of great length.

courtship. The behavior of animals aimed at attracting a mate.

cremaster. A spine-like process at the end of a pupa used for attachment.

crepuscular. Active primarily at dawn and dusk.

crochets. Hooked spines at tip of the prolegs of caterpillars.

crypsis. An animal's natural camouflage, allowing it to conceal itself from predators.

cytoplasm. The fluid inside a cell.

derived. Describes a character that differs from the ancestral condition.

diapause. A state of dormancy.

disruptive coloration. A form of crypsis in which a high-contrast pattern breaks up an animal's outline, fooling predators.

distal. Away from the center or base, toward the tip or end, as of an appendage.

distal margin. The end of the wing away from its attachment with the body.

diurnal. Active during the daytime.

dorsal. Referring to the back, top, or upper side

dorsolateral. Referring to the upper part of the side close to the dorsum or to both the dorsal and the lateral areas.

dorsoventral. From the dorsal surface (top) to the ventral surface (bottom).

dorsum. The back, top, or upper side.

endemic (noun: endemism). Describes a species whose natural distribution is limited to a specific geographical area.

epiphyte. A plant that grows on another plant but is not parasitic.

escaping mimicry. A type of crypsis in which an animal resembles another species that is too hard (e.g., flying too fast or erratically) for predators to catch.

eversible. Capable of being everted, or turned outward or inside out.

exoskeleton. A skeletal or supportive structure on the outside of the body.

extrafloral nectary. Nectar-producing tissue located on a plant outside the flower (e.g., on the leaves).

eyespot. A color pattern that resembles an eye.

falcate. Hooked or curved like a sickle.

family. In taxonomy, a group of genera descending from the same ancestor.

frass. Excrement of insects.

frass chain. A long, thin structure caterpillars build by sticking bits of frass together with silk.

frenulum. A spine or bristle in the moth hind wing that inserts in a pocket in the forewing, locking the wings together.

galea. The outer lobe of the maxilla (a mouthpart) in insects.

gregarious. Living in groups.

herbaceous. Describes soft, nonwoody plants.

hill-topping. Male butterfly behavior of establishing and protecting a territory on the top of a hill.

homoneurous. Having the vein arrangement of the forewing identical to the arrangement on the hind wing.

host plant. The specific plant on which an insect can feed; in butterflies and moths it is the plant on which the larva feeds.

hyperparasitoid. A parasitoid whose host is another parasitoid.

inflorescence. Cluster of flowers.

infraorder. A taxonomic category below suborder.

instar. A stage of an insect between successive molts.

intertropical convergence zone. A low-pressure belt in the equatorial zone where the south and north trade winds converge (known to sailors as "the doldrums").

labial palps. A pair of small feelerlike structures arising from the labium.

labium. One of the structures that make up an insect's mouthparts; the "lower lip."

LAREBUB. Laboratory of Research on Butterfly Breeding at the Biology School of the University of Costa Rica.

lateral line. A longitudinal line on the side of the body.

lek. Gathering of males for group display to attract females.

mandible. Jaws or paired chewing mouthparts.

melanism. An unusual darkening of color owing to increased amount of black pigment.

metabolite. A substance produced as a result of the metabolism of an organism.

metathorax. The most posterior segment of the thorax.

microhabitat. A small habitat somewhat different from the surrounding environment.

mimetic ring. A group of unrelated species that obtain a benefit by sharing morphological similarities through mimicry.

mimicry. A situation in which one organism resembles another organism, often to its benefit. A species may mimic the color or body form of another species or even its odor.

molt. To shed the exoskeleton as part of the growth process, changing from one instar to the next.

monophagous. Feeding from a single specific food source.

montane. In or of mountains.

morph. A regularly occurring form (such as a color) of a species that varies from other members of the species.

morphology. The study of the form of organisms.

Müllerian mimicry. Describes a group of unrelated noxious organisms that share physical characteristics to obtain protection against their predators.

mutualistic relationship. A symbiotic relationship in which all species involved obtain a benefit from the others.

myrmecophily. A close mutualistic relationship with ants.

natural succession. The process by which an ecosystem naturally develops over time, such as from a field to a forest.

necrosis. Death of body tissue.

nectary. A structure on a plant that produces nectar, a carbohydrate-rich solution.

Neogene. The geological period from 23 to 2.5 million years ago, including the Miocene and Pliocene epochs.

Neotropics. The tropical region of the Americas.

ocellus. A simple eye that mainly distinguishes light intensity.

olfactory. Concerning the sense of smell.

oligophagous. Feeding from a limited group of plants.

order. A taxonomic subdivision of a class containing a group of related families.

orogenic. The formation of mountain ranges by upward displacement of the earth's crust.

osmeterium. A fleshy, eversible, Y-shaped tubular gland located in the first thorax segment of a caterpillar.

oviposit/oviposition. To lay eggs.

páramo. A high-altitude biome of Central and South America with vegetation composed mainly of low shrubs and grasses.

parasitoid. An organism that lives in a close parasitic association with another organism. Unlike a parasite, however, a parasitoid always kills its host (parasites only sometimes kill their host, if present in large numbers).

petiole. Leafstalk.

pheromone. A chemical substance released by an animal that serves to influence the behavior or physiology of other members of the same species. In butterflies and moths, pheromones may be used to attract sexual partners or send out an alarm signal.

photosynthesis. The process by which plants use sunlight, water, and carbon dioxide to create oxygen and energy in the form of sugar.

phylogeny. The study of the history of the lines of evolution in a group of organisms.

plesiomorphic. Describes a species character that looks primitive, similar to its form in its ancestors.

polychromatic. The occurrence of several different coloration forms in a population.

polymorphic/polymorphism. The occurrence of several different morphs or forms in a population.

polyphagous. Feeding on a diversity of plant species.

posterior. Toward the rear or the tail.

pre-montane. Describing the zone immediately below the montane zone.

primary consumers. In a food chain or web, the animals that eat the primary producers (plants).

proboscis. Elongated, often tubular mouthparts used for sucking up food.

prolegs. Fleshy structures in the ventral area of the caterpillar's abdomen used for clinging and locomotion.

prothorax (adjective: prothoracic). The anterior-most of the three thoracic segments.

proximal. Nearer to the body or to the base (as of an appendage).

pubescence. A downy covering of short fine hairs.

puddling. The butterfly behavior of drinking moisture or water from the ground, often on sand or riverbanks.

pupa. In insects undergoing complete metamorphosis, the motion-diminished immature stage during which the larval body is broken down and the adult body forms.

rain shadow. A region having little rainfall because it is sheltered from prevailing rain-bearing winds by mountains.

reproductive diapause. A period of arrested reproductive behavior.

reticulated. Having a netlike or networklike appearance or pattern.

retinaculum. A hollow fold in the ventral part of the moth forewing that receives the frenulum, locking the two wings together.

retractile. Capable of being pushed out and drawn back in (like a cat's claws).

rhizome. A rootlike, usually horizontal underground plant stem that produces shoots above and roots below.

riparian. Describes the area of wet land next to a river or stream.

savanna. A grassy plain with few trees in tropical and subtropical regions.

sclerotized. Hardened.

scoli (singular: scolus). Branched projections with setae at tip of each branch.

secondary compounds. Molecules produced to obtain a specific benefit but that are not directly involved in the organism's growth and development.

setae (singular: seta). Bristles or hairlike processes, such as those on many caterpillars.

sexual dimorphism. Refers to the sexes of the same species having different morphological characteristics.

skeletonize. To reduce a leaf to its veins by feeding on all the live tissue.

skipper. Lepidopteran that belongs to the family Hesperiidae.

spatulated. Spoon-shaped.

species. The basic unit of living organisms, determined mostly by the capacity to produce fertile offspring.

species density. The number of species in a standardized area, volume, or weight.

species richness. The number of species in a region.

spermatophore. A gelatinous sperm capsule passed from male to female during copulation.

sphingophyly. Pollination carried out by moths of the family Sphingidae.

sphragis. A mating plug deposited in the genital opening of a female by a male during mating to prevent the female from mating with other males.

spiracles. Small openings in the insect outer skeleton through which air enters the respiratory system. Generally, most body segments have a spiracle on each side.

stemmata (singular: stemma). The lateral simple eyes of insect larvae.

stridulatory. Concerning the ability to make a noise by rubbing two structures or surfaces together.

subfamily. A taxonomic category below family and above genus.

suborder. A taxonomic category below order and above family.

superfamily. A group of closely related families.

tarsus (plural: tarsi). The terminal, jointed part of the leg, usually with four segments and paired claws.

taxonomy. The practice and science of categorization or classification of life forms.

tegula. A small scale-like structure overlying the base of the forewing; a "shoulder pad."

territorial. Relating to a territory and the methods by which an animal establishes and protects it from incursions by others of its species.

thermoregulation. A process that allows the body of an organism to maintain its core internal temperature.

thorax (adjective: thoracic). The middle part of an insect body that holds the wings and legs.

tibia. The middle elongate segment of the insect leg between the femur and the tarsus.

tornus. The posterior corner of the wing.

tubercle. A knotlike, wartlike, or rounded protuberance.

tympanic/tympanal organ. A hearing structure consisting of a vibrating membrane.

type specimen. A specimen designated when a species is first described to serve as a reference.

univoltine. Undergoing an annual cycle, producing only one generation in a season; often the single brood of eggs or pupae undergoes diapause in winter or dry season.

venation. In Lepidoptera, the system of thickened lines in the wings that provides structural support and fluid transport.

ventral. Referring to the belly or underside of the body.

ventrolateral. Referring to the lower side of the body close to the ventral area or to both the ventral and lateral areas.

vermiform. Wormlike or worm-shaped.

vesicles. Small, fluid-filled sacs; in larvae, may hold substances that can cause burning or stinging on contact.

volatile compounds. Various organic chemicals released as gases, largely by plants.

wing pad. The area in the butterfly pupa of the newly forming wings.

BIBLIOGRAPHY

Ackery, P. R., R. Jong, and R. I. Vane-Wright. 1999. The butterflies: Hedyloidea, Hesperioidea and Papilionoidea. In N. P. Kristensen (ed.), *Lepidoptera, Moths and Butterflies*, vol. 1: *Evolution, Systematics and Biogeography*, 263–300. *Handbuch der Zoologie*, vol. 4, pt. 35. Berlin: Walter de Gruyter.

Aiello, A. 1979. Life history and behavior of the case-bearer *Phereoeca allutella* (Lepidoptera: Tineidae). *Psyche: J. Entomol.* 86 (2–3): 125–36.

Aiello, A. 1981. Life history of *Dismorphia amphiona beroe* (Lepidoptera: Pieridae: Dismorphiinae) in Panama. *Psyche: J. Entomol.* 87(3–4): 171–75. doi:10.1155/1980/38348.

Allen, P. 2010. Group size effects on survivorship and adult development in the gregarious larvae of *Euselasia chrysippe* (Lepidoptera, Riodinidae). *Insect. Soc.* 57: 199–204.

Allen, P. 2012. Survival patterns under Costa Rican field conditions of the gregarious caterpillar *Euselasia chrysippe* (Lepidoptera: Riodinidae), a potential biological control agent of *Miconia calvescens* (Melastomataceae) in Hawaii. *J. Res. Lepid.* 45: 77–84.

Alvarez Garcia, D. M., J. A. Díaz Pérez, and Á. Amarillo-Suárez. 2015. *Hylesia continua* (Walker, 1865) (Lepidoptera: Saturniidae) en una localidad del norte de Colombia: Dimorfismo en pupas y lepidopterismo. *Acta Zoológica Mexicana*, n.s. 31(2): 327–30.

Arias, C., C. Salazar, C. Rosales, M. Kronforst, M. Linares, E. Bermingham, and O. McMillan. 2014. Phylogeography of *Heliconius cydno* and its closest relatives: Disentangling their origin and diversification. *Molecular Ecology* 23: 4137–52.

Austin, G. 1998. Hesperiidae of Rondonia, Brazil: Notes on *Talides* Hübner (Lepidoptera: Hesperiidae: Hesperiinae). *Tropical Lepidoptera*, 9 (Suppl. 2): 26-32.

Austin, G., J. Brock, and O. Mielke. 1993. Ants, birds, and skippers. *Tropical Lepidoptera* 4, suppl. 2.

Barth, F. 1991. *Insects and Flowers: The Biology of a Partnership*. Princeton, NJ: Princeton University Press.

Barro, A., and R. Núñez. 2011. Diversidad endemismo y conservación. In *Lepidopteros de Cuba*, ed. A. Barro and R. Núñez. Finland: UPC Print.

Basset, Y., H. Barrios, S. Segar, R. Srygley, A. Aiello, A. Warren, et al. 2015. The butterflies of Barro Colorado Island, Panama: Local extinction since the 1930s. *PLoS One* 10 (8): e0136623. doi:10.1371/journal.pone.0136623.

Barreto Espindola, C. 2000. Biologia de *Oxydia vesulia* (Cramer, 1779) (Lepidoptera: Geometridae). *Floresta e Ambiente* 7 (1): 80–87.

Battisti, A., G. Holm, G. Fagrell, and S. Larsson. 2011. Urticating hairs in arthropods: Their nature and medical significance. *Annual Review of Entomology* 56: 203–20.

Beccaloni, G. W., A. L. Viloria, S. K. Hall, G. S. Robinson. 2008. *Catalogue of the Hostplants of the Neotropical Butterflies*. Vol. 8. Zaragoza, Spain: S.E.A.

Becker, V. O., and S. E. Miller. 2002. The large moths of Guana Island, British Virgin Islands: A survey of efficient colonizers (Sphingidae, Notodontidae, Noctuidae, Arctiidae, Geometridae, Hyblaeidae, Cossidae). *Journal of the Lepidoptera Society* 56: 9–44.

Bezzerides, A, and T. Eisner. 2002. Apportionment of nuptial alkaloidal gifts by a multiply-mated female moth (*Utetheisa ornatrix*): Eggs individually receive alkaloid from more than one male source. *Chemoecology* 12: 213–18.

Blau, W. 1980. Notes on the natural history of *Papilio polyxenes stabilis* (Papilionidae) in Costa Rica. *Journal of the Lepidopterists' Society* 34 (3): 321–24.

Boyle, B. L., N. Matasci, D. Mozzherin, T. Rees, G. C. Barbosa, R. Kumar Sajja, and B. J. Enquist. 2021. Taxonomic Name Resolution Service, v. 5.0. Botanical Information and Ecology Network. https://tnrs.biendata.org.

Bolaños, R., V. Watson, and J. Tosi. 2005. *Mapa ecológico de Costa Rica (zonas de vida) según el sistema de clasificación de zonas de vida del mundo de L. R. Holdridge, escala 1:750 000*. San José, Costa Rica: Centro Científico Tropical.

Borror, D., C. Triplehorn, and N. Johnson. 1992. *An Introduction to the Study of Insects*. 6th ed. Philadelphia: Saunders College Pub.

Braby, M. and K. Nishida. 2007. The immature stages, larval food plants and biology of neotropical mistletoe butterflies. I. The *Hesperocharis* group (Pieridae: Anthocharidini). *Journal of the Lepidopterists' Society* 61(4): 181–95.

Braby, M., and K. Nishida. 2010. The immature stages, larval food plants and biology of Neotropical mistletoe butterflies (Lepidoptera: Pieridae). II. The *Catasticta* group (Pierini: Aporiina). *Journal of Natural History* 44 (29–30): 1831–1928. doi:10.1080/00222931003633227.

Brower, A. 2010. Alleviating the taxonomic impediment of DNA barcoding and setting a bad precedent: Names for ten species of 'Astraptes fulgerator' (Lepidoptera: Hesperiidae: Eudaminae) with DNA-based diagnoses. *Systematics and Biodiversity* 8 (4): 485–91.

Brower, L. P., J.V.C. Brower, and C. T. Collins. 1963. Experimental studies of mimicry. 7. Relative palatability and Müllerian mimicry among butterflies of the subfamily Heliconiinae. *Zoologica* (NY) 48: 65–84.

Brown, F. M., and B. Heineman. 1972. *Jamaica and Its Butterflies*. London: E. W. Classey.

Brown, J. W. 1990. Sphingidae (Lepidoptera) of Isla del Coco, Costa Rica, with remarks on the macrolepidoptera fauna. *Brenesia* 33: 81–84.

Burns, J., J. Janzen, W. Hallwachs, M. Hajibabaei, and P. Hebert. 2009. Genitalia, DNA barcodes, and life histories synonymize *Telles* with *Thracides*—A genus in which *Telles arcalaus* looks out of place (Hesperiidae: Hesperiinae). *Journal of the Lepidopterists' Society* 63 (3): 141–53.

Caldwell, P. M., and R. L. Kluge. 1993. Failure of the introduction of *Actinote anteas* (Lep.: Acraeidae) from Costa Rica as a biological control candidate for *Chromolaena odorata* (Asteraceae) in South Africa. *Entomophaga* 38: 475–78.

Callaghan, C. J., and G. Lamas. 2004. Riodinidae. In G. Lamas (ed.), *Atlas of the Neotropical Lepidoptera*. Part 4A. *Hesperioidea–Papilionoidea*, 192–201. Gainesville, FL: Scientific Publishers.

Callahan, P. S. 1965. Far infra-red emission and detection by night-flying moths. *Nature* 207 (4989): 1172–73.

Calvo, R. 2004. Parasitoidismo por dípteros en larvas de *Caligo atreus* (Lepidoptera: Nymphalidae) en Cartago, Costa Rica. *Rev. Biol. Trop. (Int. J. Trop. Biol.)* 52 (4): 915–17.

Canet, M. 1986. Algunos aspectos del comportamiento, ciclo de vida, parasitismo y depredación de Caligo Memnos (Lepidoptera: Nymphalidae). Thesis, Biology Faculty, Universidad de Costa Rica.

Cave, R., and N. Acosta. 1999. *Telenomus remus* Nixon: Un parasitoide en el control biológico del gusano cogollero, *Spodoptera frugiperda* (Smith). *Ceiba* 40 (2): 215–27.

Cave, R., and R. Cordero. 1999. Parasitoides de *Leptophobia aripa* Boisduval (Lepidoptera: Pieridae) en repollo y brócoli en Honduras. *Ceiba* 40 (1): 51–55.

Chacón, I. 1986. Historia natural de *Papilio cleotas archytas* (Lepidoptera, Papilionidae) en Costa Rica. *Brenesia* 25–26: 215–20.

Chacón, I. 2001. *Hades noctula*. Biodiversity of Costa Rica. Accessed March 2020; website no longer active. http://www.crbio.cr:8080/neoportal-web/species/Hades%20noctula.

Chacón, I., and J. Montero. 2007. *Mariposas de Costa Rica*. Heredia, Costa Rica: Editorial INBio.

Chacón, I., J. Montero-Ramírez, D. Janzen, W. Hallwachs, P. Blandin, C. Bristow, and M. Hajibabaei. 2012. A new species of *Opsiphanes* Doubleday [1849] from Costa Rica (Nymphalidae: Morphinae: Brassolini), as revealed by its DNA barcodes and habitus. *Bull. Allyn Mus.* 166: 1–15.

Chacón, I. A., D. H. Janzen, W. Hallwachs, T. Dapkey, D. Harvey, and N. V. Grishin. 2017. Six new cryptic species of *Xylodonta* Becker, 2014 (Notodontidae: Nystaleinae) from Costa Rica. *Trop. Lepid. Res.* 27 (1): 33–58.

Chacón, I., P. J. DeVries, and C. Penz. 2018. Description of a new subspecies of *Cunizza hirlanda* (Pieridae: Anthocharidini) from Costa Rica. *J. Lep. Soc.* 72 (2): 121–26.

Chai, P. 1986. Field observations and feeding experiments on the responses of rufous-tailed jacamars (*Galbula ruficauda*) to free-flying butterflies in a tropical rainforest. *Bio. J. Linn. Soc.* 29 (3): 161–89.

Chai, P. 1990. Relationships between visual characteristics of rainforest butterflies and responses of a specialized insectivorous bird. In M. Wicksten (ed.), *Adaptive Coloration in Invertebrates: Proceedings of a Symposium Sponsored by the American Society of Zoologists*, 31–60. Galveston: Texas A&M University.

Cock, M. J. W. 2015. Observations on the biology of skipper butterflies in Trinidad, West Indies: *Urbanus, Astraptes* and *Narcosius* (Hesperiidae: Eudaminae). *Living World, Journal of the Trinidad and Tobago Field Naturalists' Club*, (2015): 1–14.

Coen, E. 1983. Climate. In D. Janzen (ed.), *Costa Rican Natural History*. Chicago: University of Chicago Press.

Collins, C. W., and S. F. Potts. 1932. Attractants for the flying gypsy moths as an aid in locating new infestations. *USDA Tech. Bull.* 336: 1–43.

Córdoba-Alfaro, J., and D. Gómez. 2017. Early stages of *Morpho amathonte* (Lepidoptera: Nymphalidae, Morphinae) and its variation on the Pacific coast of Costa Rica. *Revista Peruana de Biología* 24 (2): 151–54.

Corro, P. 2018. Panama moths: Notes on the life history of *Gonodonta incurva* (Sepp, [1840]) (Erebidae, Calpinae). *Int. J. Avian and Wildlife Biol.* 3 (6): 405[]7.

Coto, D. 1987. Noticias del servicio de diagnóstico. *Boletin Informativo de Manejo Interado de Plagas* 3: 5.

Coto, D., and J. Saunders. 2004. *Insectos plaga de cultivos perennes con énfasis en frutales en América Central*. Serie técnica 52. Turrialba, Costa Rica: Centro Agronómico Tropical de Investigación y Enseñanza.

Cott, H. B. 1940. *Adaptive Coloration in Animals*. London: Methuen.

Danaher, M. W., C. Ward, L. W.. Zettler, and C. Covell. 2019. Pollinia removal and suspected pollination of the endangered ghost orchid, *Dendrophylax lindenii* (Orchidaceae) by various hawk moths (Lepidoptera: Sphingidae): Another mystery dispelled. *Florida Entomologist* 102, (4): 671–83.

Desmier de Chenon, R., A. Sipayung, and A. Sudharto. 2002. A new biological agent, *Actinote anteas*, introduced into Indonesia from South America for the control of *Chromolaena odorata*. In C. Zachariades, R. Muniappan, and L.W. Strathie (eds.), *Proceedings of the Fifth International Workshop on Biological Control and Management of Chromolaena odorata*, 170–76. Pretoria, South Africa: ARC-PPRI.

DeVries, P. J. 1978. Apparent lek behavior in Costa Rican *Perrhybris pyrra* (Lep.:Pieridae) from the Osa Peninsula. *J. Res. Lepid.* 17: 142–44.

DeVries, P. J. 1983a. *Phoebis philea*. In D. Janzen (ed.), *Costa Rican Natural History*. Chicago: University of Chicago Press.

DeVries P. J. 1983b. *Catasticta teutila* (Paracaída, Teutilla). In D. Janzen (ed.), *Costa Rican Natural History*. Chicago: University of Chicago Press.

DeVries, P. J. 1987. *The Butterflies of Costa Rica and Their Natural History: Papilionidae, Pieridae, and Nymphalidae*. Princeton, NJ: Princeton University Press.

DeVries, P. J. 1997. *The Butterflies of Costa Rica and Their Natural History*. Vol. 2: *Riodinidae*. Princeton, NJ: Princeton University Press.

DeVries, P. J., J. Schull, and N. Greig. 1987. Synchronous nocturnal activity and gregarious roosting in the neotropical skipper butterfly *Calaenorrhinus fritzgaertneri* (Lepidoptera: Hesperiidae). *Zool. J. Linn. Soc.* 89: 89–103.

DeVries P. J., and R. Dudley. 1990. Morphometrics, airspeed, thermoregulation, and lipid reserves of migrating *Urania fulgens* (Uraniidae) moths in natural free flight. *Physiol. Zool.* 63: 235–51

DeVries, P. J., I. Chacon, and D. Murray. 1992. Toward a better understanding of host use and biodiversity in riodinid butterflies (Lepidoptera). *J. Res. Lep.* 31 (1–2): 103–26.

Dudley, R., and P. J. DeVries. 1990. Flight physiology of migrating *Urania fulgens* (Uraniidae) moths: Kinematics and aerodynamics of natural free flight. *J. Comp. Physiol.* A (1990) 167: 145–54.

Dudley, R., R. B. Srygley, E. G. Oliveira, and P. J. DeVries. 2002. Flight speeds, lipid reserves, and predation of the migratory Neotropical moth *Urania fulgens* (Uraniidae). *Biotropica* 34: 452–58.

Durán, J., G. Fagua, J. Robles, and E. Gil. 2012. Sequestration of Aristolochic Acid I from *Aristolochia pilosa* by *Mapeta xanthomelas* Walker, 1863. *J. Chem. Ecol.* 38: 1285–88.

Epstein, M. E. 1996. Revision and phylogeny of the Limacodid-group families, with evolutionary studies on slug caterpillars (Lepidoptera: Zigaenoidea). *Smithsonian Contributions to Zoology* 582: 1–102.

Espeland, M., J. Breinholt,K. Willmott, A. Warren, R. Vila, E. Toussaint et al. 2018. A comprehensive and dated phylogenomic analysis of butterflies. *Current Biology* 28: 1–9. https://doi.org/10.1016/j.cub.2018.01.061.

Estrada, C., and C. Jiggins. 2002. Patterns of pollen feeding and habitat preference among *Heliconius* species. *Ecological Entomology* 27: 448–56.

Flores-Vindas, E. 1999. *La planta: estructura y funsión.* Cartago, Costa Rica: Libro Universitario Regional.

Fuentes-Quintanar, J. H., E. O. Martínez-Luque, and H. Álvarez-García. 2019. Asociación de larvas de *Oxydia vesulia*

Cramer, 1779 (Lepidoptera: Geometridae) con parasitoides microgastrinos (Hymenoptera: Braconidae). *Entomología Mexicana* 6: 87–90.

Fleming, A., D. Wood, D. Janzen, W. Hallwachs, and M. Smith. 2015. Seven new species of *Spathidexia* Townsend (Diptera: Tachinidae) reared from caterpillars in Area de Conservación Guanacaste, Costa Rica. *Biodivers. Data J.* 3: e4597. doi:10.3897/BDJ.3.e4597

Fleming, A.J. , D.M. Wood, M. A. Smith, W. Hallwachs, and D. H. Janzen. 2018. Revision of the Mesoamerican species of *Calolydella* Townsend (Diptera: Tachinidae) and descriptions of twenty-three new species reared from caterpillars in Area de Conservación Guanacaste, northwestern Costa Rica. *Biodivers. Data J.* 6: e11223.

Garwood K., B. Huertas, I. C. Ríos-Málaver, J. G. Jaramillo. 2022. *Mariposas de Colombia lista de chequeo / Butterflies of Colombia Checklist (Lepidoptera: Papilionoidea).* 2nd ed. BioButterfly Database. http://www.butterflycatalogs.com.

Gilbert, L. 1972. Pollen feeding and reproductive biology of *Heliconius* butterflies. *PNAS* 69 (6): 1403–7.

Greeney, H., and M. Jones. 2003. Shelter building in the Hesperiidae: A classification scheme for larval shelters. *J. Res. Lepidoptera* 37: 27–36.

Haber, W. A. 1983. *Hylocereus costaricensis* (pitahaya silvestre, wild pitahaya). In D. Janzen (ed.), *Costa Rican Natural History.* Chicago: University of Chicago Press.

Haber, W. A., and R. Stevenson. 2004. Diversity, migration, and conservation of butterflies in northern Costa Rica. In G. Frankie, A. Mata, and B. Vinson (eds.), *Biodiversity Conservation in Costa Rica: Learning the Lessons in a Seasonal Dry Forest.* Berkeley and Los Angeles: University of California Press.

Hammel, B. E., M. H. Grayum, C. Herrera, and N. Zamora (eds.). 2003a. *Manual de plantas de Costa Rica.* Vol. 2. Heredia, Costa Rica: Missouri Botanical Garden/INBio.

Hammel, B. E., M. H. Grayum, C. Herrera, and N. Zamora (eds.). 2003b. *Manual de plantas de Costa Rica.*Vol. 3. Heredia, Costa Rica: Missouri Botanical Garden/INBio.

Hammel, B. E., M. H. Grayum, C. Herrera, and N. Zamora (eds.). 2010. *Manual de plantas de Costa Rica.*Vol. 5. Heredia, Costa Rica: Missouri Botanical Garden/INBio.

Hammel, B. E., M .H. Grayum, C. Herrera, and N. Zamora (eds.). 2007. *Manual de plantas de Costa Rica.* Vol. 6. Heredia, Costa Rica: Missouri Botanical Garden/INBio.

Hammel, B. E., M. H. Grayum, C. Herrera, and N. Zamora (eds.). 2014. *Manual de plantas de Costa Rica.*Vol. 7. Heredia, Costa Rica: Missouri Botanical Garden/INBio.

Hammel, B. E., M. H. Grayum, C. Herrera, N. Zamora (eds.). 2015. *Manual de plantas de Costa Rica.* Vol 8. Heredia, Costa Rica: Missouri Botanical Garden/INBio.

Hebert, P., E. Penton, J. Burns, D. Janzen, and W. Hallwachs. 2004. Ten species in one: DNA barcoding reveals cryptic species in the neotropical skipper butterfly *Astraptes fulgerator*. *PNAS*, 101 (41): 14812–17

Hedelin, H., and J. Rydell. 2007. Daily habitat shifts by the neotropical butterfly *Manataria maculata* (Nymphalidae: Satyrinae) is driven by predation. *J. Lep. Soc.* 61 (2): 67–71.

Heikkilä, M., L. Kaila, M. Mutanen,C. Peña, and N. Wahlberg. 2011. Cretaceous origin and repeated tertiary diversification of the redefined butterflies. *Proc. R. Soc. B.* 279 (1731). doi:10.1098/rspb.2011.1430.

Henderson, C. 2002. *Butterflies, Moths, and Other Invertebrates of Costa Rica.* Austin: University of Texas Press.

Henning, G., R. Terblanche, and J. Ball. 2009. South African red data book: Butterflies. SAMBI Biodiversity Series 13. Pretoria: South Africa National Biodiversity Institute.

Heppner, J. B. 1998. Classification of Lepidoptera. Part 1. Introduction. *Holartic Lepidoptera* 5 (suppl.1): 1–85.

Holdridge, L. R. 1967. *Life Zone Ecology.* Rev. ed. San José, Costa Rica: Tropical Science Center.

Hurlbert, S. H. 1971. The nonconcept of species diversity: A critique and alternative parameters. *Ecology* 52: 577–86.

Janzen, D. 1982. Guía para la identificación de mariposas nocturnas de la familia Saturniidae del Parque Nacional Santa Rosa, Guanacaste, Costa Rica. *Brenesia* 19–20: 255–99.

Janzen, D. 1983. Insects. In D. Janzen (ed.), *Costa Rican Natural History.* Chicago: University of Chicago Press.

Janzen, D. H. 1984. Weather-related color polymorphism of *Rothschildia lebeau* (Saturniidae). *Bull. E.S.A.*: 16–20.

Janzen, D. H., and W. Hallwachs. 2019. Dynamic database for an inventory of the macrocaterpillar fauna, and its food plants and parasitoids, of Area de Conservación Guanacaste (ACG), northwestern Costa Rica (00-SRNP-8939, 09-SRNP-67854, 97-SRNP-66489, 06-SRNP-17415, 84-SRNP-1614, 06-SRNP-12981, 07-SRNP-2081, 17-SRNP-30870, 17-SRNP-70436, 98-SRNP-3206, 05-SRNP-35240, 99-SRNP-1186, 06-SRNP-42199, 13-SRNP-70898, 83-SRNP-560, 83-SRNP-1218, 95-SRNP-7265, 09-SRNP-65536, 97-SRNP-3245, 08-SRNP-71927, 08-SRNP-40102, 07-SRNP-42155, 07-SRNP-42032, 17-SRNP-32588, 95-SRNP-4760, 05-SRNP-4432, 08-SRNP-31839, 04-SRNP-22248, 08-SRNP-31609, 08-SRNP-42207, 80-SRNP-200, 84-SRNP-603, 08-SRNP-30693, DHJ462310, 17-SRNP-20592, 06-SRNP-46888, 09-SRNP-35341, 12-SRNP-40500, 07-SRNP-58220, 11-SRNP-30007, 15-SRNP-70633, 07-SRNP-45461, 08-SRNP-71928, 17-SRNP-71358, 04-SRNP-56422, 08-SRNP-31825, 05-SRNP-59184, 06-SRNP-9938, 08-SRNP-31902, 11-SRNP-65635, 07-SRNP-46175, 08-SRNP-55392, 16-SRNP-31601, 08-SRNP-55456, 84-SRNP-1597, 08-SRNP-56565, 95-SRNP-4612), 07-SRNP-60873, 07-SRNP-60784, 08-SRNP-20256, 07-SRNP-4898, 04-SRNP-55192, 04-SRNP-55451, 07-SRNP-59430, 08-SRNP-16396, 07-SRNP-57237, 13-SRNP-55505). http://janzen.sas.upenn.edu.

Jiggins, C. 2017. *The Ecology and Evolution of* Heliconius *Butterflies.* Oxford and New York: Oxford University Press.

Jiménez, V. 2016. Revisión taxonómica de la familia Aristolochiaceae en Costa Rica y Flora Vascular de las Sabanas Miravalles, Volcán Miravalles, Costa Rica. Master's thesis, School of Biology, University of Costa Rica.

Kappelle, M. 2005. Insectos de los páramos de Costa Rica. In M. Kappelle, and S. P. Horn (eds.), *Páramos de Costa Rica.* Heredia, Costa Rica: Editorial INBio.

Kawahara, A., and J. Breinholt. 2014. Phylogenomics provides strong evidence for relationships of butterflies and moths. *Proc. R. Soc. B* 281: 20140970. http://dx.doi.org/10.1098/rspb.2014.0970.

Kellert, S. R. 1993. Values and perceptions of invertebrates. *Conserv. Biol.* 7: 845–55.

Kendall, R. 1970. A day-flying moth (Pericopidae) new to Texas and the United States. *J. Lep. Soc.*, 24 (4): 301–3.

Kendall, R. 1980. Larval food plants and life history notes for eight moths from Texas and Mexico. *J. Lep. Soc.* 30 (4): 264–71.

Kern, W. H. Jr. 2015. Featured creatures: *Epicorsia oedipodalis* (Guenée, 1854) (Lepidoptera: Ditrysia: Pyraloidea: Pyralidae: Pyraustinae). University of Florida, Institute of Food and Agriculture Sciences. http://entnemdept.ufl.edu/creatures/ORN/fiddlewood_leafroller.htm.

Krasnoff, S. B, , and D. E. Dussourd. 1989. Dihydropyrrolizine attractants for Arctiid moths that visit plants containing pyrrolizidine alkaloids. *J. Chem. Ecol.* 15: 47–60.

Kristensen, N., M. Scoble, and O. Karsholt. 2007. Lepidoptera phylogeny and systematics: The state of inventorying moth and butterfly diversity. *Zootaxa* 1668: 699–747.

Koh, L., and N. Sodhi. 2004. Importance of reserves, fragments, and parks for butterfly conservation in a tropical urban landscape. *Ecol. Appl.* 14: 1695–1708.

Lamas, G. 1997. Comparing the butterflies faunas of Pakitza and Tambopata, Madre de Dios, Peru, or Why is Peru such a megadiverse country? In H. Ulrich (ed.), *Tropical Biodiversity and Systematics*, 165–68. Bonn: Zoologisches Forschungsinstitut und Museum Alexander Koenig.

Lamas, G. 2000. Lepidopteros neotropicales. In F. Martín Piera, J. Morrone, and A. Melic (eds.), *Hacia un proyecto Cyted para el inventario y estimacion de la diversidad entomológica en Iberoamérica: PrIBES 2000.* m3m-Monografías Tercer Milenio, Vol. 1. Zaragoza: Sociedad Entomológica Aragonesa.

Lamas, G. (ed.). 2004. *Atlas of the Neotropical Lepidoptera.* Part 4A. *Hesperioidea–*

Papilionoidea, 192–201. Gainesville, FL: Scientific Publishers.

Leite, L. A. R., A. V. L. Freitas, E. P. Barbosa, M. Casagrande, and O. H. H. Mielke. 2014. Immature stages of nine species of genus *Dynamine* Hübner, [1819]: Morphology and natural history (Lepidoptera: Nymphalidae, Biblidinae). *SHILAP Revta. Lepid.* 42 (165): 27–55.

Lemaire C. 1978. *Les Attacidae Americains: Attacinae.* Neuilly-sur-Seine: Édition C. Lemaire.

Lemaire C. 1980. *Les Attacidae Americains: Arsenurinae.* Neuilly-sur-Seine: Édition C. Lemaire.

Lemaire C. 1987 Les Attacidae Americains: Ceratocampinae. Neuilly-sur-Seine: Édition C. Lemaire.

Lewinsohn, T., V. Freitas, and P. Prado. 2005. Conservation of terrestrial invertebrates and their habitats in Brazil. *Conserv. Biol.* 19 (3): 640–45.

Ley de biodiversidad y su reglamento. 1st. ed. San José, Costa Rica: Editorial Investigaciones Jurídicas.

Llorente-Bousquets, J., I. Vargas Fernández, A. Luis-Martínez, M. Trujano-Ortega, B. Hernández-Mejía, and A. Warren. 2014. Biodiversidad de Lepidóptera en México. *Revta. Mexicana de Biodiversidad*, suppl. 85: S353–71.

Llorente-Bousquets, J., S. Nieves-Uribe, A. Flores-Gallardo, B. C. Hernández-Mejía, and J. Castro-Gerardino. 2018. Chorionic sculpture of eggs in the subfamily Dismorphiinae (Lepidoptera: Papilionoidea: Pieridae). *Zootaxa* 4429 (2): 201–46. doi:10.11646/zootaxa.4429.2.1.

Maclauring, J., and K. Sterrenly. 2008. *What Is Biodiversity?* Chicago: University of Chicago Press.

Malicky, H. 1970. New aspects on the association between Lycaenid larvae (Lycaenidae) and ants (Formicidae, Hymenoptera). *J. Lep. Soc.* 24 (3): 190–202.

Mallet, J. 1986. Gregarious roosting and home range in *Heliconius* butterflies. *Natl. Geo. Res.* 2 (2): 198–215.

Martínez, M., M. Peichoto, M. Piriz, A. Zapata, and O. Salomón. 2019. Erucismo, etiología, epidemiología y aspectos clínicos en San Ignacio, Misiones, Argentina. *Revta. Venez. Salud Pública* 7 (2): 25–34.

Méndez, V., and J. Monge-Nájera. 2010. *Costa Rica: Historia Natural*. San José, Costa Rica: EUNED.

Mexzón, R., C. Chinchilla, and R. Rodríguez. 2003. El gusano canasta, *Oiketicus kirbyi* Lands Guilding (Lepidoptera: Psychidae), plaga de la palma aceitera. *ASD Oil Palm Paper* (Costa Rica) 25: 24–28.

Mielke, O.H.H. 2004. Morphinae. In G. Lamas (ed.), *Atlas of the Neotropical Lepidoptera*, Part 4A, *Hesperioidea–Papilionoidea*, 192–201. Gainesville, FL: Scientific Publishers.

Miller, J., D. Janzen, and W. Hallwachs. 2006. *100 Caterpillars*. Cambridge, MA: Belknap Press of Harvard University Press.

Miller, J., D. Janzen, and W. Hallwachs. 2007. *100 Butterflies and Moths*. Cambridge, MA: Belknap Press of Harvard University Press.

Missouri Botanical Garden. 2021. *Tropicos*. Missouri Botanical Garden. http://www. tropicos.org. Accessed 5 March 2021.

Mohanraj, P., and K. Veenakumari. 2014. Preimaginal stages and natural history of two endemic subspecies of *Polyura* Billberg (Lepidoptera: Nymphalidae: Charaxinae) from the Andaman Islands. *Proc. Natl. Acad. Sci., India, Sect. B Biol. Sci.* 84 (2): 265–73.

Monge, M. 2018. Guía para la identificación de las principales plagas y enfermedades en el cultivo de piña. Universidad de Costa Rica-MAG. Costa Rica. 44p.

Monge-Nájera, J. 1992. Clicking butterflies, *Hamadryas*, of Panama: Their biology and classification (Lepidoptera, Nymphalidae). In D. Quintero and A. Aiello (eds.), *Insects of Panama and Mesoamerica: Selected Studies*, 567–72. Oxford: Oxford University.

Monge-Nájera, J., F. Hernández, M. I. González, J. Soley, J. Araya, and S. Zolla. 1998. Spatial distribution, territoriality and sound production by tropical cryptic butterflies (*Hamadryas*, Lepidoptera: Nymphalidae): Implications for the 'Industrial Melanism' debate. *Rev. Biol. Trop.* 48: 297–329.

Monroe, J. L. 2016. *The Large Sulphurs of the Americas*. Gainesville, FL: International Biodiversity Foundation.

Montero, R. 2007. *Manual para el manejo de mariposarios*. Heredia, Costa Rica: Editorial INBio.

Moraes, S. S., and M. Duarte. 2014. Phylogeny of Neotropical Castniinae (Lepidoptera: Cossoidea: Castniidae): Testing the hypothesis of the mimics as a monophyletic group and implications for the arrangement of the genera. *Zool. J. Linn. Soc.* doi:10.1111/zoj.12102.

Murillo-Hiller, L. R., and K. Nishida. 2003. Life history of *Manataria maculata* (Lepidoptera: Satyrinae) from Costa Rica. *Rev. Biol. Trop.* 51 (2): 463–70.

Murillo-Hiller, L. R. 2008. Notas sobre el comportamiento y la migración de *Urania fulgens* (Lepidoptera: Uraniidae) en Costa Rica. *Acta Zoológica Mexicana* n.s 24 (1): 239–41.

Murillo-Hiller, L. R. 2008. Clave dicotómica para la identificación de las familias de mariposas (Rhopalocera) pertenecientes a las superfamilias Papilionoidea y Hesperioidea. Métodos en Ecología y Sistemática 3 (2): 6–11.

Murillo-Hiller, L. R. 2009. Early stages and natural history of *Cithaerias p. pireta* (Satyrinae) from Costa Rica. *J. Lep. Soc.* 63 (3): 169–72.

Murillo-Hiller, L. R. 2016. *Mariposas y Polillas de Costa Rica*. San José: School of Biology, University of Costa Rica.

Murillo-Hiller, L. R., and N. Canet. 2018. Early stages and natural history of *Morpho menelaus amathonte* Deyrolle, 1869 and *Morpho helenor marinita* Butler, 1872 (Nymphalidae: Morphinae) from Costa Rica. *J. Lep. Soc.* 72 (1): 74–80.

Murillo-Hiller, L. R., O. Segura-Bermúdez, J. D. Barquero, and F. Bolaños. 2019. The skipper butterflies (Lepidoptera: Hesperiidae) of the Reserva Ecológica Leonelo Oviedo, San José, Costa Rica. *Rev. Biol. Trop.* 67 (2), suppl.: S228–48.

Muyshondt, A. 1973a. Notes on the life cycle and natural history of butterflies of El Salvador. I A. *Catonephele numilia esite* (Nymphalidae-Catonephelinae). *J. NY Entomol. Soc.* 81 (3): 164–74.

Muyshondt, A. 1973b. Notes on the life cycle and natural history of butterflies of

El Salvador. II A. *Epiphile adrasta adrasta* (Nymphalidae-Catonephelinae). *J. NY Entomol. Soc.* 81 (4): 214–23.

Muyshondt, A. 1975. Notes on the life cycle and natural history of butterflies of El Salvador. VI A. *Diaethria astala* Guérin. (Nymphalidae-Callicorinae). *J. NY Entomol. Soc.* 83 (1): 10–18.

Muyshondt, A., and. A. Muyshondt Jr. 1978. Notes on the life cycle and natural history of butterflies of El Salvador. II C. *Smyrna blomfildia* and *S. karwinskii* (Nymphalidae: Coloburini). *J. Lep. Soc.* 32 (3): 160–74

New, T. E. 1993. Introduction to the biology and conservation of the Lycaenidae. In T. R. New (ed.), *Conservation Biology of Lycaenidae*, 1–21. Switzerland: International Union for Conservation of Nature and Natural Resources.

Nield, A. 2008. *The Butterflies of Venezuela.* Part 2. *Nymphalidae II (Acaeinae, Libytheinae, Nymphalinae, Ithomiinae, Morphinae).* London: Meridian Publications.

Nishida, K. 2010. Description of the immature stages and life history of *Euselasia* (Lepidoptera: Riodinidae) on *Miconia* (Melastomataceae) in Costa Rica. *Zootaxa* 2466: 1–74.

Nishida, K., I. Nakamura, and C. O. Morales. 2009. Plants and butterflies of a small urban preserve in the Central Valley of Costa Rica. *Rev. Biol. Trop.* 57 (1): 31–67.

Orellana, A. 2000. Palatabilidad, mimetismo y defensas químicas en polillas Dioptinae (Lepidoptera, Notodontidae). Thesis. Departament of Biology, University of the Andes.

Paniagua, M., and L. R. Murillo-Hiller. 2015. Ciclo de vida de la mariposa *Morpho helenor narcissus* en Costa Rica. *Boletin INAGROP* 5. (1): 3–4.

Penz, C. 1998. Early stages of *Myscelia cyaniris cyaniris* (Doubleday) from Panama (Nymphalidae, Nymphalinae). *J. Lep. Soc.* 52 (3): 338–41.

Penz, C. 2008. Phylogenetic revision of *Eryphanis* Boisduval, with a description of a new species from Ecuador (Lepidoptera, Nymphalidae). *Insecta Mundi* 35: 1–25.

Peña, C., and M. Espeland. 2015. Diversity dynamics in Nymphalidae butterflies: Effect of phylogenetic uncertainty on diversification rate shift estimates. *PLoS One* 10 (4): e0120928. doi:10.1371/journal.pone.0120928.

Phillips-Rodríguez, E., and J. Powell. 2007. Phylogenetic relationships, systematics, and biology of the species of *Amorbia* Clemens (Lepidoptera: Tortricidae: Sparganothini). *Zootaxa* 1670: 1–109.

Pinheiro, C. 1996. Palatability and escaping ability in Neotropical butterflies: Tests with wild kingbirds (*Tyrannus melancholicus*, Tyrannidae). *Biol. J. Linn. Soc.* 59: 351–65.

Pinheiro, C., and A. Freitas. 2014. Some possible cases of escape mimicry in Neotropical butterflies. *Neotrop. Entomol.* 43: 393–98.

Pinheiro, C., A. Freitas, V. Campos, P. J. DeVries, and C. Penz. 2016. Both palatable and unpalatable butterflies use bright colors to signal difficulty of capture to predators. *Neotrop. Entomol.* 45: 107–13. doi:10.1007/s13744-015-0359-5.

Piovesan, M., M. M. Casagrande, and O.H.H. Mielke. 2022. Systematics of *Opsiphanes* Doubleday, [1849] (Lepidoptera: Nymphalidae, Satyrinae, Brassolini): An integrative approach. *Zootaxa* 5216 (1): 1–278.

Plants of the World Online. 2021. Facilitated by the Royal Botanic Gardens, Kew. http://www.plantsoftheworldonline.org. Accessed March 5, 2021.

Pohl, G., R. Cannings, J.-F. Landry, D. Holden, and G. Scudder. 2015. Check list of the Lepidoptera of British Columbia, Canada. *Entomol. Soc. British Columbia*, occ. paper 3.

Prance, G. 1994. A comparison of the efficacy of higher taxa and species numbers in the assessment of biodiversity in the Neotropics. *Phil. Trans. R. Soc. Lond. B* 345: 89.

Rhainds, M., D. Davis, and P. W. Price. 2009. Bionomics of bagworms (Lepidoptera: Psychidae). *Annu. Rev. Entomol.* 54: 209–26.

Ribeiro, R. C., I. O. Carvalho, G. K. Souza, H. A. Fouad, and W. P. Lemos. 2012. *Thracides phidon* (Cramer) (Lepidoptera: Hesperiidae: Hesperiinae): Novo registro em plantios comerciais de *Heliconia* spp. na região Amazônica do Brasil. *EntomoBrasilis* 5 (1): 82–83.

Robbins, R. K. 2004. Introduction to the checklist of Eumaeini (Lycaenidae). In G. Lamas (ed.), *Atlas of the Neotropical Lepidoptera*, Part 4A, *Hesperioidea–Papilionoidea*, 192–201. Gainesville, FL: Scientific Publishers.

Robbins, R. K. 2010. The "upside down" systematics of hairstreak butterflies (Lycaenidae) that eat pineapple and other Bromeliaceae. *Studies on Neotrop. Fauna and Environ.* 45 (1): 21–37.

Robbins, R. K., and G. Lamas. 2004. Lycaenidae. In G. Lamas (ed.), *Atlas of the Neotropical Lepidoptera*, Part 4A, *Hesperioidea–Papilionoidea*, 192–201. Gainesville, FL: Scientific Publishers.

Robbins, R, and M. Duarte. 2005. Phylogenetic analysis of *Cyanophrys* Clench, a synopsis of its species, and the potentially threatened *C. bertha* (Jones) (Lycaenidae: Theclinae: Eumaeini). *Proc. Entomol. Soc. Wash.* 107 (2): 398–416.

Rojas-Ugalde, C., and L. R. Murillo-Hiller. 2021. A description of the immature stages and natural history of *Doxocopa laurentia cherubina* (C. Felder & R. Felder, 1867) with comparison to *Doxocopa laure laure* (Drury, 1773) (Nymphalidae: Apaturinae). *J. Lep. Soc.* 75 (2): 153–57.

Rydell, J., S. Kaerma, H. Hedelin, and N. Skals. 2003. Evasive response to ultrasound by the crepuscular butterfly *Manataria maculata*. *Naturwissenschaften* 90: 80–83.

Samayoa, A. C., and R. Cave. 2008. Catálogo de las especies de Sphingidae (Lepidoptera) en Honduras. *Ceiba* 49 (1): 103–17.

Sánchez, J. 2002. Aves del Parque Nacional Tapantí. 1st ed. Heredia, Costa Rica: Editorial INBio.

Sandoval, L., C. O. Morales, J. D. Ramírez-Fernández, P. Hanson, L. R. Murillo-Hiller, and G. Barrantes. 2019. The forgotten habitats in conservation: Early successional vegetation. *Rev. Biol. Trop.* 67 (2), suppl.: S36–52.

Saunders, J., D. Coto, and A. King. 1998. Plagas invertebradas de cultivos anuales alimenticios en América Central. 2nd ed. Turrialba, Costa Rica: CATIE.

Schwartz, A. 1989. *The Butterflies of Hispaniola*. Gainesville, FL: University of Florida Press.

Schlegel, J., G. Breuer, and R. Rupf. 2015. Local insects as flagship species to promote nature conservation? A survey among primary school children on their attitudes toward invertebrates. *Anthrozoös* 28: 229–45.

Scoble, M. J. 1992. *The Lepidoptera: Form, Function and Diversity*. New York: Oxford University Press.

Seraphim, N., L. Kaminski, P. J. DeVries, C. Penz, C. Callaghan, N. Wahlberg, K. Silva-Brandão, and A. Freitas. 2018. Molecular phylogeny and higher systematics of the metalmark butterflies (Lepidoptera: Riodinidae). *System. Entomol.* 43: 407–25.

Sermeño-Chicas, J. M., and F. A. Parada-Berríos. 2013. Insectos asociados al mamey (*Mammea americana* L.) en El Salvador. *Bioma* 11: 50–59.

Sermeño-Chicas, J. M., D. Pérez, F. A. Parada-Berríos, R. Menjívar, and R. Estrada. 2014. Guía de artrópodos asociados al árbol de ojushte (*Brosium alicastrum* Swartz) en El Salvador. San Salvador: Universidad de El Salvador.

Silva, N., M. Duarte, E. Araújo, and H. C. Morais. 2014. Larval biology of anthophagous Eumaeini (Lepidoptera: Lycaenidae, Theclinae) in the cerrado of central Brazil. *J. Insect Sci.* 14: 184. doi:10.1093/jisesa/ieu046.

Silberglied, R. E., A. Aiello, and G. Lamas. 1979. Neotropical butterflies of the genus *Anartia*: systematics, life histories and general biology (Lepidoptera: Nymphalidae). *Psyche: J. Entomol.* 86 (2–3): 219–60. doi:10.1155/1979/50172.

Smith, B., J. McCormack, A. Cuervo, M. Hickerson, A. Aleixo, C. Cadena, et al. 2014. The drivers of tropical speciation. *Nature* 515: 406–9. doi:10.1038/nature13687.

Smith, N. G. 1983. Host plant toxicity and migration in the dayflying moth *Urania*. *Florida Entomol.* 66: 76–85.

Smith, N. G. 1992. Reproductive behavior and ecology of *Urania* (Lepidoptera: Uraniidae) moths and of their larval food plants, *Omphalea* spp. (Euphorbiaceae). In

D. A. Quintero and A. Aiello (eds.), *Insects of Panama and Mesoamerica: Selected Studies*, 576–93. Oxford: Oxford University Press.

Stevenson, R., and W. A. Haber. 1996. Time budgets and the crepuscular migration activity of a tropical butterfly, *Manataria maculata* (Satyrinae). *Bull. Ecol. Soc. Am.* 77 (suppl. 3, part 2): 424.

Soga, M., and K. J. Gaston. 2016. Extinction of experience: The loss of human-nature interactions. *Fron. Ecol. Environ.* 14: 94–101.

Solis, M. A., J. E. Hayden, F. Vargas, F. Gonzalez, C. Sanabria, and G. Connor. 2019. A new species of *Sufetula* Walker (Lepidoptera: Crambidae) feeding on the roots of pineapple, *Ananas comosus* (L.) (Bromeliaceae), from Costa Rica. *Proc. Entomol. Soc. Wash.* 121 (3): 497–510.

Stehr, F. 1987. Order Lepidoptera. In F. Stehr (ed.), *Immature Insects*. Dubuque, IA: Kendall/Hunt Publishing.

Steinhauser, S. 1981. A revision of the *Proteus* group of the genus *Urbanus* Hubner Lepidoptera: Hesperiidae. *Bull. Allyn Mus.* 62.

Tarmann, G., and E. Drouet. 2015. Zygaenidae from French Guiana, with a key to the narrow-winged Zygaenidae genera of the new world. *J. Lep. Soc.* 69 (3): 209–35.

Tella, R. 1955. Dados bionomicos de Utetheisa ornatrix (L.,1758) (Lepidoptera, Arctiidae). *Bragantia: Bol. Técn. Inst. Agronôm. São Paulo* 14 (11): 109–15.

Turner, J. 1971. Experiments on the demography of tropical butterflies. II. Longevity and home-range behaviour in *Heliconius erato. Biotropica* 3(1): 21–31.

Tyler, H., K. Brown, and K. Wilson. 1994. *Swallowtail Butterflies of the Americas: A Study in Biological Dynamics, Ecological Diversity, Biosystematics and Conservation.* Gainesville, FL: Scientific Publishers.

USDA, NRCS. 2021. The Plants Database. http://plants.usda.gov. Accessed March 5, 2021.

Van Dam, W., and G. Wilde. 1977. Biology of the Bean Leafroller *Urbanus proteus* (Lepidoptera: Hesperidae). *J. Kans. Entomol. Soc.* 50 (1): 157–60.

Vane-Wright, D. 2015. *Butterflies: A Complete Guide to Their Biology and Behavior.* Ithaca, NY: Comstock Publishing Associates.

Van Nieukerken, E. J., L. Kaila, I. J. Kitching, N. P. Kristensen, D. C. Lees, J. Minet, et al., 2011. Order Lepidoptera Linnaeus, 1758. In Zhang, Z.-Q. (ed.), *Animal Biodiversity: An Outline Of Higher-Level Classification and Survey of Taxonomic Richness.* Zootaxa 3148. Auckland, NZ: Magnolia Press.

Vargas, E. 2011. *Guía para la identificación y manejo integrado de plagas en piña.* Costa Rica: ProAgroin and Repcar.

Vega, G., and P. Gloor. 2001. Lista preliminar de mariposas diurnas (Hesperioidea: Papilionoidea) de la Zona Protectora El Rodeo, Ciudad Colón, Costa Rica. *Brenesia* 55–56: 101–22.

Villalobos, A. 2015. Entomofauna, herbivory, and chemical cues in different zones of the Naranjo Beach mangrove in the north Pacific coast of Costa Rica. MS thesis. Georg-August-Universität Göttingen.

Viloria, A. 2000. Mariposas ropalocera de Venezuela. In F. Martín Piera, J. Morrone, and A. Melic (eds.), *Hacia un proyecto Cyted para el inventario y estimacion de la diversidad entomológica en Iberoamérica: PrIBES 2000.* m3m-Monografías Tercer Milenio, Vol. 1. Zaragoza: Sociedad Entomológica Aragonesa.

Willmott, K. 2003. *The Genus Adelpha: Its Systematics, Biology and Biogeography (Lepidoptera: Nymphalidae: Limenitidinae).* Gainesville, FL: Scientific Publishers.

Willmott, K., L. M. Constantino, and J. P. W. Hall. 2001. A review of *Colobura* (Lepidoptera: Nymphalidae) with comments on larval and adult ecology and description of a sibling species. *Ann. Entomol. Soc. Am.* 94 (2): 185–96.

Wolfe, J. M., A. C. Daley, D. A. Legg, and G. D. Edgecombe. 2016. Fossil calibrations for the arthropod tree of life. *Earth-Sci. Rev.* 160: 43–110.

Young, A. 1984. Natural history notes for *Taygetis andromeda* (Cramer) (Satyridae) in eastern Costa Rica. *J. Lep. Soc.* 38 (2): 102–13.

Zahiri, R., J. D. Holloway, I. J. Kitching, J. D. Lafontaine, M. Mutanen, and N. Wahlberg. 2011. Molecular phylogenetics of Erebidae (Lepidoptera, Noctuoidea). *Syst. Entomol.* 37: 102–24.

Zahiri, R., J. D. Lafontaine, J. D. Holloway, I. J. Kitching, B. C. Schmidt, L. Kaila, and N. Wahlberg. 2013. Major lineages of Nolidae (Lepidoptera, Noctuoidea) elucidated by molecular phylogenetics. *Cladistics* 29: 337–59.

Zhang, M., T. W. Cao, K. Jin, Z. Ren, Y. Guo, J. Shi, Y. Zhong, and E. Mai. 2008. Estimating divergence times among subfamilies in Nymphalidae. *Chinese Sci. Bull.* 53 (17): 2652–58.

CREDITS

Unless otherwise indicated below, photographs in this book were taken by the author (LRHM).

The initials appended to many of the photographs in this book indicate the name of the following photographers:

AC = A. Cascante; AV = A. Villalobos; BH = B. Higgins; CB = C. Bolaños; CC = C. Chaves; CH = C. Hidalgo; CRU = C. Rojas Ugalde; EVC = E. Vargas Carrillo; HBl = H. Blanco; HBo = H. Bockler; HZ = H. Ziegler; JA = J. Alfaro; JCn = J. Cochran; JCo = J. Corrales; JDO = J.D. Obando; JDS = J.D. Salazar; JL = J. Lobo; JMR = J. Montero Ramirez; KN = K. Nishida; KW = K. Wolfe; NR = N. Roncen; RC = R. Cubero.

Other photo credits:

front cover: *Parides photinus* (Pink-spotted Cattleheart), ondrejprosicky/Adobe Stock

p. ii: *Siproeta stelenes* (Malachite), Denis-Huot/Minden Pictures
p. vi: *Dryas iulia*, Reimar Gaertner/UIT/Getty Images
p. 32: *Acraga* sp., Kenji Nishida
p. 44: *Creonpyge creon*, Kenji Nishida
p. 62: *Protesilaus protesilaus dariensis*, L. R. Murillo Hiller
p. 76: *Itaballia demophile centralis*, L. R. Murillo Hiller
p. 98: *Arcas imperialis*, L. R. Murillo Hiller
p. 110: *Caria mantinea*, Sandra Shaull
p. 126: *Diaethria astala astala* (Astala Eighty-eight), L. R. Murillo Hiller
p. 198: *Thysania agrippina*, Kenji Nishida

INDEX OF PLANT NAMES

This index lists the scientific names for all plants mentioned in the species accounts.

INDEX OF BUTTERFLY
AND MOTH NAMES

This index contains the scientific names, family names, and, when they exist, English common names for all butterfly and moth species described in the species accounts.

NOTES

NOTES

NOTES

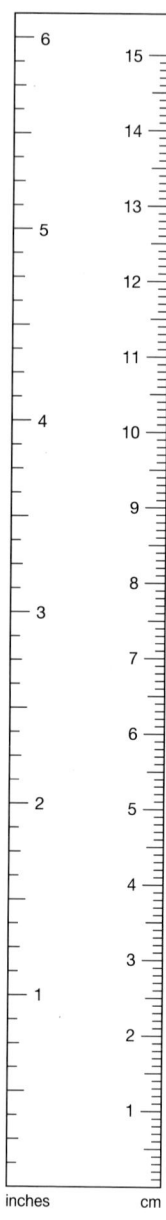

inches   cm